Lecture Notes in Computer Science   15529

Founding Editors

Gerhard Goos
Juris Hartmanis

Editorial Board Members

Elisa Bertino, USA
Wen Gao, China
Bernhard Steffen, Germany
Moti Yung, USA

Advanced Research in Computing and Software Science
Subline of Lecture Notes in Computer Science

Subline Series Editors

Giorgio Ausiello, *University of Rome 'La Sapienza', Italy*
Vladimiro Sassone, *University of Southampton, UK*

Subline Advisory Board

Susanne Albers, *TU Munich, Germany*
Benjamin C. Pierce, *University of Pennsylvania, USA*
Bernhard Steffen, *University of Dortmund, Germany*
Deng Xiaotie, *Peking University, Beijing, China*
Jeannette M. Wing, *Microsoft Research, Redmond, WA, USA*

More information about this series at https://link.springer.com/bookseries/558

Contents – Part II

Abstract Interpretation

Abstract Local Completeness: A Local Form of Abstract Non-interference 3
   Isabella Mastroeni

An Abstract Domain for Heap Commutativity 26
   Jared Pincus and Eric Koskinen

A Static Analysis of Entanglement 50
   Nicola Assolini, Alessandra Di Pierro, and Isabella Mastroeni

Synthesis

Synthesis of Parametric Locally Symmetric Protocols from Abstract
Temporal Specifications 75
   Ruoxi Zhang, Richard Trefler, and Kedar S. Namjoshi

1–2–3–Go! Policy Synthesis for Parameterized Markov Decision
Processes via Decision-Tree Learning and Generalization 97
   Muqsit Azeem, Debraj Chakraborty, Sudeep Kanav,
   Jan Křetínský, Mohammadsadegh Mohagheghi, Stefanie Mohr,
   and Maximilian Weininger

LLOR: Automated Repair of OpenMP Programs 121
   Utpal Bora, Saurabh Joshi, Gautam Muduganti,
   and Ramakrishna Upadrasta

Synthesis of Controllers for Continuous Blackbox Systems 137
   Benedikt Maderbacher, Felix Windisch, Alberto Larrauri,
   and Roderick Bloem

Applications

Automated Flaw Detection for Industrial Robot RESTful Service 163
   Yuncheng Wang, Puzhuo Liu, Yaowen Zheng, Dongliang Fang,
   Shuaizong Si, Zhiwen Pan, Weidong Zhang, and Limin Sun

Formally Verifiable Generated ASN.1/ACN Encoders and Decoders: A Case Study 185
  Mario Bucev, Samuel Chassot, Simon Felix, Filip Schramka, and Viktor Kunčak

ExpectAll: A BDD Based Approach for Link Failure Resilience in Elastic Optical Networks 208
  Gustav S. Bruhns, Martin P. Hansen, Rasmus Hebsgaard, Frederik M. W. Hyldgaard, and Jiří Srba

Constructing Trustworthy Smart Contracts 231
  Devora Chait-Roth and Kedar S. Namjoshi

Author Index 253

Invited Keynote Papers

Discovering Likely Invariants for Distributed Systems Through Runtime Monitoring and Learning

Yuan Xia[✉], Deepayan Sur, Aabha Shailesh Pingle, Jyotirmoy V. Deshmukh, Mukund Raghothaman, and Srivatsan Ravi

University of Southern California, Los Angeles, CA, USA
{yuanxia,deepayan,apingle,jdeshmuk,raghotha,srivatsr}@usc.edu

Abstract. Characterizing the set of reachable states of a distributed protocol that uses asynchronous message-passing communication is difficult due to the exponential number of possible interleavings of local executions. Any syntactic expression overapproximating the set of reachable states is an *invariant formula* of the system, and is a valuable tool that can aid programmers in understanding global program behavior. In this paper, we propose a method for obtaining a formula that approximates the set of reachable states; we call this formula a likely invariant, and we learn it using information only obtained from system executions. Our method doubles up as a way for identifying states that may not be known to be reachable (based on the best-known likely invariant) and hence may appear anomalous to the system designer. In some cases, they may be actually anomalous and may indicate a lurking (*heisenbug*). Our method has the following main steps: (1) we observe the global states of the system reached during its execution, (2) we asynchronously learn a *likely invariant* from the observed global states, (3) we monitor the learned likely invariant for the system states that do not satisfy it, and (4) if such states are found, we *revise* the likely invariant. We implement our overall methodology for a number of distributed protocols written in the Promela language and show that our technique can learn useful information about the system from just runtime executions.

1 Introduction

Distributed systems serve as the backbone of most real-world computing applications. These systems are modeled as a collection of concurrent processes that rely on local computations and asynchronous message-passing to achieve their objectives, involving multiple functions, e.g., consensus, coordination, memory coherence, decentralized computation, and database consistency. Therefore, the inherent nondeterminism can exponentially increase the number of possible execution sequences and can significantly expand the space of reachable system states, making it challenging to formally reason about system behaviors.

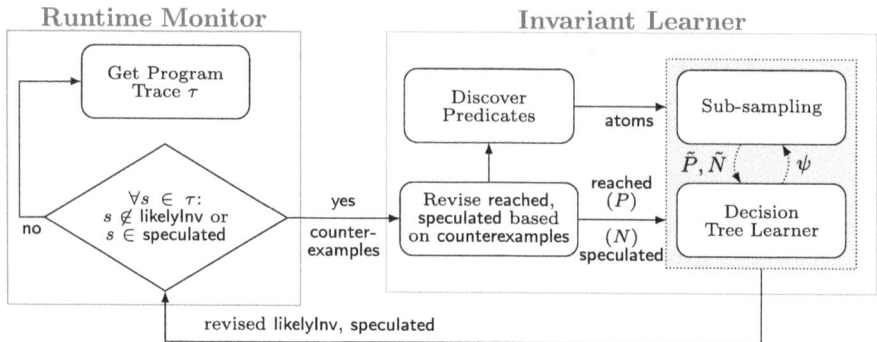

Fig. 1. High-level Algorithm 1 for learning and online monitoring likely invariants

A Boolean-valued expression over program variables that is true for every reachable state of the program is called an *invariant formula* or simply, a program *invariant*. Learning program invariants is valuable as it enhances the user understanding of the set of reachable states of the program, and can also be used to prove safety of the program w.r.t. a user-specified safety specification [2]. We define a *likely invariant* as a syntactic expression which holds true for all *observed* states of a program (in general, a subset of the reachable states[1]).

While many formal techniques [7,12,18,19,29,35,36,42,45] including model checking [2] have been effective for invariant learning, they usually struggle when applied to large-scale distributed systems. Some approaches require prior knowledge from the user; for instance, some invariant-learning techniques such as ICE-learning [18,19] require a user-provided safety property as a vital step in learning the invariant. Other invariant synthesis techniques make use of model checkers, SMT solvers, theorem provers, or constrained Horn clause solvers [36,41,47,51] to verify if a synthesized invariant is valid. Unfortunately, this restricts the use of invariant synthesis to programs for which the above tools are tractable. These limitations motivate our use of *dynamic invariant generation*, where we approximate reachable states during execution, examples include approaches based on Daikon [11,25]. Given a sequence of program states, Daikon uses predefined expression templates to synthesize invariant expressions derived from these templates. However, its reliance on specific syntactic expressions may result in expressions that are too conservative (i.e., with large error in overapproximating the reachable state set).

To address these challenges, this paper introduces the LIDO framework (Likely Invariant Discovery through Online monitoring) shown in Fig. 1. LIDO has two main components: a runtime monitor that continuously monitors program executions for counterexample states and an invariant learning engine. We first describe the invariant learner, which is a data-driven step: we use the

---

[1] Determining if a Boolean expression holds for every reachable state is undecidable for general programs by a reduction from the halting problem for Turing machines. Likely invariants are thus a best effort solution to approximate reachable states.

observed states of the program as positive examples which must be satisfied by any candidate invariant expression. Learning such an expression usually follows the Occam's Razor principle of finding the shortest syntactic expression that explains the (positive) examples. However, as *true* is a trivial expression that is an invariant, we also need negative examples to get non-trivial expressions. To do so, we use a "guess and check" approach, i.e., we guess unreachable states (by random sampling) and use them as negative examples in the decision-tree based learning method. Checking if the speculated states is part of online validation of the invariant, and incorrectly guessed unreachable states form one kind of counterexample. Existing approaches [3,8,19–21,25,48,50,51] use formal tools to verify or refute a synthesized invariant; however, LIDO employs online monitoring to find states that should be included in the invariant (but are not). These form the other kind of counterexample states. Any counterexample state discovered by the runtime monitor leads to a *revision* of the likely invariant.

A key challenge in the decision-tree based invariant synthesis method is identifying atomic predicates that should constitute the invariant expression. In this work, we assume that the user provides predicate templates of the form $\texttt{size(a)} > c_1$, $\texttt{x}_1 + \texttt{x}_2 < c_2$, etc. Here, a, x_1, x_2 are program variables, and c_1, c_2 are symbolic constants. LIDO discovers concrete predicates from such symbolic predicates automatically.

We provide experimental evaluation of our technique for several distributed protocols written in the Promela modeling language. Through the SPIN execution environment [23], we can generate runtime traces for Promela programs. Our technique itself is language-agnostic, and with some engineering effort can be applied to programs written in other languages such as Java or P [9]. Through our experiments on Promela, we demonstrate the potential of our technique to work on more complex programming and modeling languages. As our technique for learning the invariant is a purely data-driven technique with only runtime validation, an important question is whether the likely invariants that we learn are indeed true invariants. Here, we leverage the Spin model checker that can verify Promela programs to empirically evaluate the invariants learned by LIDO. We show that *in spite of being purely data-driven*, we can learn invariants that are valid.

Contributions. To summarize, this paper makes the following contributions:

(i) A decision-tree based method to learn invariants only from observed system states and speculated unreachable states.
(ii) An online monitoring framework to validate the learned invariant and revise the candidate invariant.
(iii) A runtime technique to discover concrete predicate expressions from user-supplied template predicates.
(iv) A novel evaluation measurement for likely invariants' quality based on three metrics: soundness, tightness, and safety.
(v) Demonstrations of the practical usefulness of our approach for distributed protocols modeled in the Promela language [22].

2 Preliminaries

In this section, we formalize the terminology needed to explain our approach with Peterson's algorithm, which presents mutual exclusive access to the critical section for two concurrent processes [37] (henceforth denoted as $\mathcal{P}_{\mathsf{mutex},2}$ for brevity). We show $\mathcal{P}_{\mathsf{mutex},2}$ modeled using the Promela language [22] in Fig. 2.

```
// Peterson's solution to the mutual exclusion problem - 1981
bool turn, flag[2];
byte ncrit;
active [2] proctype user() {
    assert(_pid == 0 || _pid == 1);
again:
    flag[_pid] = 1;
    turn = _pid;
    (flag[1 - _pid] == 0 || turn == 1 - _pid);
    ncrit++;
    assert(ncrit == 1); // critical section
    ncrit--;
    flag[_pid] = 0;
    goto again }
```

Fig. 2. Peterson's mutual exclusion algorithm for 2 processes in Promela

2.1 Program Structure and Modeling Assumptions

Program variables, the program state, and program execution traces are all standard notions, we define them formally so that we can precisely state the problem we wish to solve.

Definition 1 (Variables, State). *A type t refers to a finite or an infinite set. A program variable v of type t is a symbolic name that takes values from t. We use* $\mathsf{type}(v)$ *to denote variable types. We use V to denote the set of program variables. A valuation ν maps a variable v to a specific value in* $\mathsf{type}(v)$. $\nu(V)$ *denotes the tuple containing valuations of the program variables; a program state is a specific valuation of its variables during program execution.*

The following types are included: $\mathsf{bool} = \{true, false\}$, $\mathsf{int} = \mathbb{Z}$, $\mathsf{byte} = \{0, -1, 3^2 + 1\}$, finite enumeration types which are some finite set of values, the program counter pc type that takes values in the set of line numbers of the program. We assume that the grammar of our modeling language provides syntactically accurate expressions made up of operators and program variables, and that the well-defined semantics of the language defines the valuation of an expression based on the variable valuations. Given a set of variables $\{v_1, \ldots, v_k\}$ and an expression $e(v_1, \ldots, v_k)$, we use $\nu(e)$ to denote the valuation of the expression when each v_i is substituted by $\nu(v_i)$.

Procedure calls and procedure-local variables are present in the majority of real-world programs; we assume that the program is modeled in a language that abstracts away such details. Real-world concurrent programs typically use either a shared memory or message-passing paradigm to perform concurrent operations of individual *threads* or *processes*[2]. We assume that the number of concurrent processes in the program, N, is known and fixed throughout program execution. Thus, our modeling language does not have statements to start a new process or terminate one. So, a concurrent program for us is a set of N processes, with each process being a sequence of atomic statements. We include statements such as assignments and conditional statements (which execute sequentially), and `goto` statements to alter the sequential control flow. An assignment statement is of the form $v_i \leftarrow e$, where e is an expression whose valuations must be in $\texttt{type}(v_i)$. The formal semantics of an assignment statement is that for all variables that do not appear on the LHS of the assignment, the valuation remains unchanged, while the valuation of v_i changes to $\nu(e)$.

Example 1. For $\mathcal{P}_{\mathsf{mutex},2}$, $V = \{\ell, \texttt{turn}, \texttt{flag}, \texttt{ncrit}\}$, where $\texttt{type}(\ell, turn, flag[i], ncrit) = (\texttt{pc} \times \texttt{pc}, \texttt{bool}, \texttt{bool} \times \texttt{bool}, \texttt{byte})$. The variable ℓ is a pair that maintains program counters for the two processes. The statements $\texttt{flag}[\texttt{\_pid}] = 1$ and $\texttt{turn} = \texttt{\_pid}$ are assignments. The state s_0 is

$$s_0 = \begin{pmatrix} \ell_1 & \mapsto 7, \\ \ell_2 & \mapsto 7, \\ \texttt{turn} & \mapsto 0, \\ \texttt{flag}[0] & \mapsto 0, \\ \texttt{flag}[1] & \mapsto 0, \\ \texttt{ncrit} & \mapsto 0 \end{pmatrix}$$

After Process 0 (`_pid = 0`) executes two assignment statements ($\texttt{flag}[0] \leftarrow 1$ and $\texttt{turn} \leftarrow 0$), the program state changes to the following:

$$s_2 = \begin{pmatrix} \ell_1 & \mapsto 9, \\ \ell_2 & \mapsto 7, \\ \texttt{turn} & \mapsto 0, \\ \texttt{flag}[0] & \mapsto 1, \\ \texttt{flag}[1] & \mapsto 0, \\ \texttt{ncrit} & \mapsto 0 \end{pmatrix}$$

The execution semantics of a program are typically explained using the notion of a labeled transition system.

Definition 2 (Labelled Transition System (LTS) for a concurrent program). *A labelled transition system is a tuple (S, L, T, Init) where S is a set of*

---

[2] In most modern programming languages, processes and threads have been used to mean different abstract units of concurrent operation. For the purpose of this paper, we assume that we are using a modeling language such as Promela or P that abstracts away from these finer details.

system states, $T \subseteq S \times L \times S$ is the transition relation, L is the set of program statements that serve as labels for elements in T, and Init *is a set of initial program states as* Init $\subseteq S$.

Labels identify the program statement that caused the transition, and transitions describe how a system changes from one state to another. Converting a program to its corresponding LTS is a well-defined process (see [2]); where transition is added based on the effect of the corresponding program statement on the program variables (including the program counter). Assignment statements involve updating the program counter and the variables on the LHS. For conditional statements, a transition is added to the next state only if the valuation corresponding to the current state satisfies the condition. For goto statements, only the program counter is updated. Computing the LTS for a concurrent program is a bit more tricky. We typically use an interleaved model of concurrency, so from every state, we consider N next states corresponding to each of the N concurrent processes executing. We assume that each of the sequential processes involved in the computation is *deterministic*, and the only source of nondeterminism is context switches due to an external scheduler[3]. We also remark that an LTS is finite if the type of each program variable is finite, otherwise an LTS can have an infinite set of states.

Definition 3 (Reachable States). *For a program $\mathcal{P}$ as a labeled transition system, its set of reachable states* Reach$(\mathcal{P}, \mathsf{Init})$ *is defined as the set of all states that can be reached from an initial state $s_0 \in S$ through the execution of the program statements. Formally,* Reach$(\mathcal{P}, \mathsf{Init})$ *is the smallest set R that satisfies the following:*

1. $s_0 \in \mathsf{Init} \implies s \in R$, *and,*
2. $s \in R \wedge (s, s') \in T \implies s' \in R$.

Example 2. We remark that the state s_2 in Example 1 is reachable from s_0, since: $s_0 \xrightarrow{flag[\_pid]=1;turn=\_pid} s_2$. States where $ncrit = 2$ are unreachable from s_0.

Definition 4 (Program Trace). *A program trace σ of length $k = \mathsf{len}(\sigma)$ is a sequence of states $s_0, s_1, \ldots, s_{k-1}$, s.t., $s_0 \in \mathsf{Init}$, and for all $j \in [1, k-1]$, $(s_{j-1}, s_j) \in T$.*

---

[3] Our techniques can also handle nondeterminism in the sequential processes. If the number of nondeterministic choices available is fixed and known *a priori*, then the procedure in Fig. 1 will converge. If the nondeterminism is unbounded then the convergence of the procedure depends on the generalizability of the learning procedure and the kind of counterexamples obtained by the testing procedure.

Example 3. A possible trace σ of $\mathcal{P}_{\mathsf{mutex},2}$ with $\mathsf{len}(\sigma) = 5$ is shown below.

$$\begin{pmatrix} \underbrace{\langle 7,7,0,0,1,0 \rangle}_{s_0}, \\ \underbrace{\langle 7,8,1,0,1,0 \rangle}_{s_1}, \\ \underbrace{\langle 7,9,1,0,1,0 \rangle}_{s_2}, \\ \underbrace{\langle 7,11,1,0,1,1 \rangle}_{s_3}, \\ \underbrace{\langle 7,13,1,0,0,0 \rangle}_{s_4} \end{pmatrix}$$

Remark 1. To obtain a program trace, starting from a random $s_0 \in \mathsf{Init}$, we can randomly sample a successor s_1 from all possible pairs $(s_0, s') \in T$, and repeat this procedure from each subsequent s_i. Some of the successor states for a given state s correspond to a context switch for the distributed program (as it may require a different process to execute its atomic instruction than the one that executed to reach the state s). To randomly sample the initial or successor states, we need a suitable distribution over the initial states and outgoing labelled transitions from a given state. This distribution is defined by the scheduler and assumed to be *unknown* to the program developer.

Definition 5 (Monitor). *A monitor is to observe a finite prefix of a program trace $\sigma = (s_0, s_1, \ldots, s_k)$, where k is the length of the prefix. The monitor evaluates whether the observed trace σ satisfies a given set of properties ϕ and yields a prediction.*

Runtime monitoring refers as monitoring the system's states during its execution and detecting whether the behaviors align with predefined specifications. Runtime monitoring is a popular approach for dynamic verification [28]. The *offline monitoring* [17] collects runtime information during the system's execution and stores it for later analysis. The analysis is performed offline once the system has completed execution. In contrast, *online monitoring* [17] refers to the simultaneous observation and analysis of system states as it is running. The concurrent monitor enables early detection and response, which is essential for critical systems requiring immediate corrective actions.

Definition 6 (Invariants). *An invariant I is a Boolean-valued formula over program variables and constants that is satisfied by every reachable program state.*

Example 4. For instance, in the Peterson's model, an invariant is $0 \leq ncrit \leq 1$. This invariant is useful to show that at most one process is in the critical section at any given time.

Remark 2 (About inductiveness of invariants). Invariants and *inductive invariants* are sometimes confused. An inductive invariant is defined as a set of states s.t., after executing any program statement from any state within this set, the resulting state also belongs to the same set. The exact set of reachable states of a program is an inductive invariant. However, obtaining a formula that accurately characterizes all reachable states can be challenging. Instead, it is often more practical to find a formula that over-approximates the reachable states to facilitate safety verification. While any formula that over-approximates the reachable states is considered an invariant, it may not be an inductive invariant. Many verification methods prefer inductive invariants because they can be verified using automated tools like SMT solvers. However, deriving inductive invariants using data-driven techniques requires is more complex. In our approach, we would require an execution engine that uses a labeled transition system representation of the given protocol to execute arbitrary transitions from states in the likely invariant. As the invariant expressions that we obtain are adequate to perform safety proofs, we defer learning inductive invariants to future work.

Algorithm 1: Pseduo-code for LIDO

   input : Number of points randomly sampled ℓ
             Precision threshold δ
             Depth of already constructed decision tree d
             Maximum depth of decision tree k
             Ratio of speculative negative to positive examples α
   output: likelyInv ϕ

1 reached $\leftarrow \{\}$, speculated $\leftarrow \{\}$, $\phi \leftarrow$ *false*, $n \leftarrow 0$
2 **repeat**
3      $\tau \leftarrow$ get_trace
4   ce $\leftarrow \{s \mid s \in \tau \cap [\![\neg\phi]\!]\}$ // reached state in $\neg\phi$
5   ce $\leftarrow$ ce $\cup \{s \mid s \in \tau \cap$ speculated$\}$ // reached state assumed unreachable
6   reached $\leftarrow$ reached $\cup$ ce // update reached states
7   speculated $\leftarrow$
        (speculated$\setminus$ce) $\cup$ randomSample($S \setminus$ ((speculated$\setminus$ce) $\cup$ reached), ℓ)
8   s.t. $\frac{|\text{speculated}|}{|\text{reached}|} < \alpha$
9   **if** ce $\neq \emptyset$ **then**
10   atoms $\leftarrow$ predicates_discovery(reached, template_predicates)
11         $\phi \leftarrow$ Learner(atoms, reached, speculated, d, k, δ)
12 **until** *true*
13 **return** ϕ

Problem Statement. Let $\mathcal{P}$ be a distributed program. Let Reach($\mathcal{P}$, Init) be the set of reachable states of $\mathcal{P}$. Let G denote a user-defined grammar to specifies (a possibly infinite) set of Boolean-valued expressions over program variables, and let $\mathcal{L}(G)$ denote this set. The objective of this paper is to design an runtime algorithm able to learn a program likely invariant ϕ without interrupting the system execution. The invariant has the following properties:

1. Soundness of the likely invariant:

$$((s \in \mathsf{Reach}(\mathcal{P}, \mathsf{Init})) \Rightarrow (s \in \phi)) \tag{1}$$

2. Tightness of the likely invariant:

$$\phi = \underset{\phi \in \mathcal{L}(G)}{\arg\min} \ | \{s \mid s \notin \mathsf{Reach}(\mathcal{P}, \mathsf{Init}) \wedge s \in \phi\} | \tag{2}$$

3 Data-Driven Invariant Generation

In this section, we present the overall algorithm for synthesizing likely invariants in an *online* learning setting in Algorithm 1. We use the notation $[\![\phi]\!]$ to represent the set of states that satisfy ϕ. Initially, the likely invariant is set to *false*, so in Line 4, the set ce contains all states encountered in the sampled trace T, which are then added to the set reached in Line 6. The next step involves sampling a set of states speculated to be unreachable, represented by the set speculated. We sample ℓ states from the state space S (excluding those already in reached or speculated) and add them to speculated as long as the ratio $\frac{|\mathsf{speculated}|}{|\mathsf{reached}|}$ remains below a user-specified threshold α (Line 8). In the first iteration, since ce is non-empty, the algorithm invokes the Learner procedure with the sets reached (positive examples), speculated (negative examples), and hyper-parameters d, k, and δ (elaborated in the next section) (Line 11). The main loop of the algorithm (Lines 2-12) iteratively checks if the likely invariant from the previous iteration survives. This is done by sampling a new trace T (Line 3) and verifying whether T contains any states not included in ϕ (Line 4) or any previously speculated unreachable states (Line 5). If either condition is met, a revision is triggered in Line 11. As shown in Fig. 1, we have a monitor and a generator to keep monitoring the system states and generate positive states P and negative states N. The states are then forwarded to Learner to further learn/revise the likelyInv.

Discovering Concrete Predicates. Before each revision, we first discover a dynamic set of concrete predicates (Line 10). The predicates_discovery function assumes a grammar (similar to the one shown in Table 1) with predicate templates. It constructs a dynamic set of concrete predicates based on the observed states of the system; essentially, it replaces parameters in the predicate templates with variable values in the set of observed states. We remark that for any invariant expression to be tight, one the concrete predicates thus obtained must be present in the final synthesized predicate, and thus it is enough to only consider as many concrete predicates as the number of observed states. Thus, though there may be a very large number of predicates associated with a given set of variables, we can restrict the set of concrete atomic predicates to a finite number.

Example 5. We now give an example run of the algorithm on $\mathcal{P}_{\mathsf{mutex},2}$. Recall that the state of $\mathcal{P}_{\mathsf{mutex},2}$ is a valuation of $(\ell_1, \ell_2, \mathtt{turn}, \mathtt{flag}[0], \mathtt{flag}[1], \mathtt{ncrit})$.

The trace of valuations of these state variables from Example 3 is a possible sampled trace of $\mathcal{P}_{\mathsf{mutex},2}$ obtained by using RecordStates($\mathcal{P}_{\mathsf{mutex},2}$), reproduced here for ease of exposition:

$$\langle 7,7,0,0,1,0\rangle, \langle 7,8,1,0,1,0\rangle, \langle 7,9,1,0,1,0\rangle, \langle 7,11,1,0,1,1\rangle, \langle 7,13,1,0,0,0\rangle, \langle 7,7,0,0,0,0\rangle$$

In Line 8, we speculatively add following states to speculated:

$$\mathsf{speculated} = \{\langle 8,7,0,1,0,0\rangle, \langle 9,7,0,1,0,0\rangle\} \tag{3}$$

We then use the positive (reached) and negative (speculated) examples to learn a candidate likely invariant ϕ that over-approximates reached but has minimal overlap with speculated. Suppose we learn the likely invariant $\phi_1 \equiv \mathtt{flag}[0] = 0$. We can check that all states in reached satisfy ϕ_1 and none of the states in speculated satisfy it. In the next iteration, suppose we sample the following trace next:

$$\langle 7,7,0,1,0,0\rangle, \langle 8,7,0,1,0,0\rangle, \langle \mathbf{9,7,0,1,0,0}\rangle, \langle 11,7,0,1,0,1\rangle, \langle 13,7,0,0,0,0\rangle, \langle 7,7,0,0,1,0\rangle$$

Clearly, the states in this trace refute the likely invariant ϕ_1 because $\mathtt{flag}[0] \neq 0$, which means that these states will appear in the set ce. We note that the second state in the set speculated (shown in (3)) was speculated to be unreachable but is actually reached in the trace (shown in bold). Thus, this state also gets added to ce, and removed from speculated. With the revised speculated and reached, we can re-learn the likely invariant. □

Table 1. Grammar for predicate templates; v is an arbitrary numeric variable, a an array variable, and p a parameter. Here, i is an integer.

$ArrayExpr\ (ae)$	::=	$a \mid subset(a)$
$ArrayArithExprs\ (aae)$	::=	$index(a,i) \mid \lambda(a,f)$
$ArrayFunc\ (f)$	::=	$len(a) \mid sum(a) \mid min(a) \mid max(a)$
$ArithExprs\ (e)$	::=	$v \mid e \diamond e \mid aae \diamond aae \mid aae \diamond e \mid e \diamond aae$
$Operator\ (\diamond)$	::=	$+ \mid - \mid \times \mid \div \mid mod$
$Comparator\ (\circ)$	::=	$= \mid > \mid < \mid \leq \mid \geq \mid \neq$
$Predicate\ Template$	::=	$e \circ p$

4 Decision Tree Learning for Invariants

In this section, we describe the details of Learner as presented in Algorithms 2, 3. The Learner algorithm combines decision tree learning, syntax-guided synthesis, and a novel subsampling technique to enable efficient inference of likely invariants at runtime.

Syntax-Guided Synthesis. Our method for learning likely invariants is motivated by syntax-guided synthesis (SyGuS) [1]. We assume that a grammar G is provided, which specifies Boolean-valued formulas constructed from user-defined parametric atomic predicates: $\phi ::= \mathsf{atom}(\mathbf{p}) \mid \neg \phi \mid \phi \wedge \phi \mid \phi \vee \phi$. In this context, $\mathsf{atom}(\mathbf{p})$ denotes Boolean-valued expressions involving program variables and parameters, referred to as parametric atomic predicates. The parameters $\mathbf{p}$ in atom act as placeholders for constant values of corresponding types. Replacing a parametric predicate with a suitable constant results in a concrete atomic predicate. Following predicate_discovery, we obtain a set of concrete atomic predicates, represented by atoms, which correspond to a finite collection of instantiated atomic predicates. The pre-defined parametric atomic predicates are sufficient for exploring concrete atoms, as they are derived from combinations of global variables and potential grammar rules. The grammar for atomic predicates that we use for invariant synthesis is provided in Table 1.

Example 6. Consider the following grammar for parameterized atom symbols.

$$\mathsf{atom}(c) ::= (x \leq c) \mid (x \geq c) \mid (y \leq c) \mid (y \geq c)$$

Here, the program variables V is the set $\{x, y\}$ (say of type byte), and c is a parameter of type byte. Substituting c with values, e.g., 2, -3, etc., gives atomic predicates $x \leq 2$, $y \geq -3$, etc. □

Next, we present the concept of a signature for a well-formed formula in G. We assume that we have an ordered set of positive examples $P = \langle e_1, \ldots, e_{|P|} \rangle$ and an ordered set of negative examples $N = \langle f_1, \ldots, f_{|N|} \rangle$.

Definition 7 (Formula signature). *Given ordered sets P and N, the signature of a formula ϕ is a $(|P| + |N|)$-bit vector σ^ϕ, where for each $i \in [1, |P|]$, $\sigma_i^\phi = 1$ iff $e_i \models \phi$ and 0 otherwise, and for $i \in [|P| + 1, |P| + |N|]$, $\sigma_i^\phi = 1$ iff $f_{i-|P|} \not\models \phi$ and 0 otherwise.*

Assuming the sets P and N are concatenated, the bit at a given index in σ is 1 iff the corresponding example in the concatenated set satisfies ϕ. It is important to recognize that, with the formula signatures for specific atomic predicates, we can easily derive the signatures for more complex formulas in the grammar by recursively applying the following rules, where bw_op represents the bitwise application of the corresponding logical operation.

$$\sigma^{\neg\phi} = \mathsf{bw\_not}(\sigma^\phi),\ \sigma^{\phi_1 \wedge \phi_2} = \mathsf{bw\_and}(\sigma^{\phi_1}, \sigma^{\phi_2}),\ \sigma^{\phi_1 \vee \phi_2} = \mathsf{bw\_or}(\sigma^{\phi_1}, \sigma^{\phi_2})$$

Using the formula signature, we can also calculate the precision and recall of a formula. In the definitions below, we first define precision and recall, and then provide the expressions over σ. Precision and recall play a role in our invariant learning and the stopping criteria of subsampling, which will be explained in the following subsections.

$$\text{precision}(\phi, P, N) = \frac{|P \cap [\![\phi]\!]|}{|(P \cup N) \cap [\![\phi]\!]|}$$
$$= \frac{\left|\{i \mid i \le |P| \wedge \sigma_i^\phi = 1\}\right|}{\text{sum}(\sigma^\phi)} \qquad (4)$$

$$\text{recall}(\phi, P, N) = \frac{|P \cap [\![\phi]\!]|}{|P|}$$
$$= \frac{\left|\{i \mid i \le |P| \wedge \sigma_i^\phi = 1\}\right|}{|P|} \qquad (5)$$

Learner is presented in two parts: a decision tree learner and a sub-sampling procedure to improve the scalability and generalizability of decision-tree learning. Previous work such as [46] frequently employs enumerative solvers for learning expressions. Although these solvers perform effectively in practice for learning expressions over more complex types, we have found that they do not scale as effectively when it comes to learning Boolean-valued functions. In contrast, the decision tree model is particularly effective to tackle the combinatorial aspect of learning Boolean combinations of atomic predicates.

Algorithm 2: DecisionTreeLearner(atoms, P, N, d, k, δ)

input :
- atoms: a set of atoms
- P : set of positive examples, N : set of negative examples
- d : depth of already constructed decision tree
- k : maximum allowed depth for the decision tree

output: likelyInv ϕ

1 $\delta \leftarrow$ MinPrecisionThreshold
2 **if** $d < k$ **then**
3 $\quad$ atom $= \arg\max_{\text{atom}\in\text{atoms}} (\text{IG}(\text{atom}, P, N))$
4 $\quad$ $P' \leftarrow P \cap [\![\text{atom}]\!]$, $N' \leftarrow N \cap [\![\neg\text{atom}]\!]$
5 $\quad$ atoms$' =$ atoms $\setminus \{\text{atom}\}$
6 $\quad$ $\phi \leftarrow$ (atom $\wedge$ DecisionTreeLearner(atoms$', P', N', d+1, k, \delta$) $\vee$
7 $\quad\quad$ ($\neg$atom $\wedge$ DecisionTreeLearner(atoms$', P \setminus P', N \setminus N', d+1, k, \delta$))
8 **if** (precision(ϕ, P, N) $> \delta \wedge$ recall(ϕ, P, N) $= 1$) **then return** ϕ
9 **else** report error: k is too low

Decision Tree Learner. Algorithm 2 presents the decision tree learning technique. Each internal node in our decision tree model represents an atom with its formula signature. Given our speculative sampling technique, ensuring that our decision tree has an appropriate depth bound is critical to avoid over-fitting (if the depth is too high) or over-generalizing (if the depth is too low). The key steps are shown in Algorithm 2. The DecisionTreeLearner procedure is a recursive algorithm; each recursive instance is invoked with sets of states P and N, and it constructs a sub-tree that best partitions states in P and N. The subtree is then returned to the caller, where it is added as a child to the partial tree being constructed by the caller. The parameter d equals the depth of the decision tree constructed by the caller. Each recursive call finds the *best* atom to use as the root of the sub-tree – we explain how we define the best atom below. All examples from P that satisfy atom are removed to get P', and those from N not satisfying atom are removed to get N' (Line 4). The predicate atom is itself removed from the set atoms to get atoms' (Line 5). Then DecisionTreeLearner is recursively invoked on the sets P' and N' (to form the left sub-tree) and on the sets $P \setminus P'$ and $N \setminus N'$ (to form the right sub-tree); the recursive invocations are with an incremented value of d (Line 7). The likely invariant expression is constructed using tail recursion. If the learned likely invariant lacks sufficient precision (i.e., too many negative states are included in the likely invariant), then either the maximum allowed depth k is too low, or the choice of atoms is insufficient to learn a good invariant, and the procedure fails. We remark that we use a precision threshold of less than 1 because states in N are not known to be truly unreachable, but are speculative. Therefore, it is reasonable to allow a certain number of speculative negative states to be included in the invariant.

Algorithm 3: Learner(P, N, δ)

    input : P : set of positive examples
             N: set of negative examples
             δ : precision threshold
    output: likelyInv ψ
1 $\tilde{P}, \tilde{N}$ = subSample(P, N)
2 **while** $\tilde{P} \neq P \lor \tilde{N} \neq N$ **do**
3   ψ = DecisionTreeLearner($\tilde{P}, \tilde{N}, 0, k,$ atoms)
4   **if** precision(ψ, P, N) $> \delta \land$ recall(ψ, P, N) $= 1$ **then return** ψ
5   **else** $\tilde{P} = \tilde{P} \cup$ subSample(P), $\tilde{N} = \tilde{N} \cup$ subSample(N)
6 increase tree depth k

To select the best atom in each recursive call, we choose the atom with the highest *information gain* over the sets P and N (denoted IG(atom, P, N)) in Line 3. This is a commonly used metric in decision tree algorithms [31]. To define information gain, we make use of Shannon entropy. Given sets P and N, consider the set $P \cup N$. The probability of a randomly drawn state being in P is $p_P = \frac{|P|}{|P \cup N|}$ and being in N is $p_N = \frac{|N|}{|P \cup N|}$. Then the Shannon entropy is defined as:

$$H(P, N) = -p_P \log_2(p_P) - p_N \log_2(p_N) \qquad (6)$$

Lower entropy signifies a higher level of separation between positive and negative points in the dataset. In other words, when the dataset is more homogeneous and primarily consists of one class, the entropy will be lower. Conversely, higher entropy signifies a greater level of randomness (or uncertainty) in the dataset, characterized by a more equal distribution of class labels. In our context, this means that a more balanced set of points is being separated. Information Gain with Shannon Entropy (as proposed in [52]) is then defined as follows:

$$\mathsf{IG}(P, N, \mathsf{atom}) = H(P, N) - p_P H(P \cap [\![\mathsf{atom}]\!], N \cap [\![\mathsf{atom}]\!]) \\ - p_N H(P \cap [\![\neg\mathsf{atom}]\!], N \cap [\![\neg\mathsf{atom}]\!]) \quad (7)$$

Subsampling for Efficiency. Now we explain how the overall Learner procedure works. In Line 1 of Algorithm 3, we use subSample to sample states from the positive examples P and negative examples N. Recall that Learner is invoked with P = reached and N = speculated. The main idea is to only use the subsets $\tilde{P}, \tilde{N}$ of P, N to invoke DecisionTreeLearner (Line 3), which returns ψ. Once we identify a candidate ψ, we check if its recall is 1 (i.e., all reached states are included in the likely invariant) and if its precision is above a pre-defined threshold δ. If yes, the algorithm returns ψ as the candidate likelyInv (Line 4). If it fails to find a likely invariant with desired precision/recall, subSample gathers more samples and invokes DecisionTreeLearner (Line 5).

Grammar. The figure illustrates the template-free grammar $\mathcal{G}$ designed for atomic predicates to express invariants. This grammar provides an extensive range of constructs for logical and arithmetic expressions, array manipulations, and comparison operations. The grammar is also compatible with standard SMT solvers, enabling users to verify invariants when necessary. By limiting expressiveness to linear integer arithmetic and simple array manipulations, the grammar **G** ensures that invariants can be synthesized within a feasible computational framework while still capturing the essential properties needed by programmers for practical use.

5 Experimental Evaluation

To demonstrate the feasibility of our framework and the quality of the learned likely invariants, we evaluated LIDO for distributed protocols, which are modeled in the Promela language. Although our framework can accommodate systems in other languages such as P [9], TLA+ [34], and others, we selected Promela for two main reasons: firstly, numerous descriptions of distributed protocols are readily accessible in Promela. Secondly, Promela programs are compatible with the Spin model checker [23] its capability to sample finite-length execution traces of programs. As a comparative baseline, we employed Daikon [11] a widely used dynamic invariant generation tool that is known for its ability to identify likely invariants based on runtime traces. Daikon can serve as a front-end invariant synthesis in the LIDO framework as well. In our experiments, Daikon is used to substitute Learner, while the operational framework LIDO remained unchanged.

Python Implementation. The implementation of the LIDO framework is with two concurrent Python processes, one to continuously monitor system states in real-time using the trace generation functionality of the Spin model checker, while the other to asynchronously synthesize likely invariants based on the monitored program traces. The two processes: the Runtime Monitor process monitors global program states. It does so by collecting a trace of the global system state of a fixed length. It then checks if any state in the trace violates the invariant or invalidates the speculated negative examples. If true, then the likely invariant needs to be revised, and the process writes the counterexample state(s) on a shared channel. The Invariant Learner process subscribes to this channel, and whenever a new counterexample is published to this channel, it invokes the invariant learning procedure. This process terminates when a likely invariant is synthesized, and writes the new likely invariant and the updated set of speculated unreachable states to the shared channel.

In Python, the ready method is used to determine if a subsequent read on the shared communication channel would block; we use this to check if the invariant is synthesized before issuing a get on the shared channel to obtain the updated likely invariant and the set of speculated unreachable states. Both processes are guaranteed to run concurrently in two separate Python interpreters, thus avoiding potential performance degradation caused by Python's Global Interpreter Lock (GIL). The Runtime Monitor process is not expected to stop; however, the Invariant Learner process does terminte

5.1 Benchmarks and Measurements

To investigate our research goals, we formulated three key questions:

- **RQ1**: Is the previous leading method, Daikon, capable of producing high-quality likely invariants during the runtime monitoring of real-world distributed systems?
- **RQ2**: How does LIDO compare to Daikon in terms of the quality of likely invariants generated during online monitoring of distributed systems?
- **RQ3**: Is LIDO able to effectively learn likely invariants in large-scale distributed systems at runtime with minimal overhead?

We assess the quality of a learned likely invariant using three distinct metrics. The first metric measures the tightness of the invariant by counting the total number of states that satisfy it. This can be estimated through model counting, which helps determine if the invariant is overly broad. A maximum bound can also provide insights into the total number of states, especially when the invariant implies an infinite number.

The second and third metrics focus on the ability of a posteriori verification tool to validate that the synthesized invariant aligns with a user-defined safety property. The soundness of an invariant reflects its accuracy in representing the distribution of reachable states without being overly aggressive. However, soundness alone does not guarantee utility; for example, a broadly defined invariant like *True* may still be considered sound. Furthermore, the safety aspect of an invariant serves as an important indicator of its practical value, as it suggests that the system being analyzed remains safe.

Our benchmarks include 13 distributed systems modeled in Promela, drawn from resources related to Spin and other distributed systems literature. These systems primarily involve conditional invariants that require more intricate combinations of atomic expressions than those typically available in Daikon's templates. Notably, our approach to invariant generation successfully identified a bug in one of Spin's official systems. Additionally, we effectively generated invariants for a large-scale, real-world smart contract system modeled in Spin [49], which were subsequently verified to ensure the system's safety properties were maintained.

5.2 Results for Quality Evaluation

In order to address research questions **RQ1** and **RQ2**, we assessed the performance of LIDO through the Learner procedure, which incorporates both positive and negative examples during the learning process. Daikon* was employed as our baseline, integrating it into our monitoring framework as the inference engine to derive invariant expressions. For the ablation study, we ensured that both Daikon* and LIDO generated invariants within the same execution time, simulating conditions from Spin. The likely invariants produced by both methods were subsequently validated using the Spin model checker. The safety of these invariants was confirmed through an SMT solver, which checked if the intersection of the invariant ϕ with a user-defined set of unsafe states was empty. To maintain a consistent comparison, both Daikon* and LIDO utilized identical execution traces from Spin. This allowed for a fair evaluation of their respective performances and effectiveness.

Table 2 presents the results of our quality assessment for the learned invariant. In all cases, the likely invariants generated by LIDO with Learner were confirmed as true invariants. This trend was similarly observed with Daikon, which is known for producing overly conservative invariants. However, due to speculative synthesis, LIDO achieved precise invariants than those generated by Daikon, while maintaining a high level of soundness by monitoring system states. Notably, in each case study, the synthesized likely invariant also aligned with the system's safety property, as indicated by ✓ in Table 2. In contrast, the likely invariants derived from the combination of LIDO and Daikon were only able to confirm system safety in one case study and failed in others (marked by ✗), further highlighting Daikon's tendency to generate overly conservative invariant expressions. These empirical findings suggest that, with sufficient monitoring time, LIDO can develop a robust understanding of system behavior.

We also modeled a smart contract for the Ethereum commodity market in Spin to simulate execution and verify compliance with specifications. This model plays a crucial role in enhancing the credibility of smart contracts by detecting vulnerabilities, as errors can lead to significant financial losses due to their immutable nature once deployed on the blockchain. Unlike static analysis tools such as OYENTE, Osiris, and Gasper, which focus on pre-execution analysis, our approach with LIDO allows for dynamic monitoring and synthesis of system behavior in real-time. For example, we identified a likely invariant

$fa\_Acc + su\_Acc + t\_Acc + sca\_Acc + scb\_Acc = invar0$, which pertains to the total account balance and is recognized as a key safety property.

Additionally, LIDO discovered the likely invariant $((seen \leq n \wedge tour \leq max) \vee (tour > max \wedge seen \leq n))$ for the *salesman* system [23], which was verified by Spin. However, the official specification stated $seen < n \vee tour > max$, which Spin did not confirm as a true invariant. This discrepancy revealed a flaw in the official specification. By adjusting the assertion to $seen \leq n \vee tour > max$, the system was successfully verified by Spin. Thus, through our tool LIDO, we not only identified this bug but also provided a resolution.

5.3 Results for Evaluating the Scalability and Efficiency

Learning likely invariants efficiently during runtime is critical for large-scale distributed systems, which often operate continuously and cannot afford to be paused for traditional offline verification methods. Given the complexity of nondeterminism inherent in distributed systems, the key question is whether LIDO can effectively handle large-scale environments while minimizing its impact on system performance. **RQ3** explores whether LIDO can balance accuracy and efficiency with minimal disruption to the system while providing meaningful invariants. To benchmark the *efficiency* of LIDO, we measured the overhead of tracking state change log information on Spin, which is the only interruption event on Spin systems. We quantify the performance impact, or overhead incurred by monitoring system of 10000 states. The overhead is defined in terms of resource usage-such as CPU, memory, and I/O operations-that the monitoring

Table 2. Likely Invariant Quality Evaluation

System	Tightness		Soundness		Safety	
	LIDO	Daikon*	LIDO	Daikon*	LIDO	Daikon*
Peterson (Binary processes) [23]	✓	✗	✓	✓	✓	✗
Peterson (N processes) [23]	✓	✗	✓	✓	✓	✗
Bakery [23]	✓	✗	✓	✓	✓	✗
Hajek [23]	✓	✗	✓	✓	✓	✗
Manna Pnuelli [23]	✓	✗	✓	✓	✓	✗
Traffic Lights [13]	✓	✗	✓	✓	✓	✗
Producer Consumer [4]	✓	✗	✓	✓	✓	✗
Alternative Bit Protocol [23]	✓	✗	✓	✓	✓	✗
Leader Election [23]	✓	✗	✓	✓	✓	✓
UPPAAL Train/Gate [23]	✓	✗	✓	✓	✓	✗
Salesman [23]	✓	✗	✓	✓	✓	✗
Distributed Lock Server [30]	✓	✗	✓	✓	✓	✗
Smart Contract (ETH) [49]	✓	✗	✓	✓	✓	✗

process consumes while recording and storing state information. By evaluating this overhead, we seek to determine whether the monitoring system can effectively scale to larger state spaces without introducing significant performance degradation. Based on the results, the overhead ranges between 0.008 to 0.5 s, which indicates that the monitoring system can efficiently handle detailed logging and state observation without significantly affecting the system's overall performance or responsiveness. The outcomes for each distributed system are outlined in Table 3. With an execution time (T_r) of at most or around 1 s, and an approximate ratio of observed to reachable states $\frac{|V|}{|R|} \approx 0$ on large-scale systems, our tool scales effectively. Notably, even with the significantly low ratio, our tool can still infer likelyInv. The data presented in Table 3 affirm the validity of **RQ3**. It is important to note that while we were able to obtain likely invariants that were sound invariants, LIDO does not guarantee soundness as we *do not use* a model checker in the learning process. This approach aligns with traditional dynamic techniques, such as Daikon, avoiding the use of model checking. Model checking typically encounters the common issues of reduced generality and scalability.

Table 3. **Evaluation results on large-scale distributed systems** Overhead: the interruption time of system execution due to the monitoring event; T_r: average execution time for each revision.

Distributed Program	LoC	Shared Vars.	No. of Reachable States (R)	$\frac{|Visited|}{|R|}$	Overhead(s) (log info) /10000 states	$T_r(s)$
Peterson(Binary Processes) [23]	20	3	16	1	0.111	0.012
Bakery [23]	24	2	8	1	0.125	0.084
Manna Pnueli [23]	29	3	18	0.61	0.092	0.015
Hajek [23]	68	2	256	0.13	0.115	0.015
Traffic Lights [13]	33	2	200	0.065	0.178	0.004
Producer Consumer [4]	37	4	1.03M	0.00044	0.135	0.049
Peterson(N Processes) [23]	45	3	19.5M	0.00018	0.088	0.070
Alternative Bit Protocol [23]	42	3	∞	≈ 0	0.154	0.013
Leader Election [23]	127	3	26K	0.0024	0.008	0.015
UPPAAL train/gate [23]	78	7	16.8M	0.000024	0.192	0.021
Salesman [23]	54	6	∞	≈ 0	0.012	0.033
Distributed Lock Server [30]	100	4	12.2K	0.015	0.503	0.024
Smart Contract(ETH) [49]	962	21	∞	≈ 0	0.172	0.022

6 Related Work and Conclusions

Runtime Monitoring. Runtime monitoring that leverages invariant inference techniques supports various applications, from software verification to error detection and enhancing system reliability. One of the pioneering and impactful tools in this domain is Daikon, which dynamically identifies likely invariants from program execution traces. Daikon has been extensively utilized for debugging and improving testing and verification processes by generating invariants that describe system behavior [5,48]. While Daikon has been integrated into a runtime monitoring framework, as discussed in previous work [5], our research indicates that the invariants produced by Daikon are often overly conservative.

Alongside Daikon, DIDUCE [20] also significantly influenced the field by employing template enumeration and checking invariants by mapping state information to bits, where unchanged bits during execution are treated as invariants. Building on these foundational efforts, Artemis [16] was introduced as an acceleration technique for dynamic monitoring, specifically applied to DIDUCE in C programs, enhancing its efficiency. Recent studies have also focused on inferring FSM (Finite State Machine) models [10,40] to develop effective methodologies and frameworks for evaluating specification miners, while ours focuses on logic expressions. [6] Another study developed an SVM-based model through runtime monitoring in cyber-physical systems. Despite the contributions of these early approaches, they often face challenges related to the quality and generality of the invariants produced. Our work aims to address these issues, improving both the generality and quality of dynamic invariant generation.

Dynamic Invariant Synthesis. The main advantages of these works are scalability and generality, while the main limitation is its sensitivity to the initial pool of templates and its inability to learn interesting and non-trivial invariants, including properties with disjunction [11]. A few invariant generators [25,26] build on Daikon. ContExt [26] combines static analysis of program properties and dynamic analysis by Daikon to generate disjunctive constraints. The work in [25] proposes an LLVM-based code instrumentation frontend on top of Daikon to achieve invariant inference on multithreaded programs. Some solvers focus on specific invariant types, such as LinearAirbitrary [52] for linear inequalities and algebraic equation invariants [43]. Other dynamic generation tools include DIDUCE [20], DySy [8], Agitator [3], and Iodine [21], but they are often limited to certain languages, programs, or invariant types. Existing dynamic generation tools compromise applicability across diverse scenarios, lagging behind static techniques. Our tool aims to retain generality like Daikon while producing higher-quality invariants through a data-driven method, complementing existing approaches by enhancing their strengths and addressing limitations.

Model Checker Assisted Invariant Synthesis. Early attempts at automatic invariant generation used first-order theorem provers like Vampire [24,38], limited by their scalability. Subsequent work includes approaches guaranteeing provably correct invariants and data-driven methods for *likely invariants*. Counter-example guided invariant generation (CEGIR) combines inductive synthesis and model checker verification, exemplified by ICE [18], ICE-DT [19], and FreqHorn [14,15]. DistAI [51] and DuoAI [50] observe system executions but are guided by a target property verified with IVy [32], potentially causing infinite loops when the property does not hold. Techniques using theorem provers and model checkers can be accelerated through additional verifier information [33] or machine learning, such as Code2Inv [44], counter-example guided neural synthesis [39], and ACHAR [27]. However, the cost of model checking and the rapid growth of the space of possible invariants limit the complexity of systems to which these techniques can be applied. On the other hand, our approach in LIDO can thus be freely applied to complex systems.

7 Conclusion

Fundamental research in runtime monitoring has paved the way for innovations in dynamic invariant generation and automated program verification. Our approach, LIDO, is an automatic and practical framework for learning likely invariants for distributed protocols. We utilize counterexamples and speculative negative states to guide the invariant learning process. We also dynamically discover new atomic predicates from the observed states; essentially replacing the parameters in the chosen grammar by values matching the observed states.

Our framework successfully learns likely invariants, preserving the with same level of scalability as widely used tools such as Daikon, without being restricted to any particular programming language, without relying on exhaustive tools such as model checkers, or necessitating prior knowledge of the system's safety properties. Our method can provide three valuable outcomes for system developers: (1) a true overapproximation of the reachable states, which yields a valid invariant learned purely from program traces and validated only through online monitoring, (2) a summary of the most commonly observed states, where any violation highlights rarely encountered system behaviors or anomalies, and (3) potential verification of system safety. In future work, we will explore replacing the decision tree learning method with more advanced generative learning models.

References

1. Alur, R., et al.: Syntax-guided synthesis. In: 2013 Formal Methods in Computer-Aided Design, pp. 1–8 (2013). https://api.semanticscholar.org/CorpusID:6705760
2. Baier, C., Katoen, J.P.: Principles of Model Checking. Representation and Mind Series. The MIT Press, Cambridge (2008)
3. Boshernitsan, M., Doong, R., Savoia, A.: From Daikon to agitator: lessons and challenges in building a commercial tool for developer testing. In: Proceedings of the 2006 International Symposium on Software Testing and Analysis, pp. 169–180. Association for Computing Machinery (2006). https://doi.org/10.1145/1146238.1146258
4. Byrd, G., Flynn, M.: Producer-consumer communication in distributed shared memory multiprocessors. Proc. IEEE **87**(3), 456–466 (1999). https://doi.org/10.1109/5.747866
5. Chen, Y., Ying, M., Liu, D., Alim, A., Chen, F., Chen, M.H.: Effective online software anomaly detection. In: Proceedings of the 26th ACM SIGSOFT International Symposium on Software Testing and Analysis. ISSTA 2017, pp. 136–146. Association for Computing Machinery, New York, NY, USA (2017).https://doi.org/10.1145/3092703.3092730
6. Chen, Y., Poskitt, C.M., Sun, J.: Learning from mutants: using code mutation to learn and monitor invariants of a cyber-physical system . In: 2018 IEEE Symposium on Security and Privacy (SP), pp. 648–660. IEEE Computer Society, Los Alamitos, CA, USA (2018). https://doi.org/10.1109/SP.2018.00016

7. Clarke, E., Grumberg, O., Jha, S., Lu, Y., Veith, H.: Counterexample-guided abstraction refinement for symbolic model checking. J. ACM **50**(5), 752–794 (2003). https://doi.org/10.1145/876638.876643
8. Csallner, C., Tillmann, N., Smaragdakis, Y.: DYSY: dynamic symbolic execution for invariant inference. In: Proceedings of the 30th International Conference on Software Engineering. ICSE '08, pp. 281–290. Association for Computing Machinery, New York, NY, USA (2008). https://doi.org/10.1145/1368088.1368127
9. Desai, A., Gupta, V., Jackson, E., Qadeer, S., Rajamani, S., Zufferey, D.: P: safe asynchronous event-driven programming. ACM SIGPLAN Not. **48**(6), 321–332 (2013)
10. Dianlin, W., Ziying, D., Donghong, L., Xiaoguang, M.: Automatic online specification mining. In: Proceedings 2011 International Conference on Transportation, Mechanical, and Electrical Engineering (TMEE), pp. 253–258 (2011). https://doi.org/10.1109/TMEE.2011.6199191
11. Ernst, M., et al.: The daikon system for dynamic detection of likely invariants. Sci. Comput. Program. **69**, 35–45 (2007). https://doi.org/10.1016/j.scico.2007.01.015
12. Ezudheen, P., Neider, D., D'Souza, D., Garg, P., Madhusudan, P.: Horn-ice learning for synthesizing invariants and contracts. Proc. ACM Program. Lang. **2**(OOPSLA) (2018). https://doi.org/10.1145/3276501
13. Faye, S., Chaudet, C., Demeure, I.: A distributed algorithm for adaptive traffic lights control. In: 2012 15th International IEEE Conference on Intelligent Transportation Systems (2012)
14. Fedyukovich, G., Bodík, R.: Accelerating syntax-guided invariant synthesis. In: Beyer, D., Huisman, M. (eds.) TACAS 2018. LNCS, vol. 10805, pp. 251–269. Springer, Cham (2018). https://doi.org/10.1007/978-3-319-89960-2_14
15. Fedyukovich, G., Kaufman, S.J., Bodík, R.: Learning inductive invariants by sampling from frequency distributions. Form. Methods Syst. Des. **56**(1–3), 154–177 (2020). https://doi.org/10.1007/s10703-020-00349-x
16. Fei, L., Midkiff, S.P.: Artemis: practical runtime monitoring of applications for execution anomalies. SIGPLAN Not. **41**(6), 84–95 (2006). https://doi.org/10.1145/1133255.1133992
17. Gao, L., Lu, M., Li, L., Pan, C.: A survey of software runtime monitoring. In: 2017 8th IEEE International Conference on Software Engineering and Service Science (ICSESS), pp. 308–313 (2017). https://doi.org/10.1109/ICSESS.2017.8342921
18. Garg, P., Löding, C., Madhusudan, P., Neider, D.: ICE: a robust framework for learning invariants. In: Biere, A., Bloem, R. (eds.) CAV 2014. LNCS, vol. 8559, pp. 69–87. Springer, Cham (2014). https://doi.org/10.1007/978-3-319-08867-9_5
19. Garg, P., Neider, D., Madhusudan, P., Roth, D.: Learning invariants using decision trees and implication counterexamples. SIGPLAN Not. **51**(1), 499–512 (2016). https://doi.org/10.1145/2914770.2837664
20. Hangal, S., Lam, M.: Tracking down software bugs using automatic anomaly detection. In: Proceedings of the 24th International Conference on Software Engineering. ICSE 2002, pp. 291–301 (2002). https://doi.org/10.1145/581376.581377
21. Hangal, S., Narayanan, S., Chandra, N., Chakravorty, S.: Iodine: a tool to automatically infer dynamic invariants for hardware designs. In: Proceedings of 42nd Design Automation Conference, 2005, pp. 775–778 (2005). https://doi.org/10.1109/DAC.2005.193920
22. Holzmann, G.J., Lieberman, W.S.: Design and Validation of Computer Protocols, vol. 512. Prentice Hall, Englewood Cliffs (1991)

23. Holzmann, G.: The model checker spin. IEEE Trans. Softw. Eng. **23**(5), 279–295 (1997). https://doi.org/10.1109/32.588521
24. Kovács, L., Voronkov, A.: First-order theorem proving and VAMPIRE. In: Sharygina, N., Veith, H. (eds.) CAV 2013. LNCS, vol. 8044, pp. 1–35. Springer, Heidelberg (2013). https://doi.org/10.1007/978-3-642-39799-8_1
25. Kusano, M., Chattopadhyay, A., Wang, C.: Dynamic generation of likely invariants for multithreaded programs. In: 2015 IEEE/ACM 37th IEEE International Conference on Software Engineering, vol. 1, pp. 835–846 (2015). https://doi.org/10.1109/ICSE.2015.95
26. Kuzmina, N., Paul, J., Gamboa, R., Caldwell, J.: Extending dynamic constraint detection with disjunctive constraints. In: Proceedings of the 2008 International Workshop on Dynamic Analysis: held in Conjunction with the ACM SIGSOFT International Symposium on Software Testing and Analysis (ISSTA 2008). WODA '08, pp. 57–63. Association for Computing Machinery, New York, NY, USA (2008). https://doi.org/10.1145/1401827.1401839
27. Lahiri, S., Roy, S.: Almost correct invariants: synthesizing inductive invariants by fuzzing proofs. In: Proceedings of the 31st ACM SIGSOFT International Symposium on Software Testing and Analysis. ISSTA 2022, pp. 352–364. Association for Computing Machinery, New York, NY, USA (2022). https://doi.org/10.1145/3533767.3534381
28. Leucker, M., Schallhart, C.: A brief account of runtime verification. J. Logic Algebr. Program. **78**(5), 293–303 (2009). https://doi.org/10.1016/j.jlap.2008.08.004
29. Li, J., Sun, J., Li, L., Le, Q.L., Lin, S.W.: Automatic loop-invariant generation and refinement through selective sampling. In: Proceedings of the 32nd IEEE/ACM International Conference on Automated Software Engineering. ASE '17, pp. 782–792. IEEE Press (2017)
30. Ma, H., Goel, A., Jeannin, J.B., Kapritsos, M., Kasikci, B., Sakallah, K.A.: I4: incremental inference of inductive invariants for verification of distributed protocols. In: Proceedings of the 27th ACM Symposium on Operating Systems Principles, pp. 370–384 (2019)
31. Maimon, O.Z., Rokach, L.: Data Mining with Decision Trees: Theory and Applications, vol. 81. World Scientific (2014). https://doi.org/10.1142/9097
32. McMillan, K.L., Padon, O.: Ivy: a multi-modal verification tool for distributed algorithms. In: Lahiri, S.K., Wang, C. (eds.) CAV 2020. LNCS, vol. 12225, pp. 190–202. Springer, Cham (2020). https://doi.org/10.1007/978-3-030-53291-8_12
33. Neider, D., Madhusudan, P., Saha, S., Garg, P., Park, D.: A learning-based approach to synthesizing invariants for incomplete verification engines. J. Autom. Reason. **64**(7), 1523–1552 (2020). https://doi.org/10.1007/s10817-020-09570-z
34. Newcombe, C., Rath, T., Zhang, F., Munteanu, B., Brooker, M., Deardeuff, M.: How amazon web services uses formal methods. Commun. ACM **58**(4), 66–73 (2015). https://doi.org/10.1145/2699417
35. Padhi, S., Sharma, R., Millstein, T.: Data-driven precondition inference with learned features, pp. 42–56 (2016). https://doi.org/10.1145/2908080.2908099
36. Padon, O., McMillan, K.L., Panda, A., Sagiv, M., Shoham, S.: Ivy: safety verification by interactive generalization. In: Proceedings of the 37th ACM SIGPLAN Conference on Programming Language Design and Implementation, pp. 614–630 (2016)
37. Peterson, G.L.: Myths about the mutual exclusion problem. Inf. Process. Lett. **12**, 115–116 (1981). https://api.semanticscholar.org/CorpusID:45492619

38. Pnueli, A., Ruah, S., Zuck, L.D.: Automatic deductive verification with invisible invariants. In: Margaria, T., Yi, W. (eds.) TACAS 2001. LNCS, vol. 2031, pp. 82–97. Springer, Cham (2001). https://doi.org/10.1007/3-540-45319-9_7, http://dblp.uni-trier.de/db/conf/tacas/tacas2001.html#PnueliRZ01
39. Polgreen, E., Abboud, R., Kroening, D.: Counterexample guided neural synthesis. arXiv (2020). https://api.semanticscholar.org/CorpusID:210920369
40. Pradel, M., Bichsel, P., Gross, T.R.: A framework for the evaluation of specification miners based on finite state machines. In: 2010 IEEE International Conference on Software Maintenance, pp. 1–10. IEEE (2010). https://api.semanticscholar.org/CorpusID:6673177
41. Riley, D., Fedyukovich, G.: Multi-phase invariant synthesis. In: Proceedings of the 30th ACM Joint European Software Engineering Conference and Symposium on the Foundations of Software Engineering, pp. 607–619 (2022)
42. Ryan, G., Wong, J., Yao, J., Gu, R., Jana, S.S.: Cln2INV: learning loop invariants with continuous logic networks. arXiv (2019). https://api.semanticscholar.org/CorpusID:202749930
43. Sharma, R., Gupta, S., Hariharan, B., Aiken, A., Liang, P., Nori, A.V.: A data driven approach for algebraic loop invariants. In: Felleisen, M., Gardner, P. (eds.) ESOP 2013. LNCS, vol. 7792, pp. 574–592. Springer, Heidelberg (2013). https://doi.org/10.1007/978-3-642-37036-6_31
44. Si, X., Dai, H., Raghothaman, M., Naik, M., Song, L.: Learning loop invariants for program verification. In: Neural Information Processing Systems (2018). https://api.semanticscholar.org/CorpusID:53319040
45. Si, X., Naik, A., Dai, H., Naik, M., Song, L.: Code2Inv: a deep learning framework for program verification. In: Lahiri, S.K., Wang, C. (eds.) CAV 2020. LNCS, vol. 12225, pp. 151–164. Springer, Cham (2020). https://doi.org/10.1007/978-3-030-53291-8_9
46. Udupa, A., Raghavan, A., Deshmukh, J.V., Mador-Haim, S., Martin, M.M., Alur, R.: Transit: specifying protocols with concolic snippets. ACM SIGPLAN Not. **48**(6), 287–296 (2013)
47. Wang, J., Wang, C.: Learning to synthesize relational invariants. In: Proceedings of the 37th IEEE/ACM International Conference on Automated Software Engineering, pp. 1–12 (2022)
48. Bo, W., Sirui, L., Jiang, J.J., Xiong, Y.: Survey of dynamic analysis based program invariant synthesis techniques. J. Softw. **31**(6), 1681–1702 (2020)
49. Yang, Z., Dai, M., Guo, J.: Formal modeling and verification of smart contracts with spin. Electronics, 3091 (2022). https://doi.org/10.3390/electronics11193091
50. Yao, J., Tao, R., Gu, R., Nieh, J.: {DuoAI}: Fast, automated inference of inductive invariants for verifying distributed protocols. In: 16th USENIX Symposium on Operating Systems Design and Implementation (OSDI 22), pp. 485–501 (2022)
51. Yao, J., Tao, R., Gu, R., Nieh, J., Jana, S., Ryan, G.: DistAI: data-driven automated invariant learning for distributed protocols. In: 15th USENIX Symposium on Operating Systems Design and Implementation (OSDI 21), pp. 405–421. USENIX Association (2021). https://www.usenix.org/conference/osdi21/presentation/yao
52. Zhu, H., Magill, S., Jagannathan, S.: A data-driven CHC solver. SIGPLAN Not. **53**(4), 707–721 (2018). https://doi.org/10.1145/3296979.3192416

Verification and Model Checking

Space-Efficient Model-Checking of Higher-Order Recursion Schemes

Florian Bruse[1,2]

[1] University of Kassel, Kassel, Germany
florian.bruse@uni-kassel.de
[2] TUM School of Computation, Information and Technology,
Technical University of Munich, Munich, Germany

Abstract. Model checking trees generated by Higher-Order Recursion Schemes (HORS) of order k against Alternating Parity Tree-Automata (APT) is known to be a k-EXPTIME-complete problem (Ong'06). We exhibit a natural fragment of HORS, called tail-recursive HORS, and a restricted APT model, called bounded-alternation APT, such that the problem of model checking trees generated by order-k tail-recursive HORS against bounded-alternation APT is $k{-}1$-EXPSPACE-complete. The upper bound is achieved by converting the problem into an alternating reachability game, the lower one via reduction from a tiling problem.

Keywords: Higher-Order Recursion Schemes · Simply-Typed Lambda-Calculus · Krivine Machine · Modal Mu-Calculus

1 Introduction

Higher-Order Recursion Schemes (HORS) are higher-order grammars that generate trees. They are employed in higher-order model-checking, i.e., the verification of functional programs with higher-order functions. The program is translated into a recursion scheme, while the property to be verified is translated into an Alternating Parity Tree-Automaton (APT). The problem reduces to checking whether the APT accepts the tree generated by the HORS.

The problem and its precursors have been studied extensively, starting from Rabin's Theorem [24] and work of Damm [7] and Courcelle [6], with partial results in [13,14]. Ong's breakthrough result [23] shows that model-checking order-k HORS against APT is not only decidable, but complete for k-EXPTIME. Later, practical model checkers for the HORS model-checking problem were developed using several distinct approaches [2,3,15,22].

Ong's result inspired several other characterizations of the HORS model-checking problem, e.g., type-based [18], using collapsible pushdown automata [10] and many more. A particularly interesting characterization is that of Salvati and Walukiewicz [25,26], using the λY calculus instead of HORS, and Krivine's abstract machine [19] to define the semantics. Rather than a grammar with several nonterminals, as in the case of HORS, the λY calculus defines a single term with several recursive subterms, not unlike the modal μ-calculus, but with only one kind of fixpoint operator. Krivine's machine is then adapted to compute the semantics of such a term, i.e., a tree. This characterization is helpful as it makes visible the reduction steps that construct the tree.

While practical aspects of model-checking HORS against APT have been studied well, there still are gaps in the theoretical characterization. Kobayashi and Ong [17] exhibit fragments that have a model-checking problem whose complexity is a full exponential less than that of the standard instances. This is obtained by restricting the automaton model, i.e., the expressive power of the verification device. Beyond this result, the literature contains no further results. A natural question is whether there is a fragment of the HORS model-checking problem, i.e., natural classes of HORS and APT, such that the problem becomes more space-efficient. A fragment with a model-checking problem in, or even complete for, k-EXPSPACE would naturally fit between those that are complete for k-EXPTIME and $(k+1)$-EXPTIME, resp., constituting a balance between improved complexity of the latter and expressive power of the automaton.

There are reasons to believe that such a fragment exists, since there are links from the problem of HORS model checking to that of HFL model checking. HFL [29] is an extension of the modal μ-calculus by a simply-typed λ-calculus. The formulas of this calculus are model checked over a labeled transition system. HFL's model-checking problem is also naturally stratified by type order, and the order-k version is known to be k-EXPTIME complete [1]. Moreover, there is a family of natural fragments, called the tail-recursive fragment (of order k) s.t. the model-checking problem is $(k-1)$-EXPSPACE complete [5,20]. Additionally, there is a pair of translations that map the order-k model-checking problem for HORS to that for HFL, and vice versa [16]. Hence, a $(k-1)$-EXPSPACE complete fragment for the HORS model-checking problem should exist as the image and preimage of that for HFL under the translations.

In this paper, we characterize this fragment in a natural way, adapting the notion of tail recursion to HORS. We do this using the characterization via the λY calculus [25,26], as it naturally comes with the notion of a syntax tree. Additionally, we define a restricted class of APT, called bounded-alternation APT (baAPT), that have a simplified level of boolean alternation, but no trivial boolean alternation as those in [17]. We show that model-checking the trees generated by an order-k tail-recursive λY term can be done in $(k-1)$-EXPSPACE, using a characterization of the problem as a finite parity game. We also show that the problem is hard for $(k-1)$-EXPSPACE by encoding a tiling problem into it. This follows a pattern similar to the one in [5,16], encoding large numbers into trees in order to capture the tiling problem.

The paper is structured as follows: In Sect. 2, we recall the notions of HORS, the λY calculus, and we give an abridged version of the main points in [25,26]. In Sect. 3, we define the tail-recursive fragment, baAPT, and the associated model-checking problem. We establish the upper bound in Sect. 4 and the lower bound in Sect. 5. We conclude in Sect. 6.

2 Preliminaries

We write $[n]$ for the set $\{0, \ldots, n-1\}$. Define $2_0^n = n$ and $2_{k+1}^n = 2^{2_k^n}$.

2.1 Trees and Parity Automata

A *tree* is a set $T \subseteq \mathbb{N}^*$ that is *prefix-closed* and *left-closed*, i.e., if $vi \in T$ for $v \in \mathbb{N}^*$ then $v \in T$ and, moreover, $v(i-1) \in T$ if $i > 0$. A *tree alphabet* is a finite, nonempty Σ together with an arity function $ar \colon \Sigma \to \mathbb{N}$. We often identify the alphabet with Σ if ar is clear from context. Σ^i is the set of i-ary symbols in Σ. A Σ-tree for some finite, nonempty Σ (and its arity function ar) is a pair (T, ℓ) where T is a tree and $\ell \colon T \mapsto \Sigma$ is a *labeling function*, such that, for all $v \in T$, if $ar(\ell(t(v))) = k$ then $vk \notin T$ but $v(k-1) \in T$ if $k > 0$. We often simply write T to refer to Σ-trees, with the labeling function ℓ to be present implicitly.

Parity Games. A *parity game* $\mathcal{G} = (V, V_\exists, E, v_0, \Omega)$ is a two-player game between $\exists$ and $\forall$ where (V, E) is a directed graph, $V_\exists \subseteq V$ is the set of nodes belonging to $\exists$, $v_0 \in V$ is the starting position, and $\Omega \colon V \mapsto \mathbb{N}$ is the *priority function* which has finite range. $V_\forall$ is $V \setminus V_\exists$, and we often give parity games implicitly, i.e., by stating the successors, ownership and priorities at a given position.

A *play* in $\mathcal{G}$ is a sequence of nodes that starts in v_0 and respects E. A finite play is extended by the player owning the last node by picking a valid successor; they lose the game if no such successor exists. A *maximal* play is either infinite or one where a player is stuck in the last node. Finite maximal plays are won by the player not stuck. An infinite play $v_0, \ldots$ is won by $\exists$ iff the highest number that occurs infinitely often in the sequence $\Omega(v_0), \ldots$ is even.

Positional strategies are defined as usual. A player *wins* a game iff they have a (positional) strategy for it, which can be computed in time $\mathcal{O}(p \cdot |E| \cdot |V|^{\lfloor p/2 \rfloor})$ [12] if the game is finite and has at most p priorities.

Let $\mathcal{G} = (V, V_\exists, E, v_0, \Omega)$ be a finite parity game, i.e., one where $|V|$ is finite. Then there is some minimal n s.t. $\{0, \ldots, n\} \supseteq codom(\Omega)$. A *signature* [28] $\sigma \colon \{0, \ldots, n\} \to \{0, \ldots, |V|\}$ assigns to each priority a value between 0 and $|V|$. If $m \in codom(\Omega)$ and $\sigma(m) > 0$, define $\sigma[\downarrow m]$ as $\sigma[\downarrow m](k) = \sigma(k)$ if $m < k \le n$, $\sigma[\downarrow m](k) = \sigma(k) - 1$ if $k = m$ and $\sigma[\downarrow m](k) = |V|$ if $k < m$. Let $\sigma_\top$ be defined via $\sigma_\top(k) = |V|$ for all $0 \le k \le n$.

An *alternating reachability game* is played by the same rules as a parity game, but there are no infinite plays, whence any play necessarily ends with one player getting stuck and the other player winning the game. Hence, there are also no priorities. Any finite parity game $\mathcal{G}$ as above can be converted into an alternating reachability game $\mathcal{G}_\sigma$ that has positions of the form (v, σ) where $v \in V$ and σ is a signature. The starting position is $(v_0, \sigma_\top)$. Positions (v, σ) with $\sigma(\Omega(v)) = 0$ have no successors and belong to $V_\exists$ (i.e., they are winning for $\forall$) iff $\Omega(v)$ is even. Positions (v, σ) with $\sigma(\Omega(v)) \ne 0$ have as successors all $(v', \sigma[\downarrow \Omega(v)])$ such that $(v, v') \in E$. Such a position belongs to $V_\exists$ iff v does.

Proposition 1 ([28]). *Let $\mathcal{G}$ be a finite parity game. $\exists$ wins $\mathcal{G}$ iff she wins $\mathcal{G}_\sigma$.*

We also call such a game with signatures *finitary* since, as an alternating reachability game, there is a finite upper bound on the length of plays (as opposed to *finite* parity games, where plays can be infinite).

Tree Automata. Given some set S, let $\mathcal{B}^+(S)$ denote the set derived via the grammar $\varphi ::= s \mid \top \mid \bot \mid \varphi \vee \varphi \mid \varphi \wedge \varphi$, where $s \in S$.

An *alternating parity tree-automaton* (APT) is an $\mathcal{A} = (Q, \Sigma, \delta, q_0, \Lambda)$ where Q is a nonempty, finite set of *states*, Σ is a tree alphabet, $\delta = \bigcup_{0 \le i \le n} \delta^i$ where $n = \max\{ar(a) \mid a \in \Sigma\}$ and $\delta^i \colon Q \times \Sigma^i \to \mathcal{B}^+(Q^i)$ (where Q^i is the set of i-tuples from Q) is the *transition function*, $q_0 \in Q$ is the *starting state*, and $\Lambda \colon Q \mapsto \mathbb{N} \setminus \{0\}$ is the *priority function*. The assumption that the lowest possible priority appearing in the priority function is at least 1 unless stated otherwise is for technical reasons. The *size* of an APT is the number of its states.

A *run* of $\mathcal{A}$ on a Σ-tree T is a parity game $A(\mathcal{A}, T)$, called the *acceptance game*, with positions from $T \times (\bigcup_{0 \le i \le n} \mathcal{B}^+(Q^i))$ with n the maximal arity of Σ, and starting position (ϵ, q_0). From a position of the from (v, q), the unique successor position is $(v, \delta(q, \ell(v)))$, positions of the form $(v, \varphi_1 \vee \varphi_2)$ and $(v, \varphi_1 \wedge \varphi_2)$ have successors (v, φ_1) and (v, φ_2), and a position of the from $(v, (q_0, \ldots, q_j))$ has successors of the form $(v_0, q_0), \ldots, (vj, q_j)$. Positions of the form $(v, \varphi_1 \wedge \varphi_2)$ and $(v, (q_0, \ldots, q_j))$ and $(v, \top)$ belong to $\forall$, all other positions belong to $\exists$. Moreover, $\Omega(v, q) = \Lambda(q)$ and $\Omega(v, \varphi) = 1$ otherwise. Then $\mathcal{A}$ accepts T if $\exists$ wins $A(\mathcal{A}, T)$, and $\mathcal{A}$ accepts T *from state* q iff $\mathcal{A}_q$ accepts T where $\mathcal{A}_q$ is the APT obtained by changing the starting state of $\mathcal{A}$ to q. We assume w.l.o.g. that each APT contains states $q_\top, q_\bot$ with $\delta(q_\top, a) = \top, \delta(q_\bot, a) = \bot$ for all $a \in \Sigma$, whence $\mathcal{A}$ will accept any (sub)tree from $q_\top$ and reject any tree from $q_\bot$.

Given some APT $\mathcal{A} = (Q, \Sigma, \delta, q_0, \Lambda)$, we define its *complement* automaton $\overline{\mathcal{A}} = (\overline{Q}, \Sigma, \overline{\delta}, \overline{q_0}, \overline{\Lambda})$ via $\overline{Q} = \{\overline{q} \mid q \in Q\}$, with $\overline{\delta}$ defined via $\overline{\delta}(\overline{q}, a) = \overline{\delta(q, a)}$ and $\overline{(q_0, \ldots, q_j)} = \bigvee_{0 \le i \le j} (\overline{q_\bot}, \ldots, \overline{q_i}, \ldots, \overline{q_\bot})$, $\overline{\varphi_1 \vee \varphi_2} = \overline{\varphi_1} \wedge \overline{\varphi_2}$, $\overline{\varphi_1 \wedge \varphi_2} = \overline{\varphi_1} \vee \overline{\varphi_2}$, $\overline{\top} = \bot, \overline{\bot} = \top$ and $\overline{\Lambda}(\overline{q}) = \Lambda(q) + 1$. Note that $\overline{q_\bot}$ plays the role of $q_\top$ in $\overline{\mathcal{A}}$ and vice versa. Clearly $\mathcal{A}$ accepts T iff $\overline{\mathcal{A}}$ rejects T, since a winning strategy for $\exists$ in $A(\mathcal{A}, T)$ is one for $\forall$ in $A(\overline{\mathcal{A}}, T)$ and vice versa.

Proposition 2. *Let $\mathcal{A}$ be an APT, and let $\overline{\mathcal{A}}$ be its complement automaton. Then $\mathcal{A}$ accepts a tree T iff $\overline{\mathcal{A}}$ does reject it. Both automata have the same size.*

2.2 Higher-Order Recursion Schemes

Instead of the classical definition of Higher-Order Recursion Schemes (HORS, cf. e.g., [14]), we use a version going back to Damm [7] and re-introduced by Walukiewicz and Salvati [26], using the λY-calculus for the syntax, while using the Krivine Machine [19], adjusted as per [25] to canonical form, for the semantics. The presentation here follows [26] in part.

Types. The set of *types* is defined inductively via $\tau ::= \bullet \mid \tau \to \tau$ where $\bullet$ is the type of trees, $\bullet \to \bullet$ is the type of functions that consume a tree and yield a tree, etc. We write $\tau^i \to \bullet$ for the type $\tau \to \cdots \tau \to \bullet$ with i many copies of τ. The *order* of a type is defined via $ord(\bullet) = 0$ and $ord(\tau_1 \to \tau_2) = \max\{1 + ord(\tau_1), ord(\tau_2)\}$.

Terms. A *tree signature* Σ is a set of *tree constructors*, each of type $\bullet^i \to \bullet$ for some i. Let $\mathcal{V}$ be a set of typed λ *variables*, let $\mathcal{F}$ be a set of typed Y *variables* or recursion variables. Lower case letters $a, b, \ldots$ denote tree constructors, lower case letters $x, y, \ldots$ denote λ variables, upper case letters $F, G, \ldots$ denote Y variables. $(x\colon \tau)$, $(F\colon \tau)$ etc. indicates the type of the respective variable; we omit the type annotation if it is not important or clear from context.

Given $\Sigma, \mathcal{V}, \mathcal{F}$, the set of λY terms is defined inductively: a tree constructor of type $\bullet^i \to \bullet$ is a term of type $\bullet^i \to \bullet$, a lambda variable $x\colon \tau$ and a Y variable $F\colon \tau$ are terms of type τ. Given $x\colon \tau_1$ and a term t of type τ_2, $\lambda(x\colon \tau).\, t$ is a term of type $\tau_1 \to \tau_2$. Given terms t_1, t_2 of type $\tau_2 \to \tau_1$ and τ_2, resp., $(t_1\, t_2)$ is a term of type τ_1. If $F\colon \tau$ is a Y variable and t is a term of type τ, then $Y(F\colon \tau).\, t$ is a term of type τ. We write $type(t)$ to denote the type of some term t. Every term is a *subterm* of itself, t is a subterm of $\lambda x.\, t$ and of $YF.\, t$, and t_1 and t_2 are subterms of $(t_1\, t_2)$. Application associates to the left and binds stronger than λ abstraction and Y. We contract consecutive abstractions into the form $\lambda x, y, z.\, t$.

Free and *bound variables* are defined as usual, with x and F being bound in $\lambda x.\, t$ and $YF.\, t$. The set of free variables of t is $free(t)$, and $t[t'/x]$ denotes capture-avoiding substitution of t' for all free occurrences of x in t, and similarly for $t[t'/F]$, assuming that the types match. We also assume that each variable is bound at most once in a term whence, for closed terms, there is a function $term$ that maps each Y variable F to $term(F)$, defined as t if the unique binding of F is $YF.\, t$. The *order* of a term is that of the highest order of any of its subterms, and the size of a term is the number of its distinct subterms.

Semantics. Besides α-conversion, i.e., variable renaming, the λY calculus has β-reduction and δ-reduction. β-reduction $\to_\beta$ reduces $((\lambda x.\, t)\, t')$, where x and t' have the same type, to $t[t'/x]$. δ-reduction $\to_\delta$ reduces $YF.\, t$ to t and F to $term(F)$. This definition is slightly different from the usual one where Y is a genuine operator, but more useful when defining the semantics of a λY term. The reflexive transitive closure of $\to_\beta \cup \to_\delta$ is $\to^*_{\beta\delta}$. Since both β-reduction and δ-reduction maintain the type of a term, so does $\to^*_{\beta\delta}$. It is standard that this relation is confluent. A closed term of type $\bullet$ is in *weak head normal form* if it is of the form $a\, t_0 \cdots t_j$. Not every λY term reduces to a head normal form, consider e.g., $Y(F\colon \bullet).\, F$.

Another reduction rule often used is η-reduction. It reduces $\lambda x.\, t\, x$ to t if x does not appear freely in t. Since this rule is somewhat awkward w.r.t. typing, we assume w.l.o.g. that all terms we work with are in so-called η-*long form* [11], which requires that every term not of type $\bullet$ is either a λ abstraction or applied to some argument term.

Let Σ be a tree signature and let ω be a new nullary symbol, i.e., of type •. We define the *Böhm tree* $BT(t)$ of a closed term of type • as follows: If t does not reduce via $\to^*_{\beta\delta}$ to a weak head normal form, then $BT(t) = \omega$, i.e., the tree has just one node. Otherwise, let $a\ t_0 \cdots t_j$ be a weak head normal form of t. The root of $BT(t)$ is labeled by a, and its successors are, in order, the roots of $BT(t_0), \ldots, BT(t_j)$. Due to confluence, this definition is sound.

The HORS model-checking problem (defined via λY terms) is the following: Given a closed term t of type • over Σ and an APT $\mathcal{A}$ with alphabet $\Sigma \cup \{\omega\colon •\}$, does $\mathcal{A}$ accept $BT(t)$? For terms of order $k \geq 0$, this is known to be a k-EXPTIME complete problem [23].

2.3 The Krivine Machine

The Krivine machine [19] computes weak head normal forms of λY terms using call-by-name semantics. Given some closed λY term t_0 of type •, its constituent parts are *closures* and *environments*, which are defined by mutual recursion. A closure $C = (t, e)$ is a pair of a subterm of t_0 and an environment. Its type is that of t. An environment e is either the *empty environment* e_0 or a pair $(e', x \mapsto C)$ of another environment called its *parent environment*, and a binding of the form $x \mapsto C$. It is tacitly understood that environments form a tree-like structure with e_0 at the root, i.e., there are no circular dependencies between environments, that environments always bind variables to closures of the same type as the variable, and, finally, that the environment part of a closure contains definitions for all free variables of its subterm, either directly or through predecessor environments. The intuition here is that a closure denotes a term, obtained by recursively substituting free variables of its term component by the values indicated by the respective environment. This justifies the notation $e(x)$, which is defined as C if $e = (e', x \mapsto C)$ and as $e'(x)$ if $e = (e', y \mapsto C)$ with $x \neq y$.

A *configuration* of a Krivine machine is a pair $((t, e), \gamma)$, where (t, e) is a closure of type $\tau_1 \to \cdots \to \tau_n \to •$ (the *primary closure*) and $\gamma = C_1, \ldots, C_n$ is a *stack* of closures of types $\tau_1, \ldots, \tau_n$, read from top to bottom. Hence, such a configuration represents a term of type •, with the arguments to the primary closure stored on the stack. The initial configuration for a term t is $((t, e_0), \epsilon)$ where ϵ denotes the empty stack.

The transitions of the machine are deterministic and depend on the form of the term in the primary closure. Let $((t, e), \gamma)$ with $\gamma = C_1, \ldots, C_n$ be a configuration.

- If $t = \lambda x.\ t'$, then the machine transitions to $((t', e'), \gamma')$ where the new stack is $\gamma' = C_2, \ldots, C_n$ and where $e' = (e, x \mapsto C_1)$.
- If $t = t_1\ t_2$ with $t_1 \neq a$, then the machine transitions to $((t_1, e), \gamma')$ where $\gamma' = (t_2, e), C_1, \ldots, C_n$.
- If $t = YF.\ t'$, then the machine transitions to $((t', e), \gamma)$.
- If $t = F$, then the machine transitions to $((term(F), e), \gamma)$.
- If $t = x$, then the machine transitions to $(e(x), \gamma)$.
- If $t = a\ t_0 \cdots t_j$, then the machine halts.

It is not hard to see that, if a Krivine machine stops, necessarily in a configuration of the form $((a\ t_0 \cdots t_j, e), \epsilon)$, the term represented by its configuration is in weak head normal form. Hence, the machine can be used to define a tree. Formally, we define $KT((t,e))$ as the tree with one node labeled by ω if the machine does not halt from $((t,e), \epsilon)$, and if it does halt in $((a\ t_0 \cdots t_j, e'), \epsilon)$ as the tree whose root is labeled by a, and whose subtrees are, in order, the trees $KT((t_0, e)), \ldots, KT((t_j, e))$. It is then immediate [19,26] that $KT(t) = BT(t)$. Hence, the model-checking problem for HORS (presented as λY terms) can equivalently be explained in terms of the Krivine machine.

We define the order of a Krivine machine as the order of the term that generated it, and its size as the size of said term.

2.4 The Krivine Machine Acceptance Game

Given a Krivine Machine $\mathcal{M}$ and an APT $\mathcal{A}$, Salvati and Walukiewicz [26] present a (generally infinite) parity game $\mathcal{K}(\mathcal{A}, \mathcal{M})$ such that $\exists$ wins $\mathcal{K}(\mathcal{A}, \mathcal{M})$ iff $\mathcal{A}$ accepts $KT(\mathcal{M})$. We present it here in order to build intuition. Let $\mathcal{A} = (Q, \Sigma, \delta, q_0, \Lambda)$. The game contains positions of the two forms $(q, (t, e), \gamma)$ and $(q, \varphi, (a\ t_0 \ldots t_j, e), \epsilon)$ where $q \in Q$, $\varphi \in \bigcup_{0 \leq i \leq n} \mathcal{B}^+(Q^i)$ and n is the maximal arity in the tree alphabet in question, (t, e) is a closure, and γ is a stack of closures. The starting position is $(q_0, (t_0, e_0), \epsilon)$ where q_0 is the starting state of $\mathcal{A}$ and $((t_0, e_0), \epsilon)$ is the starting configuration of $\mathcal{M}$. The successors of a position of the form $(q, (t, e), \gamma)$ with $\gamma = C_1, \ldots, C_n$ depend on the form of t:

- A node with $t = \lambda x.\ t'$ has one successor $(q, (t', e'), C_2, \ldots, C_n)$ where $e' = (e, x \mapsto C_1)$.
- A node with $t = t_1\ t_2$ with $t_1 \neq a$ has one successor $(q, (t_1, e), (t_2, e) \cdot \gamma)$.
- A node with $t = YF$. t' has one successor $(q, (t', e), \gamma)$.
- A node with $t = F$ has one successor $(q, (term(F), e), \gamma)$.
- A node with $t = x$ has one successor $(q, e(x), \gamma)$.
- A node with $t = a\ t_0 \cdots t_j$ and, hence, with $\gamma = \epsilon$, has a unique successor $(q, \delta(q, a), (a\ t_0 \cdots t_j, e), \epsilon)$.

Moreover, nodes of the form $(q, \varphi, (a\ t_0 \cdots t_j), \epsilon)$ have successors as follows:

- A node with $\varphi = \varphi_1 \vee \varphi_2$ or $\varphi_1 \wedge \varphi_2$ has successors $(q, \varphi_i, (a\ t_0 \cdots t_j), \epsilon)$ for $i \in \{1, 2\}$.
- A node with $\varphi = (q_0, \ldots, q_j)$ has successors $(q, q_i, (t_i, e), \epsilon)$ for $i \in \{0, \ldots, j\}$.

The only positions not in $V_\exists$ are those of the form $(q, \varphi_1 \wedge \varphi_2, (a\ t_0 \cdots t_j), \epsilon)$ and $(q, (q_0, \ldots, q_j), (a\ t_0 \cdots t_j, e), \epsilon)$. The priority labeling of $(q, \varphi, (a\ t_0 \cdots t_j), \epsilon)$ is $\Omega(q)$. Recall that the lowest priority assigned by Λ is at least 1. The definition of the parity game is completed by assigning priority 1 to all nodes of the form $(q, (t, e), \gamma)$. The following is established in [26].

Proposition 3. $\exists$ wins $\mathcal{K}(\mathcal{A}, \mathcal{M})$ iff $\mathcal{A}$ accepts $KT(\mathcal{M})$.

Note that, since the lowest priority of $\mathcal{A}$ is 1, plays following a run of $\mathcal{M}$ on which no constant is seen eventually will only visit positions with parity 1 and, hence, will be rejecting. This mirrors diverging runs of $\mathcal{M}$ which would generate a Böhm tree of the form ω. For the remainder of the paper, we will drop ω and assume that all HORS define trees where ω does not occur. This restriction can easily be lifted by re-adding ω at the appropriate places.

A Finite Parity Game. Salvati and Walukiewicz follow up the presentation of $\mathcal{K}(\mathcal{A}, \mathcal{M})$ by an equivalent finite parity game $\mathcal{G}(\mathcal{A}, \mathcal{M})$, i.e., a game that is won by the same player. In order to make the game finite, parts of the game are converted from a syntactic characterization into a semantic one: In a configuration of the form $(q, (t_1 t_2, e), \gamma)$, rather than moving a closure containing t_2 onto the stack, a semantic representation of (t_2, e) is moved there, including information regarding q and the expected behavior of the game w.r.t. priorities up to the point when the closure involving t_2 would have been passed to as part of variable lookup (this, of course, requires this closure to be bound to a variable first).

The idea here is this: The object (called a residual R) that is actually put on the stack instead of the closure (t_2, e) contains certain promises to player $\exists$ that she wins the game if she reaches this residual with the correct state. Reaching R means that R is eventually moved from the stack and bound to some variable in some environment, and that the play then reaches a configuration of the form $(q, (x, e'), \gamma)$ where $e(x)$ yields R.

For the moment assume that the type of t_2 is $\bullet$. Then R is a set of pairs of states and priorities. If the play actually reaches the R, i.e., if it reaches a position $(q', (x, e'), \gamma)$ where the closure (t_2, e) would have been returned by the variable lookup $e'(x)$ in $\mathcal{K}(\mathcal{A}, \mathcal{M})$, then q and the highest priority seen since R was created are compared to the set of such pairs contained in R. If the pair (q', p), where p is the highest priority seen in the play since R was created, is contained[1] in R, then $\exists$ wins. Otherwise, $\forall$ wins. This formalizes the promise that, if $(q, p) \in R$, then $\exists$ would have won the game from any successor where a position $(q', (x, e'), \gamma)$ is reached with $e'(x) = R$ and where the highest priority seen since the creation of R is p. The concept is then generalized to higher-order closures and residuals (cf. [26]).

It remains to explain how the residuals created in a configuration of the form $(q, (t_1 t_2, e), \gamma)$ are obtained. This is done via a classical pattern in such a game: $\exists$ proposes a residual, and $\forall$ can either accept it or challenge it. In the latter case the play continues by verifying the residual: $\forall$ choses a list of arguments to the residual (which are residuals itself), and the play is continued from some position $(q'', (t_2, e''), \gamma')$ where γ' is a completely new stack containing the list of arguments given by $\forall$, and $\exists$ is now challenged to show that the machine would accept from such a position. This pattern of application via alternation is also used in the study of temporal logics [1]. We now define residuals formally.

---

[1] Technically, R is manipulated during the game rather than storing the highest priority seen since the creation of R, but the effect is the same.

Definition 1 (Residuals [26]). Let $\mathcal{A} = (Q, \Sigma, \delta, q_0, \Lambda)$ and let $d = 1 + \max\{\Lambda(q) \mid q \in Q\}$. The set of residuals $\mathcal{D}_\tau$ for a given type $\tau = \tau_1 \to \cdots \to \tau_n \to \bullet$ is defined as $\mathcal{D}_{[\tau_1 \to \cdots \to \tau_n \to \bullet]} = \mathcal{D}_{[\tau_1]} \to \cdots \to \mathcal{D}_{[\tau_n]} \to \mathcal{P}(Q \times [d])$. Given a residual $R \colon \tau_1 \to \cdots \to \tau_n \to \bullet$ and some $1 \leq d' \leq d$, we define the lifting of R by d', written $R{\upharpoonright}_{d'}$ as the function

$$R{\upharpoonright}_{d'} \colon S_1, \ldots, S_n \mapsto = \{(q, d'') \in R(S_1, \ldots, S_n) \mid d'' \geq d'\}$$
$$\cup \{(q, d'') \mid \ ex. \ (q, d') \in R(S_1, \ldots, S_n), d'' \leq d'\}.$$

Note that, since in $\mathcal{G}(\mathcal{A}, \mathcal{M})$ environments do not store closures but residuals, we have to update our definition of an environment. An environment, for the purposes of $\mathcal{G}(\mathcal{A}, \mathcal{M})$, is either the empty environment e_0, or has the form $e = (e', x \mapsto R)$ where R is a residual of matching type. The other stipulations w.r.t. well-definedness of environments stay put. The *lifting* $e_0{\upharpoonright}_{d'}$ of the empty environment is the empty environment. For $e_0 \neq e = (e', x \mapsto R)$, the *lifting* $e{\upharpoonright}_{d'}$ is defined as $(e'{\upharpoonright}_{d'}, x \mapsto R{\upharpoonright}_{d'})$. The size of residuals is as follows [26].

Observation 1. *A residual in $\mathcal{D}_{[\tau]}$ has k-fold exponential size if $ord(\tau) = k$.*

We are now ready to define $\mathcal{G}(\mathcal{A}, \mathcal{M})$, which we are doing in the same style as for $\mathcal{K}(\mathcal{A}, \mathcal{M})$. It has three types of positions: those of the form $(q, (t, e), \gamma)$, those of the form $(q, \varphi, (t, e), \gamma)$ and those of the from $((q, R), (t, e), \gamma)$. In each case, γ is now a stack of residuals. For nodes of the form $(q, (t, e), \gamma)$ with $\gamma = R_1, \ldots, R_n$, the successors depend on the form of t:

- A node with $t = \lambda x.\ t'$ has one successor $(q, (t', e'), R_2, \ldots, R_n)$ where $e' = (e, x \mapsto R_1)$.
- A node with $t = YF.\ t'$ has one successor $(q, (t', e), \gamma)$.
- A node with $t = F$ has one successor $(q, (term(F), e), \gamma)$.
- A node with $t = a\ t_0 \cdots t_j$ and, hence $\gamma = \epsilon$, has a unique successor $(q, \delta(q), (a\ t_0 \cdots t_j, e), \epsilon)$.

All these nodes belong to $V_\exists$ and have priority 1.

A node of the form $(q, \varphi, (a\ t_0 \cdots t_j, e), \epsilon)$ has priority $\Omega(q)$ and successors depending on the form of φ:

- A node with $\varphi = \varphi_1 \vee \varphi_2$ or $\varphi = \varphi_1 \wedge \varphi_2$ has successors $(q, \varphi_i, (a\ t_0 \cdots t_j), e), \epsilon)$ for $i \in \{1, 2\}$. It belongs to $V_\exists$, resp. $V_\forall$.
- A node with $\varphi = (q_0, \ldots, q_j)$ belongs to $V_\forall$. For each $i \in \{0, \ldots, j\}$, the position $(q_i, (t_i, e\ {\upharpoonright}_{\Lambda(q)}), \epsilon)$ is a successor.

Moreover, applications and variable lookup are handled as follows:

- A node with $t = t_1 t_2$ and $t_1 \neq a$ and t_1 of type τ has one successor $((q, R), (t_1 t_2, e), \gamma)$ for every residual in $R \in \mathcal{D}_\tau$.
- A position of the form $((q, R), (t_1 t_2, e), \gamma)$ belongs to $V_\forall$ and has one successor of the form $(q, (t_1, e), R\ {\upharpoonright}_{\Lambda(q)} \cdot \gamma)$ and, if the type of R is $\tau_1 \to \cdots \to \tau_n \to \bullet$, one successor of the form $(q', (t_2, e\ {\upharpoonright}_{d'}), R_1, \ldots, R_n)$ for every $R_1 \in \mathcal{D}_{\tau_1}, \ldots, R_n \in \mathcal{D}_{\tau_n}$ and $(q', d') \in R\ {\upharpoonright}_{\Lambda(q)}(R_1, \ldots, R_n)$. The priority of a successor of the form $(q', (t_2, e{\upharpoonright}_{d'}), R_1, \ldots, R_n)$ is d'.

– A node of the from $(q,(x,e),R_1,\ldots,R_n)$ has no successors. It belongs to $V_\exists$ iff $(q,\Lambda(q)) \notin e(x)(R_1,\ldots,R_n)$. Hence, it is winning for $\exists$ if $(q,\Lambda(q)) \in e(x)(R_1,\ldots,R_n)$.

All nodes of the last form have priority 1.

Proposition 4 ([26]). *Let $\mathcal{A}$ be an APT and let $\mathcal{M}$ be a Krivine machine. Then $\exists$ wins $\mathcal{G}(\mathcal{A},\mathcal{M})$ iff she wins $\mathcal{K}(\mathcal{A},\mathcal{M})$. Moreover, computing the winner of $\mathcal{G}(\mathcal{A},\mathcal{M})$ is k-EXPTIME complete where k is the order of $\mathcal{M}$.*

Note that the complexity result does require some changes to the way Krivine machines work. These can be found in [25], we have incorporated them here.

Clearly, the previous result is also compatible with APT complementation:

Observation 2. $\exists$ *wins* $\mathcal{G}(\mathcal{A},\mathcal{M})$ *iff* $\forall$ *wins* $\mathcal{G}(\overline{\mathcal{A}},\mathcal{M})$.

3 Tail Recursion

Tail recursion in the context of programming refers to recursive definitions of functions in which the recursive call happens last, i.e., in which the result of a call to a recursive function is not manipulated any further. The classical definition of the factorial function via $\mathsf{fac}(1) = 1, \mathsf{fac}(n+1) = n \cdot \mathsf{fac}(n)$ is not tail recursive, since the result of the recursive call $\mathsf{fac}(n)$ is multiplied by n to yield the result of $\mathsf{fac}(n+1)$. Hence, the recursive call appears in an *argument position*. Conversely, a definition of the greatest common divisor via $\gcd(a,b) = a$ if $b = 0$ and $\gcd(a,b) = \gcd(b, a \bmod b)$ is tail recursive.

The concept of tail recursion can also be defined in the context of higher-order verification, cf. [4,5,20]. A defining feature of tail recursion in this context is that it reduces problems characteristic of time to those characteristic of space, but with one less exponential.

We introduce the concept to the world of HORS, presented in the form of λY terms. The concept is easily translated into the classical form of presentation. A notable difference from classical tail recursion in the context of programming is that tail recursion in the area of higher-order verification not only makes restrictions w.r.t. recursion in argument positions, but also restricts boolean alternation. This is not surprising since space-bounded problems with unrestricted boolean alternation become alternating space-bounded problems, i.e., time-bounded problems with one added exponential of complexity. In the context of model-checking the trees generated by HORS, restricting boolean alternation means tightening the definition of APT to some form that contains less inherent alternation than the classical form. There, the actual transition at some tree node is chosen nondeterministically, but the automaton has to accept on all subsequent paths, which constitutes universal nondeterminism.

Fix some alphabet Σ of tree constructors. For technical reasons that will become clear later, we partition it into two sets Σ_r and Σ_u where all tree constructors in Σ_r are binary.

Bounded-Alternation Parity Automata. Let $\mathcal{A} = (Q, \Sigma, \delta, q_0, \Lambda)$ be an APT such that Q is partitioned into sets $\{q_\top, q_\bot\} \cup Q_\mathsf{u}, Q_1, \ldots, Q_m$ for some m.

Let $q \in Q_j$ for some $1 \leq j \leq m$ and let $a \in \Sigma^2$. We say that $\mathcal{A}$ is *branching* at q and a if $\delta(q,a) = (q_1, q_\top) \vee (q_\top, q_2)$ with $q_1, q_2 \in Q_j \cup \{q_\top, q_\bot\}$ or $\delta(q,a) = (q_1, q_2)$ with $q_2 \in Q_j \cup \{q_\top, q_\bot\}$ and $q_1 \in Q_{j-1} \cup \cdots \cup Q_1 \cup \{q_\top, q_\bot\}$.

We say that it is *universal* at q and a if $\delta(q,a) = (q_1, q_2)$ with both $q_1, q_2 \in Q_j \cup \{q_\top, q_\bot\}$, or $\delta(q,a) = (q_1, q_\top) \vee (q_\top, q_2)$ with $q_1 \in Q_{j-1} \cup \cdots \cup Q_1 \cup \{q_\top, q_\bot\}$ and $q_2 \in Q_j \cup \{q_\top, q_\bot\}$.

$\mathcal{A}$ is called a *bounded-alternation APT* (baAPT) if it satisfies the following conditions:

1. Q is partitioned into sets $\{q_\top, q_\bot\}, Q_\mathsf{u}, Q_1, \ldots, Q_m$ as per above.
2. For each $1 \leq i \leq m$, each Q_i is labeled as either branching or universal. Q_u is not labeled.
3. If Q_i is labeled as branching, then for each $q \in Q_i$ and for each $a \in \Sigma_\mathsf{r}$, $\mathcal{A}$ is branching at q and a. If Q_i is labeled as universal, then for each $q \in Q_i$ and for each $a \in \Sigma_\mathsf{r}$, $\mathcal{A}$ is universal at q and a.
4. For each $q \in Q$ and $a \in \Sigma_\mathsf{u}^i$, we have that $\delta(q,a) \in \mathcal{B}^+(Q_\mathsf{u}^i)$.
5. For each $q \in Q_\mathsf{u}$ and $a \in \Sigma_\mathsf{r}^i$, we have that $\delta(q,a) \in \mathcal{B}^+((Q \setminus Q_\mathsf{u})^i)$.

The intuition here is as follows: The acceptance problem of a baAPT in which each state is branching or in which each state is universal can be reduced to a one-player game. The acceptance problem for such a baAPT with empty set Q_u is a bounded-alternation game. For general baAPT, the reduction to a bounded-alternation game is still possible if offending states are in a set of lower index and there is an a priori bound on the number of times a play passes through states in Q_u. The following definition of tail-recursive HORS yields this bound.

In Proposition 2 we have seen that APT are easy to complement. This holds for baAPT, too, but requires a little more work.

Lemma 1. *Let $\mathcal{A}$ ba a baAPT and let $\overline{\mathcal{A}}$ be its complement. Then $\overline{\mathcal{A}}$ can be converted into an cquivalent baAPT in linear time.*

Proof. The conversion consists of minor adaptions to the transition function. The partition of $\overline{\mathcal{A}}$ is the same as for $\mathcal{A}$, except that universal sets become branching and vice versa.

Let $q \in Q_j$. Obviously, if $\delta(q,a)$ is branching then $\overline{\delta}(\overline{q},a)$ is not necessarily universal. There are two cases: either $\delta(q,a) = (q_1, q_\top) \vee (q_\top, q_2)$ whence $\overline{\delta}(\overline{q},a) = ((\overline{q_1}, \overline{q_\bot}) \vee (\overline{q_\bot}, \overline{q_\top})) \wedge ((\overline{q_\top}, \overline{q_\bot}) \vee (\overline{q_\bot}, \overline{q_2}))$ which can be simplified to universal $(\overline{q_1}, \overline{q_2})$. Or $\delta(q,a) = (q_1, q_2)$ with $q_2 \in Q_j \cup \{q_\top, q_\bot\}$ and $q_1 \in Q_{j-1} \cup \cdots \cup Q_1 \cup \{q_\top, q_\bot\}$. Then $\overline{\delta}(\overline{q},a) = (\overline{q_1}, \overline{q_\bot}) \vee (\overline{q_\bot}, \overline{q_2})$, which is also universal. Similar reasoning holds for the complement of universal transition functions.

baAPT intuitively work similar to weak automata [21] in that their state sets are partitioned. However, instead of the *parity* of the states in a set, the boolean alternation of the automaton is restricted in the sense that all states in a set behave either like in a (restricted version of) nondeterministic automaton or the or the complement of one, with the exception of the set Q_u.

The intuition in terms of the modal μ-calculus is that a formula in negation normal form to be translated into a baAPT is one derived via the following:

- A formula that contains no conjunctions and no modal boxes (i.e., only disjunctions and modal diamonds) can be translated into a baAPT with only branching states.
- A formula that contains no disjunctions and no modal diamonds (i.e., only conjunctions and modal boxes) can be translated into a baAPT with only universal states.
- *closed* formulas of the above types may be nested, i.e., fixpoints from one sort of formula cannot appear freely in the other type.

The first two kinds correspond to the sets labeled as branching, resp. universal in a baAPT, while the nesting without mutual recursion reflects the condition of passing to a set of lower index when branching in a universal set, or vice versa, in a baAPT. It is not hard to see that the model-checking problem of such formulas of the μ-calculus constitutes a bounded-alternation model-checking problem. Note that bounded alternation for space-bounded problems can be eliminated at no cost via repeated applications of Savitch's Theorem [27].

Finally, the above restrictions can be lifted further by introducing additional subformulas that have no restrictions beyond that (I) they may occur only guarded by a modal operator parameterized by a symbol from Σ_u, and that (II) any free occurrences of fixpoint variables from the above types must occur guarded by a model operator parameterized by a symbol from Σ_r, reflecting the last two conditions in the definition of a baAPT. Note that formulas of this form do not yield a bounded-alternation model-checking problem on arbitrary trees. However, over trees generated by tail-recursive HORS (to be defined next), the problem is still bounded-alternation.

Tail-Recursive Higher-Order Recursion Schemes. Let t be a term of the λY calculus over $\Sigma_r \cup \Sigma_u$, $\mathcal{V}$ and $\mathcal{F}$. It is called *tail-recursive* if it satisfies the following conditions:

1. For all subterms of t of the form $t_1\ t_2$, the operand-side subterm t_2 has no free $\mathcal{F}$ variables.
2. For all subterms of the form $a\ t_0 \cdots t_j$ with $a \in \Sigma_u$, none of the t_i has free $\mathcal{F}$ variables.

Note that there are no restrictions w.r.t. variables in $\mathcal{V}$. The notion of tail recursiveness is not related to that of safety [14].

Definition 2. *Let Σ be partitioned into $\Sigma_r \cup \Sigma_u$. Let t be a closed tail-recursive term of type $\bullet$ in the λY calculus over Σ, and let $\mathcal{A}$ be a baAPT. The problem of tail-recursive HORS model checking is to decide whether $\mathcal{A}$ accepts $BT(t)$.*

Example 1. Let $\Sigma = \{a^{(2)}, e^{(2)}\} \cup \{u^{(1)}\}$ be an alphabet partitioned into $\Sigma_r \cup \Sigma_u$, with the superscripts indicating arities of the symbols. Consider the APT

$\mathcal{A} = (\{q_a, q_e^0, q_e^1, q_u, q_\top, q_\bot\}, \{x, u^{(2)}\}, \delta, q_a, \Lambda)$ where $\Lambda(q_e^0) = \Lambda(q_1^e) = 1$ and $\Lambda(q_a) = \Lambda(q_u) = \Lambda(q_\top) = \Lambda(q_\bot) = 2$ and

$\delta(q_a, e) = (q_e^0, q_\top) \vee (q_\top, q_a)$ $\delta(q_a, a) = (q_a, q_a)$ $\delta(q_a, u) = (q_u)$

$\delta(q_e^0, e) = (q_e^1, q_\top) \vee (q_\top, q_e^1)$ $\delta(q_e^0, a) = (q_\bot, q_\bot)$ $\delta(q_e^0, u) = (q_u)$

$\delta(q_e^1, e) = (q_e^0, q_\top) \vee (q_\top, q_e^0)$ $\delta(q_e^1, a) = (q_\bot, q_\bot)$ $\delta(q_e^1, u) = (q_\bot)$

$\delta(q_u, e) = (q_e^0)$ $\delta(q_u, a) = (q_a)$ $\delta(q_u, u) = (q_\bot)$

It is a baAPT with $\{q_a\} = Q_2$ being universal, $\{q_e^0, q_e^1\} = Q_1$ being branching, and $\{q_u\} = Q_u$. Let P be the following property of trees: for each sequence of as starting from the root, there is a subtree of es such that, on the rightmost path through that subtree, at each node starts a maximal path to the left in this subtree with an even number of es, and ending in a u. Then $\mathcal{A}$ verifies that the tree has this property recursively, i.e., that it holds again after the u.

For more examples of tail-recursive λY terms and baAPT, see Sect. 5.

Theorem 3. *The problem of tail-recursive HORS model checking (i.e., against a baAPT) for terms of order $k > 0$ is complete for $(k-1)$-EXPSPACE.*

The upper bound is by Theorem 5, the lower bound is by Theorem 6.

4 The Upper Bound

We now show that the problem of tail-recursive HORS model-checking for recursion schemes of order k is in $(k-1)$-EXPSPACE.

Let Σ be partitioned into $\Sigma_u \cup \Sigma_r$, let $\mathcal{A}$ be a baAPT and let t be a tail-recursive λY term of order $k \geq 0$, both over Σ. We are converting $\mathcal{G}(\mathcal{A}, \mathcal{M})$, where $\mathcal{M}$ is the Krivine machine for t, into a finitary reachability game $\mathcal{G}_\sigma(\mathcal{A}, \mathcal{M})$ using signatures, and then we give a bounded-alternation $(k-1)$-EXPSPACE algorithm to solve this game.

By Lemma 1, a baAPT is easily complemented. Recall that $\overline{\mathcal{A}}$ accepts some (sub)tree T from state $\overline{q}$ iff $\mathcal{A}$ rejects the same subtree from state q. Hence $\exists$ wins the acceptance game $A(\mathcal{A}, T)$ from some position iff $\forall$ wins the acceptance game $A(\overline{\mathcal{A}}, T)$ from the same position, but with q replaced by $\overline{q}$. We write q for $\overline{\overline{q}}$, i.e., we identify the complement of $\overline{\mathcal{A}}$ with $\mathcal{A}$.

Recall that, for each finite parity game $\mathcal{G}$, there is a finitary game $\mathcal{G}_\sigma$ where the winning player is not decided via the parity condition, but via alternating reachability, i.e., which player exhausts "their" signature first. In particular, there is the finitary variant $\mathcal{G}(\mathcal{A}, \mathcal{M})$ of $\mathcal{G}_\sigma(\mathcal{A}, \mathcal{M})$. Moreover, recall that, by Observation 2, $\exists$ wins $\mathcal{G}(\mathcal{A}, \mathcal{M})$ and, hence, $\mathcal{G}_\sigma(\mathcal{A}, \mathcal{M})$, iff $\forall$ wins $\mathcal{G}_\sigma(\overline{\mathcal{A}}, \mathcal{M})$ and $\mathcal{G}(\overline{\mathcal{A}}, \mathcal{M})$. This carries over to single positions, assuming the signature, environments and residuals are updated for the shift in priorities. We now define this.

Definition 3. *Let $\mathcal{A}$ be an APT and let $\mathcal{M}$ be a Krivine machine over matching Σ. Let R, e, σ be a residual, an environment and a signature in $\mathcal{G}_\sigma(\mathcal{A}, \mathcal{M})$. Their matching residual, signature and environment $R_\rightarrow, e_\rightarrow, \sigma_\rightarrow$ in $\mathcal{G}_\sigma(\overline{\mathcal{A}}, \mathcal{M})$ are defined via:*

- $R_\to : S_1, \ldots, S_n = \{(q, d+1) \mid (q,d) \in R(S_1, \ldots, S_n)\}$,
- $(e_0)_\to = e_0$
- $e_\to = (e'_\to, x \mapsto R_\to)$ if $e = (e', x \mapsto R)$,
- $\sigma_\to(p+1) = \sigma(p)$ and $\sigma_\to(0) = 0$.

Conversely, let R be a residual, e an environment and σ a signature in $\mathcal{G}_\sigma(\overline{\mathcal{A}}, \mathcal{M})$. Their matching residual, signature and environment $R_\leftarrow, e_\leftarrow, \sigma_\leftarrow$ in $\mathcal{G}_\sigma(\mathcal{A}, \mathcal{M})$ are defined via:

- $R_\leftarrow : S_1, \ldots, S_n = \{(q, d) \mid (q, d+1) \in R(S_1, \ldots, S_n)\}$,
- $(e_0)_\leftarrow = e_0$
- $e_\leftarrow = (e'_\leftarrow, x \mapsto R_\leftarrow)$ if $e = (e', x \mapsto R)$,
- $\sigma_\leftarrow(p) = \sigma(p+1)$.

Given some fixed $\mathcal{A}$, we define $R_\leftrightarrow, e_\leftrightarrow$ and $\sigma_\leftrightarrow$ to denote $R_\to, e_\to$ and $\sigma_\to$ if R, e and σ are defined for $\mathcal{G}_\sigma(\mathcal{A}, \mathcal{M})$, and $R_\leftarrow, e_\leftarrow$ and $\sigma_\leftarrow$ if these are defined for $\mathcal{G}_\sigma(\overline{\mathcal{A}}, \mathcal{M})$.

The following is not hard to see, yet tedious to verify:

Observation 4. Let $\mathcal{A}, \mathcal{M}$ be an APT and a matching Krivine machine over matching Σ. $\exists$ wins from $(q, (t, e), R_1, \ldots, R_n), \sigma)$ in $\mathcal{G}_\sigma(\mathcal{A}, \mathcal{M})$ iff $\forall$ wins from $((\overline{q}, (t, e_\leftrightarrow), R_{1\leftrightarrow}, \ldots, R_{n\leftrightarrow}), \sigma_\leftrightarrow)$ in $\mathcal{G}_\sigma(\overline{\mathcal{A}}, \mathcal{M})$.

The upper bound is obtained via procedure MC in Algorithm 1. We show its correctness first.

Lemma 2. Let Σ be partitioned into $\Sigma_u \cup \Sigma_r$, let $\mathcal{A} = (Q, \Sigma, \delta, q_0, \Lambda)$ be a baAPT and let t be a tail-recursive λY term, both over Σ. Then $MC(q_0, q_0, (t, e_0), \epsilon, \sigma_\top)$ in Algorithm 1 terminates and returns true iff $\mathcal{A}$ accepts $BT(t)$.

It remains to show the complexity result, i.e., that $MC(q_0, (t, e_0), \epsilon, \sigma_\top)$ is a k–1-EXPSPACE procedure for t of order k. Towards this, we define the *recursion depth* of a position $(q, \varphi, (t, e), R_1, \ldots, R_n, \sigma)$ and, as an auxiliary, the *operand depth* of a tail-recursive term t.

Definition 4. Let t be a tail-recursive term over $\Sigma = \Sigma_u \cup \Sigma_r$. The operand depth of t is defined inductively, depending on the form of t:

- If t is $F \in \mathcal{F}$ or $x \in \mathcal{V}$, then the operand depth of t is 0.
- If t is $\lambda x. t'$ or $YF. t'$ then the operand depth is the operand depth of t'.
- If t is $a\ t_0\ t_1$, and $a \in \Sigma_r$, then the operand depth of t is the maximum of the operand depths of t_0 and t_1.
- If t is $a\ t_0 \cdots t_j$ and $a \in \Sigma_u$, then the operand depth of t is 1 plus the maximum of the operand depths of the t_i.
- If t is $t_1\ t_2$ with $t_1 \neq a$, then the operand depth of t is the maximum of the operand depth of t_1 and one plus the operand depth of t_2.

We now define the *recursion depth* of a position $(q, \varphi, (t, e), R_1, \ldots, R_n, \sigma)$ in $\mathcal{G}_\sigma(\mathcal{A}, \mathcal{M})$. Let $\mathcal{A}$ be partitioned into $Q_u, Q_1, \ldots, Q_m$. Then the recursion depth of $(q, \varphi, (t, e), R_1, \ldots, R_n, \sigma)$ is $k \cdot m$ if $q \in Q_u$ and $k \cdot m + m'$ if $q \in Q_{m'}$.

Algorithm 1. Order-k tail-recursive model-checking in $(k-1)$-exponential space.

```
 1: procedure MC((q, φ, (t, e), R₁, ..., Rₙ, σ))
 2:   if φ = q' then
 3:     if q' = q⊤ then return true
 4:     if q' = q⊥ then return false
 5:     if σ(Ω(q)) = 0 and Ω(q) is even then return false
 6:     if σ(Ω(q)) = 0 and Ω(q) is odd then return true
 7:   case (φ, t) of
 8:     (q, λx. t'):    return MC(q, q, (t', (e, x ↦ R₁)), R₂, ..., Rₙ, σ)
 9:     (q, YF. t'):    return MC(q, q, (t', e), R₁, R₂, ..., Rₙ, σ)
10:     (q, F):         return MC(q, q, (term(F), e), R₁, ..., Rₙ, σ)
11:     (q, a t₀ ··· tⱼ): return MC(q, δ(q, a), (at₀ ··· tⱼ, e), ϵ, σ[↓ Ω(q)])
12:     ((q₁, q⊤) ∨ (q⊤, q₂), a t₀ t₁):
13:       guess i ∈ {1, 2}
14:       return MC(qᵢ, qᵢ, (tᵢ, e), ϵ, σ)
15:     ((q₀, q₁), a t₀ t₁):                                    ▷ a ∈ Σᵣ, q branching
16:       b ← MC(q₀, q₀, (t₀, e), ϵ, σ)
17:       if b = false then return false
18:       return MC(q₁, q₁, (t₁, e), ϵ, σ)
19:     ((q₀, q₁), a t₀ t₁):                                    ▷ a ∈ Σᵣ, q universal
20:       bᵢ ← MC(q̄ᵢ, q̄ᵢ, (tᵢ, e↔), ϵ, σ↔)                      ▷ i ∈ {1, 2}
21:       if b₀ = true or b₁ = true then return false
22:       return true
23:     ((q₀, q⊤) ∨ (q⊤, q₁), a t₀ t₁):                         ▷ a ∈ Σᵣ, q universal
24:       bᵢ ← MC(q̄ᵢ, q̄ᵢ, (tᵢ, e↔), ϵ, σ↔)                      ▷ i ∈ {1, 2}
25:       if b₀ = true and b₁ = true then return false
26:       return true
27:     (φ₁ ∨ φ₂, a t₀ t₁):                                     ▷ a ∈ Σᵤ
28:       bᵢ ← MC(q, φᵢ, (tᵢ, e), ϵ, σ)                         ▷ for both i ∈ {1, 2}
29:       if b₁ = b₂ = false then return false
30:       return true
31:     (φ₁ ∧ φ₂, a t₀ t₁):                                     ▷ a ∈ Σᵤ
32:       bᵢ ← MC(q, φᵢ, (tᵢ, e), ϵ, σ)                         ▷ for both i ∈ {1, 2}
33:       if b₀ = b₁ = true then return true
34:       return false
35:     ((q₀, ..., qⱼ), a t₀ ··· tⱼ):                           ▷ a ∈ Σᵤ
36:       for i ∈ {0, ..., j} do
37:         if qᵢ is branching then bᵢ ← MC(qᵢ, qᵢ, (tᵢ, e), ϵ, σ)
38:         else bᵢ ← (1 − MC(q̄ᵢ, q̄ᵢ, (tᵢ, e↔), ϵ, σ↔))
39:       if b₀ = ··· = bⱼ = true then return true
40:       return false
41:     (q', t₁t₂) with t₁ ≠ a:
42:       τ ← type(t₂)                                          ▷ τ = τ₁ → ··· τₘ → •
43:       guess R ∈ Dτ
44:       for all (R'₁, ..., R'ₘ) ∈ Dτ₁ × ··· × Dτₘ do
45:         for all (q'', d') ∈ R ↾_{Λ(q)} do
46:           b ← MC(q'', q'', (t₂, e ↾_{d'}, R'₁, ..., R'ₘ, σ⊤)
47:           if b = false then return false
48:       return MC(q', q', (t₁, e), R, R₁, ..., Rₙ, σ)
49:     (q', x):
50:       if (q', Λ(q')) ∈ e(x)(R₁, ..., Rₙ) then return true
51:       else return false
52:   end case
```

The notion of recursion depth is important as non-tail-recursive calls in Algorithm 1 reduce the recursion depth of their argument. This bounds the nesting depth of such calls and, hence, the total space needed by the algorithm. Also note that the implicit alternation in calls to complement states also happens only in situations where the recursion depth decreases. Hence, this alternation is bounded and can be eliminated by repeated applications of Savitch's Theorem [27].

Theorem 5. *Algorithm 1 runs in $(k-1)$-EXPSPACE for order-k tail-recursive λY terms and matching baAPT. Hence, the order-k tail-recursive HORS model-checking problem is in $(k-1)$-EXPSPACE.*

5 The Lower Bound

5.1 Tiling Problems

A *tiling system* is a $\mathcal{K} = (T, H, V, t_I, t_\square, t_F)$ where T is a finite set of *tiles*, $H, V \subseteq T \times T$ are the *horizontal* and *vertical matching relations*, and $t_I, t_\square, t_F$ are the *initial*, *blank* and *final tiles*. The *order-k corridor tiling problem* consumes as input a tiling system $\mathcal{T}$ and some $n \in \mathbb{N}$ encoded unarily. It asks whether there is a sequence $\rho_0, \ldots, \rho_{m-1}$, where each of the ρ_i is a T-word of length 2_k^n that satisfies the following:

- $\rho_0 = t_I t_\square \cdots t_\square$,
- for each $0 \leq i \leq m-1$, and $0 \leq j \leq 2_k^n - 2$, we have $(\rho_i(j), \rho_i(j+1)) \in H$,
- for each $0 \leq i \leq m-2$ and $0 \leq j \leq 2_k^n - 1$, we have $(\rho_i(j), \rho_{i+1}(j)) \in V$, and
- $\rho_{m-1}(0) = t_F$,

where $\rho_i(j)$ denotes the jth tile in ρ_i. We call the ith word in this solution the *ith row*.

Proposition 5 ([8,9]). *The order-k corridor tiling problem is k-EXPSPACE-hard for each $k \geq 0$.*

We ow show that the model-checking problem for order-k tail-recursive HORS against baAPT is hard for $(k-1)$-EXPSPACE. For that, we need some encoding of rows, each of which are a mapping $\rho \colon [2_k^n] \to T$. Modeling the domain of these mappings requires the manipulation of large numbers.

5.2 Encoding Large Numbers

The following constructions are based on an idea from [5]. Fix some n. Let $Q^n = \{q_{\text{yes}}^0, q'^{0}_{\text{yes}}, q_{\text{no}}^0, q'^{0}_{\text{no}}, \ldots, q_{\text{yes}}^{n-1}, q'^{n-1}_{\text{yes}}, q_{\text{no}}^{n-1}, q'^{n-1}_{\text{no}}, q_\top, q_\bot\}$. We define an automaton with these states. $q_\top$ and $q_\bot$ function as before, and, for each $0 \leq i < n$, any tree over the alphabet given below will either be accepted by q_{yes}^i and q'^{i}_{yes} and rejected by q_{no}^i and q'^{i}_{no}, or vice versa. The pairs q_{yes}^i, q'^{i}_{yes} and q_{no}^i, q'^{i}_{no} will have the same semantics for all i; the distinction is necessary for technical reasons that are explained later.

Hence, any tree T defines a set $S = \{i \mid \text{the automaton accepts } T \text{ from } q_{\text{yes}}^i\}$, which can be seen to represent the number $\sum_{i=0}^{n-1} b_i 2^i$ where $b_i = 1$ if $i \in S$. This allows us to count up to 2_1^n, and we use higher-order constructions to count further.

Manipulation of numbers, e.g., incrementation and decrementation, requires accessing bits, i.e., states, from other states. Hence, we introduce tree constructors $\mathsf{u}, \mathsf{d}, \mathsf{e}$ that signal transition to a state with higher, lower, or any index, respectively. Moreover, we add two more such constructors h and v to be used in the encoding of the tiling system later, as well as copies of these constructors to work for complementation. Finally, we add constructors that correspond to the boolean operators, and three additional constructors to help with the tiling system. Hence, let $\Sigma_\# = \{c_\vee, c'_\vee, c_\wedge, c'_\wedge, c_\neg, c_0, c_1, c_I, c_f, c_\Box, \mathsf{h}, \bar{\mathsf{h}}, \mathsf{v}, \bar{\mathsf{v}}, \mathsf{u}, \bar{\mathsf{u}}, \mathsf{d}, \bar{\mathsf{d}}, \mathsf{e}, \bar{\mathsf{e}}\}$ be a tree signature where $c_\vee, c'_\vee, c_\wedge, c'_\wedge$ have arity 2, $c_0, c_1, c_I, c_f, c_\Box$ have arity 0, and the other tree constructors have arity 1.

Consider the following APT $\mathcal{A}_\# = (Q^n, \Sigma_\#, \delta, q_{\text{yes}}^1, \Lambda)$ where for all $0 \leq i < n$, we have $\Lambda(q_{\text{yes}}^i) = \Lambda(q_{\text{no}}^i) = 2$ and $\Lambda(q'^i_{\text{yes}}) = \Lambda(q'^i_{\text{no}}) = 1$. Part of δ is given by

$\delta(q_{\text{yes}}^i, c_\vee) = (q_{\text{yes}}^i, q_\top) \vee (q_\top, q_{\text{yes}}^i) \qquad \delta(q_{\text{yes}}^i, \neg) = (q_{\text{no}}^i) \qquad \delta(q_{\text{no}}^i, \neg) = (q_{\text{yes}}^i)$

$\delta(q_{\text{yes}}^i, c'_\vee) = (q_{\text{yes}}^i, q_\top) \vee (q_\top, q'^i_{\text{yes}}) \qquad \delta(q_{\text{yes}}^i, c_1) = \top \qquad \delta(q_{\text{yes}}^i, c_0) = \bot$

$\delta(q_{\text{no}}^i, c_\vee) = (q_{\text{no}}^i, q_{\text{no}}^i) \qquad \delta(q_{\text{yes}}^i,) = \bot \qquad \delta(q_{\text{no}}^i, c_0) = \top$

$\delta(q_{\text{no}}^i, c'_\vee) = (q_{\text{no}}^i, q_{\text{no}}^i) \qquad \delta(q_{\text{yes}}^0, c_I) = \top \qquad \delta(q_{\text{yes}}^i, c_I) = \bot \text{ if } i \neq 0$

$\delta(q_{\text{no}}^i, c_\wedge) = (q_{\text{no}}^i, q_\top) \vee (q_\top, q_{\text{no}}^i) \qquad \delta(q_{\text{no}}^0, c_I) = \bot \qquad \delta(q_{\text{no}}^i, c_I) = \top \text{ if } i \neq 0$

$\delta(q_{\text{no}}^i, c'_\wedge) = (q_{\text{no}}^i, q_\top) \vee (q_\top, q'^i_{\text{no}}) \qquad \delta(q_{\text{yes}}^1, c_f) = \top \qquad \delta(q_{\text{yes}}^i, c_f) = \bot \text{ if } i \neq 1$

$\delta(q_{\text{yes}}^i, c_\wedge) = \{(q_{\text{yes}}^i, q_{\text{yes}}^i)\} \qquad \delta(q_{\text{no}}^1, c_f) = \bot \qquad \delta(q_{\text{no}}^i, c_f) = \top \text{ if } i \neq 1$

$\delta(q_{\text{yes}}^i, c'_\wedge) = \{(q_{\text{yes}}^i, q_{\text{yes}}^i)\} \qquad \delta(q_{\text{yes}}^2, c_\Box) = \top \qquad \delta(q_{\text{yes}}^i, c_f) = \bot \text{ if } i \neq 2$

$\delta(q_{\text{no}}^2, c_\Box) = \bot \qquad \delta(q_{\text{no}}^i, c_f) = \top \text{ if } i \neq 2$

The transition functions for q'^i_{yes} and q'^i_{no} are as for their unprimed counterparts. The intuition here is that $c_\vee, c_\wedge, c_\neg$ work as their first-order logic counterparts, with $c'_\vee$ and $c'_\wedge$ working almost the same. Moreover, q_{yes}^i accepts a tree rooted in c_1 and rejects one in c_0, with q_{no}^i acting in a dual manner. Since the terms defined below are rather large, for the sake of readability we will write $c_\vee, c_\wedge, c_\neg$ as $\vee, \wedge, \neg$ and, moreover, we will write $\vee$ and $\wedge$ in infix notation and with the standard FO precedence rules. We write $c'_\vee$ as $\vee'$ and $c'_\wedge$ as $\wedge'$. The conversion to the official syntax is straightforward.

Now let $H, V \subseteq [n]^2$ be given, set $U = \{(i,j) \mid n > j > i \geq 0\}$, set $D = \{(i,j) \mid 0 \leq j < i < n\}$ and set $E = \{(i,j) \mid 0 \leq i \leq j < n\}$. The second part of

δ is given as

$$\delta(q_{yes}^i, \mathsf{h}) = \{ \bigvee_{j \text{ s.t. } (i,j) \in H} (q_{yes}^j) \} \qquad \delta(q_{yes}^i, \overline{\mathsf{h}}) = \{ \bigwedge_{j \text{ s.t. } (i,j) \in H} (q_{yes}^j) \}$$

$$\delta(q_{no}^i, \mathsf{h}) = \{ \bigwedge_{j \text{ s.t. } (i,j) \in H} (q_{no}^j) \} \qquad \delta(q_{no}^i, \overline{\mathsf{h}}) = \{ \bigvee_{j \text{ s.t. } (i,j) \in H} (q_{no}^j) \}$$

The transitions for $\mathsf{v}, \overline{\mathsf{v}}, \mathsf{u}, \overline{\mathsf{u}}, \mathsf{d}, \overline{\mathsf{d}}, \mathsf{e}, \overline{\mathsf{e}}$ follow the same pattern, using the relations V, U, D and E, respectively, and the transitions for q'^i_{yes} and q'^i_{no} are as for q_{yes}^i and q_{no}^i, respectively. The intuition here is that reading h from state q_{yes}^i makes the automaton transition to any state q_{yes}^j with $(i,j) \in H$. To which such state the automaton transitions is decided by $\exists$, and if there is no such state, she loses the game. Conversely, reading $\overline{\mathsf{h}}$ makes $\forall$ pick a new index, with the same rules. From states q_{no}^i, the roles are reversed, and the same pattern appears for the other relations. In particular, reading d allows to lower the index, reading u allows to increase the index, and reading e allows to change to any index in $[n]$.

We now formalize the above definition regarding a tree encoding a number.

Definition 5. *Any $\Sigma_\#$ tree T encodes the number $\sum_{j=0}^{n-1} b_j 2^j$ where $b_j = 1$ iff $\mathcal{A}_\#$ accepts T from state q_{yes}^i.*

Now consider the following λY terms:

$$\mathtt{zero}_0 = c_0$$
$$\mathtt{isZero}_0 = \lambda(m: \tau_0).\, \overline{\mathsf{e}}\ \neg m$$
$$\mathtt{gt}_0 = \lambda(m, m': \bullet).\, \mathsf{e}\, \big((m' \wedge \neg m) \wedge \overline{\mathsf{u}}\, (\neg m \vee m')\big)$$
$$\mathtt{next}_0 = \lambda(m: \bullet).\, (m \wedge (\mathsf{d}\ \neg m)) \vee (\neg m \wedge \overline{\mathsf{d}}\ m)$$

Lemma 3.
1. *$BT(\mathtt{zero}_0)$ encodes 0.*
2. *$\mathcal{A}_\#$ accepts $BT(\mathtt{isZero}_0\, t)$ for some term t iff $BT(t)$ encodes $[0 \in 2_1^n]$.*
3. *If $BT(t_1), BT(t_2)$ encode $m_1, m_2 \in [2_1^n]$, then, for all $0 \le i < n$, we have that $BT(\mathtt{gt}_0\, t_1\, t_2)$ is accepted by $\mathcal{A}_\#$ from q_{yes}^i iff $m_1 < m_2$.*
4. *If $BT(t)$ encodes $m \in [2_1^n]$ then $BT(\mathtt{next}_0)$ encodes $m+1 \mod 2^n$.*

In order to count to numbers larger than 2_1^n, we need to employ functions that map trees to trees, etc. Let $m > 0$ be a natural number. Let $\tau_0 = \bullet$ and $\tau_1 = \bullet^i \to \bullet$ and for $k > 1$ let $\tau_k = \tau_k \to \bullet$.

Definition 6. *A closed term of type $\bullet$ is an arithmetic predicate of order 0. An otherwise closed term t of type τ_{k+1} is an arithmetic predicate of order $k+1$ if, for all terms t' of order k, either $BT(t\, t')$ is accepted by $\mathcal{A}_\#$ from q_{yes}^i for all $0 \le i < n$, or it is rejected from q_{yes}^i for all $0 \le i < n$.*

An term t of order $k > 0$ encodes the number $\sum_{j=0}^{2_k^n - 1} b_j 2^j \in [2_{k+1}^n]$ with $b_j = 1$ if $\mathcal{A}_\#$ accepts $BT(t\, t')$ from all the q_{yes}^i for $0 \le i < n$ for any order-k term t' that encodes j.

Given that, on arithmetic predicates, all the q_{yes}^i for $0 \leq i < n$ in $\mathcal{A}_\#$ behave the same, we simply say that $\mathcal{A}_\#$ accepts (the tree generated by) the predicate if it accepts said tree from one and, hence, all of them, and we say that it rejects the tree, if it rejects from one and, hence, all of them. Now consider the following λY terms:

$$\texttt{zero}_{k+1} = \lambda(b\colon \tau_k).\ c_0$$
$$\texttt{exists}_k = \lambda(p\colon \tau_{k+1}).\ \bigl(Y\ (F\colon \tau_{k+1}).\ \lambda(m\colon \tau_k).\ (p\ m) \vee' (F\ (\texttt{next}_k\ m))\bigr)\texttt{zero}_k$$
$$\texttt{forall}_k = \lambda(p\colon \tau_{k+1})\bigl(Y\ (F\colon \tau_{k+1}).\ \lambda(m\colon \tau_k).\ (p\ m) \wedge' (F\ (\texttt{next}_k\ m))\bigr)\texttt{zero}_k$$
$$\texttt{gt}_{k+1} = \lambda(m,m'\colon \tau_{k+1}).\ \texttt{exists}_k\Bigl(\lambda(i\colon \tau_k).\ ((m'\ i) \wedge \neg(m\ i)) \wedge$$
$$\bigl(\texttt{forall}_k\ \bigl(\lambda(j\colon \tau_k).\neg(\texttt{gt}_k\ i\ j) \vee \neg(m\ j) \vee (m'\ j)\bigr)\bigr)\Bigr)$$
$$\texttt{next}_{k+1} = \lambda(m\colon \tau_{k+1}, i\colon \tau_k).\ \Bigl((m\ i) \wedge \bigl(\texttt{exists}_k\bigl(\lambda(j_1\colon \tau_k).\ (\texttt{gt}_k\ i\ j_1) \wedge \neg(m\ j_1)\bigr)\bigr)\Bigr)$$
$$\vee \Bigl((\neg(m\ i)) \wedge \bigl(\texttt{forall}_k\ \bigl(\lambda(j_2\colon \tau_k).\ \neg(\texttt{gt}_k\ i\ j_2) \vee (m\ j_2)\bigr)\bigr)\Bigr)$$
$$\texttt{isZero}_{k+1} = \lambda(m\colon \tau_{k+1}).\ \texttt{forall}_k\ \bigl(\lambda(i\colon \tau_k).\ \texttt{isZero}_0\ (m\ i)\bigr)$$

Note the uses of $\vee'$ and $\wedge'$ in the definitions of $\texttt{exists}_k$ and $\texttt{forall}_k$: they serve to mark the disjunction, resp. conjunction in the recursion.

Lemma 4. 1. $\texttt{zero}_{k+1}$ encodes $0 \in 2_{k+2}^n$.
2. If t is an order-$(k+1)$ arithmetic predicate, then $BT(\texttt{exists}_{k+1}\ t)$ is accepted by $\mathcal{A}_\#$ iff there is an order-k term t' such that $\mathcal{A}_\#$ accepts $BT(p\ t')$
3. If t is an order-$(k+1)$ arithmetic predicate, then $BT(\texttt{forall}_{k+1}\ t)$ is accepted by $\mathcal{A}_\#$ if, for all order-k terms t', we have that $\mathcal{A}_\#$ accepts $BT(p\ t')$.
4. If t, t' are order-$(k+1)$ terms that encode $m, m' \in [2_{k+2}^n]$, then $\mathcal{A}_\#$ accepts $BT(\texttt{gt}_{k+1}\ t\ t')$ if $m < m'$, and it rejects $BT(\texttt{gt}_{k+1}\ t\ t')$ if $m \geq m'$.
5. If t is an order-$(k+1)$ term that encodes $m \in [2_{k+2}^n]$, then $\texttt{next}_{k+1}\ t$ encodes $m+1 \mod 2_{k+2}^n$.
6. $\mathcal{A}_\#$ accepts $BT(\texttt{isZero}_k\ t)$ iff t encodes $0 \in [2_k^n]$.

5.3 Encoding the Tiling Problem

We now move to encoding the tiling problem. Remember that a row of the order-k tiling problem is a sequence of length 2_k^n. W.l.o.g. assume that there are exactly n tiles (if not, boost their number by making identical copies). Hence, there is exactly one set $q_{yes}^i, {q'}_{yes}^i$, etc. for each tile. We encode a row of tiles as a term of type τ_{k+1}.

Definition 7. Let t be a term of type τ_{k+1}. Then t encodes a row if, for all $m \in [2_{k+1}^n]$, there is i such that for all terms t' that encode m, $\mathcal{A}_\#$ accepts $BT(t\ t')$ from q_{yes}^j iff $j = i$.

Consider the following λY terms:

$$\text{isTile} = \lambda(x\colon \bullet).\ \bar{\text{e}}\left(\neg x \vee \left((\bar{\text{u}}\ \neg x) \wedge (\bar{\text{d}}\ \neg x)\right)\right)$$
$$\text{isRow}_{k+1} = \lambda(r\colon \tau_{k+1}).\ \text{forall}_k\left(\lambda(m\colon \tau_{k-1}]).\ \text{isTile}\ (r\ m)\right)$$
$$\text{init}_k = \lambda(m\colon \tau_{k-1}).\ \left(c_I \wedge \text{isZero}_{k-1}\ m\right) \vee \left(c_\square \wedge \neg \text{isZero}_{k-1}\ m\right)$$
$$\text{isFinal}_k = \lambda(r\colon \tau_k).\ \bar{\text{e}}\left((c_f \vee \neg(r\ \text{zero}_{k-1}))\right)$$
$$\text{horiz}_k = \lambda(r\colon \tau_k).\ \text{forall}_{k-1}\Big(\lambda(m\colon \tau_{k-1}).\ \bar{\text{e}}\ \big(\neg(r\ m) \vee$$
$$\left((\text{isZero}_{k-1}\ (\text{next}_{k-1}\ m)) \vee (\text{h}\ (\text{next}_{k-1}\ m))\right)\big)\Big)$$
$$\text{vert}_k = \lambda(r_1, r_2\colon \tau_k).\ \text{forall}_{k-1}\Big(\lambda(m\colon \tau_{k-1}).\ \bar{\text{e}}\ \big(\neg(r_1\ m) \vee (\text{v}\ (r_2\ m))\big)\Big)$$

Lemma 5. *Let t be an otherwise closed term of type τ_k.*
1. *Let t define a tree. Then $\mathcal{A}_\#$ accepts $BT(\text{isTile}\ t)$ if there is exactly one i such that $\mathcal{A}_\#$ accepts $BT(t)$ from q^i_{yes}.*
2. *$\mathcal{A}_\#$ accepts $BT(\text{isRow}_{k+1}\ t)$ iff $\mathcal{A}_\#$ accepts $BT(\text{isTile}\ (t\ t'))$ for all order-(k) arithmetic predicates t'. Hence, t encodes a row.*
3. *init_k encodes the row where the 0th tile is t_I and the other tiles are $t_\square$.*
4. *Let t encode a row. Then $\mathcal{A}_\#$ accepts $BT(\text{isFinal}_k\ t)$ iff t encodes a final row, i.e., one where the 0th tile is t_F.*
5. *let t encode a row. Then $\mathcal{A}_\#$ accepts $BT(\text{horiz}_k\ t)$ iff for each $0 \leq m < 2^n_{k+1} - 1$, the mth and $m+1$st tiles match horizontally.*
6. *Let t, t' encode rows. Then $\mathcal{A}_\#$ accepts $BT(\text{vert}_k\ t\ t')$ if for each $0 \leq m \leq 2^n_{k+1}$, the $m$the tiles in the rows encoded by t and t' match vertically.*

Now consider the following λY terms:

$$S_k = R_k\ \text{init}_k$$
$$R_k = Y(F\colon \tau_{k+1}).\ \lambda(r_1\colon \tau_k).\ (\text{isFinal}\ r_1) \vee' \Big(Y(G\colon \tau_{k+1}).\ \lambda(r_2\colon \tau_k).$$
$$\left((\text{isRow}_k\ r_2) \wedge (\text{horiz}_k\ r_2) \wedge (\text{vert}_k\ r_1\ r_2) \wedge' (F\ r_2)\right) \vee' G\ (\text{next}_k\ r_2)\Big)\ r_1$$

Lemma 6. *$\mathcal{A}_\#$ accepts $BT(S_k)$ iff the order-k corridor tiling problem for the tiling system $\mathcal{K}$ has a solution.*

S_k can be seen to be tail-recursive, since the only recursive subterms are those in R_k, exists_k and forall_k, where only $\vee'$ and $\wedge'$ have subterms with free recursion variables, whence the other symbols can all be assigned to $\Sigma^u_\#$. Clearly, no free recursion variables appear in operand position.

Theorem 6. *The model-checking problem for order-k tail-recursive HORS is $(k-1)$-EXPSPACE-hard.*

The prof requires converting $\mathcal{A}_\#$ into a baAPT by introducing k many copies of each state, one for each partition set Q_i. The initial state starts in the highest set, and a decrease in index happens for the states that traverse to the left at $\vee'$ and $\wedge'$, with the exception of the $\vee'$ in R_k, which does not require a decrease.

6 Conclusion

We have exhibited a natural fragment of the HORS model-checking problem such that its complexity is characteristic of space. In particular, the order-k model-checking problem for tail-recursive λY terms against baAPT is $(k-1)$-EXPSPACE-complete.

There are several avenues for further research. One concerns some restrictions of the fragment, in particular the restriction to binary tree constructors in crucial parts of the tree alphabet. This is needed for easy complementation, but maybe this restriction can be lifted.

A clear next step is to translate the formalisms developed in this paper into some of the more popular frameworks for working with HORS. This means replacing definitions based on the Krivine Machine and the λY calculus by the traditional grammar-based presentation. This is not difficult, however adapting Algorithm 1 into a practical procedure based on intersection types or Collapsible Pushdown Automata likely requires extra work and, hence, is beyond the scope of this paper.

References

1. Axelsson, R., Lange, M., Somla, R.: The complexity of model checking higher-order fixpoint logic. Log. Methods Comput. Sci. **3**(2) (2007). https://doi.org/10.2168/LMCS-3(2:7)2007
2. Broadbent, C.H., Carayol, A., Hague, M., Serre, O.: C-shore: a collapsible approach to higher-order verification. In: Morrisett, G., Uustalu, T. (eds.) ACM SIGPLAN International Conference on Functional Programming. ICFP'13, Boston, MA, USA, 25–27 September 2013, pp. 13–24. ACM (2013). https://doi.org/10.1145/2500365.2500589
3. Broadbent, C.H., Kobayashi, N.: Saturation-based model checking of higher-order recursion schemes. In: Rocca, S.R.D. (ed.) Computer Science Logic 2013 (CSL 2013). CSL 2013, 2–5 September 2013, Torino, Italy. LIPIcs, vol. 23, pp. 129–148. Schloss Dagstuhl - Leibniz-Zentrum für Informatik (2013). https://doi.org/10.4230/LIPICS.CSL.2013.129
4. Bruse, F., Lange, M.: The tail-recursive fragment of timed recursive CTL. Inf. Comput. **294**, 105084 (2023). https://doi.org/10.1016/J.IC.2023.105084
5. Bruse, F., Lange, M., Lozes, É.: The complexity of model-checking tail-recursive higher-order fixpoint logic. Fundam. Informaticae **178**(1–2), 1–30 (2021). https://doi.org/10.3233/FI-2021-1996
6. Courcelle, B.: The monadic second-order logic of graphs IX: machines and their behaviours. Theor. Comput. Sci. **151**(1), 125–162 (1995). https://doi.org/10.1016/0304-3975(95)00049-3
7. Damm, W.: The IO- and OI-hierarchies. Theor. Comput. Sci. **20**, 95–207 (1982). https://doi.org/10.1016/0304-3975(82)90009-3
8. Demri, S., Goranko, V., Lange, M.: Temporal Logics in Computer Science: Finite-State Systems. Cambridge Tracts in Theoretical Computer Science. Cambridge University Press, Cambridge (2016). https://doi.org/10.1017/CBO9781139236119

9. van Emde Boas, P.: The convenience of tilings. In: Sorbi, A. (ed.) Complexity, Logic, and Recursion Theory, Lecture Notes in Pure and Applied Mathematics, vol. 187, pp. 331–363. Marcel Dekker, Inc., New York City (1997)
10. Hague, M., Murawski, A.S., Ong, C.L., Serre, O.: Collapsible pushdown automata and recursion schemes. In: Proceedings of the Twenty-Third Annual IEEE Symposium on Logic in Computer Science. LICS 2008, 24–27 June 2008, Pittsburgh, PA, USA, pp. 452–461. IEEE Computer Society (2008). https://doi.org/10.1109/LICS.2008.34
11. Huet, G.: Résolution d'Équations dans des Langages d'Order 1, 2., ω. Ph.D. thesis, Universite de Paris VII (1976)
12. Jurdziński, M.: Small progress measures for solving parity games. In: Reichel, H., Tison, S. (eds.) STACS 2000. LNCS, vol. 1770, pp. 290–301. Springer, Heidelberg (2000). https://doi.org/10.1007/3-540-46541-3_24
13. Knapik, T., Niwiński, D., Urzyczyn, P.: Deciding monadic theories of hyperalgebraic trees. In: Abramsky, S. (ed.) TLCA 2001. LNCS, vol. 2044, pp. 253–267. Springer, Heidelberg (2001). https://doi.org/10.1007/3-540-45413-6_21
14. Knapik, T., Niwiński, D., Urzyczyn, P.: Higher-order pushdown trees are easy. In: Nielsen, M., Engberg, U. (eds.) FoSSaCS 2002. LNCS, vol. 2303, pp. 205–222. Springer, Heidelberg (2002). https://doi.org/10.1007/3-540-45931-6_15
15. Kobayashi, N.: Model checking higher-order programs. J. ACM **60**(3), 20:1–20:62 (2013). https://doi.org/10.1145/2487241.2487246
16. Kobayashi, N., Lozes, É., Bruse, F.: On the relationship between higher-order recursion schemes and higher-order fixpoint logic. In: Castagna, G., Gordon, A.D. (eds.) Proceedings of the 44th ACM SIGPLAN Symposium on Principles of Programming Languages. POPL 2017, Paris, France, 18–20 January 2017, pp. 246–259. ACM (2017). https://doi.org/10.1145/3009837.3009854
17. Kobayashi, N., Ong, C.-H. L.: Complexity of model checking recursion schemes for fragments of the modal mu-calculus. In: Albers, S., Marchetti-Spaccamela, A., Matias, Y., Nikoletseas, S., Thomas, W. (eds.) ICALP 2009, Part II. LNCS, vol. 5556, pp. 223–234. Springer, Heidelberg (2009). https://doi.org/10.1007/978-3-642-02930-1_19
18. Kobayashi, N., Ong, C.-H. L.: A type system equivalent to the modal mu-calculus model checking of higher-order recursion schemes. In: Proceedings of the 24th Annual IEEE Symposium on Logic in Computer Science. LICS 2009, 11–14 August 2009, Los Angeles, CA, USA, pp. 179–188. IEEE Computer Society (2009). https://doi.org/10.1109/LICS.2009.29
19. Krivine, J.: A call-by-name lambda-calculus machine. High. Order Symb. Comput. **20**(3), 199–207 (2007). https://doi.org/10.1007/S10990-007-9018-9
20. Lange, M., Lozes, É.: Capturing bisimulation-invariant complexity classes with higher-order modal fixpoint logic. In: Diaz, J., Lanese, I., Sangiorgi, D. (eds.) TCS 2014. LNCS, vol. 8705, pp. 90–103. Springer, Heidelberg (2014). https://doi.org/10.1007/978-3-662-44602-7_8
21. Muller, D.E., Saoudi, A., Schupp, P.E.: Alternating automata, the weak monadic theory of trees and its complexity. Theor. Comput. Sci. **97**(2), 233–244 (1992). https://doi.org/10.1016/0304-3975(92)90076-R
22. Neatherway, R.P., Ramsay, S.J., Ong, C.L.: A traversal-based algorithm for higher-order model checking. In: Thiemann, P., Findler, R.B. (eds.) ACM SIGPLAN International Conference on Functional Programming. ICFP'12, Copenhagen, Denmark, 9–15 September 2012, pp. 353–364. ACM (2012). https://doi.org/10.1145/2364527.2364578

23. Ong, C.-H. L.: On model-checking trees generated by higher-order recursion schemes. In: 21th IEEE Symposium on Logic in Computer Science (LICS 2006), 12–15 August 2006, Seattle, WA, USA, Proceedings, pp. 81–90. IEEE Computer Society (2006). https://doi.org/10.1109/LICS.2006.38
24. Rabin, M.O.: Decidability of second-order theories and automata on infinite trees. Trans. Am. Math. Soc. **141**, 1–35 (1969)
25. Salvati, S., Walukiewicz, I.: Recursive schemes, krivine machines, and collapsible pushdown automata. In: Finkel, A., Leroux, J., Potapov, I. (eds.) RP 2012. LNCS, vol. 7550, pp. 6–20. Springer, Heidelberg (2012). https://doi.org/10.1007/978-3-642-33512-9_2
26. Salvati, S., Walukiewicz, I.: Krivine machines and higher-order schemes. Inf. Comput. **239**, 340–355 (2014). https://doi.org/10.1016/J.IC.2014.07.012
27. Savitch, W.J.: Relationships between nondeterministic and deterministic tape complexities. J. Comput. Syst. Sci. **4**(2), 177–192 (1970). https://doi.org/10.1016/S0022-0000(70)80006-X
28. Streett, R.S., Emerson, E.A.: The propositional mu-calculus is elementary. In: Paredaens, J. (ed.) ICALP 1984. LNCS, vol. 172, pp. 465–472. Springer, Heidelberg (1984). https://doi.org/10.1007/3-540-13345-3_43
29. Viswanathan, M., Viswanathan, R.: A higher order modal fixed point logic. In: Gardner, P., Yoshida, N. (eds.) CONCUR 2004. LNCS, vol. 3170, pp. 512–528. Springer, Heidelberg (2004). https://doi.org/10.1007/978-3-540-28644-8_33

Property-Agnostic Base Case Extension for Scalable Verification of Distributed Systems

Kyle Storey(✉) and Eric Mercer

Brigham Young University, Provo, UT 84602, USA
kyle.r.storey@gmail.com, egm@cs.byu.edu

Abstract. Many distributed systems require temporal properties to hold for correctness. Model checking can verify these properties on a small system but it doesn't scale for arbitrarily large systems. This work presents a new method for proving that temporal properties verified on a small system extend to an arbitrarily large system when that system has a ring topology. It uses a model checker to prove temporal properties and properties of a partial order of events in the system. It then admits the partial order properties as axioms in a theorem prover and proves a conformance relation between an arbitrary-sized ring of nodes and the model-checked base case. The conformance relation is used to prove that adding a new node to the ring does not affect the possible states of the existing nodes in the system and therefore any properties proven in the small system continue to hold in an arbitrarily large system.

We demonstrate the approach in a case study of a nontrivial distributed protocol that is used by the MyCHIP's digital currency to clear credit.

Keywords: Model checking · interactive theorem provers · formal verification · distributed systems · digital currency

1 Introduction

We present an approach to verification that combines model checking with an interactive theorem prover to verify temporal properties in an arbitrarily large distributed system used to clear credit in a digital currency called MyCHIPs [3].

We model the system with a small fixed number of nodes and verify that the required temporal properties hold. We define a partial order of events with added constraints on which events must be present, parameterized by the size of the system—collectively referred to as the *partial order*. We verify that the *partial order* is respected in the fixed-size system.

In the interactive theorem prover, we use the *partial order* to define the set of possible messages in the system. We construct an inductive proof that the set of messages observed by each node does not change with the number of participating nodes.

We use *conformance* as described by Dill [8] to show that if the set of observed messages for each node is unchanged then the set of internal states of each node is unchanged, which implies that the states considered by the model checker is the complete set of states for nodes in a system of any size and therefore the properties proved in the model checker generalize to the larger system.

MyCHIPs is a novel digital currency that facilitates the trade of goods and services between trusting partners. MyCHIPs has some similarities to time banking systems [16,20] but introduces a novel debt-clearing protocol, called a *credit lift*, to completely decentralize trade. We verify in this work that the distributed protocol to implement the credit lift preserves the security and functionality of the MyCHIPs digital currency.

A credit lift (*lift* for brevity) in the MyCHIPs protocol operates on a group of nodes arranged in a cycle such that each node is in debt to its successor in the cycle. The protocol arranges for each to forgive their predecessor's debt, and in exchange, their own debt is forgiven by their successor. Each node's debts are recorded in a *tally*, which is a record of transactions that is kept in consensus between its trading partners. The *tally* uses *CHIPs* as a unit of measure for the debt with each *CHIP* being defined to be equal to the value of one hour of an unskilled laborer's time.

The lift operates over two stages once a mutually beneficial cycle of trading nodes is discovered. First, each node *promises* to send CHIPs to their predecessor. And then second, after all have promised, each node *commits* by sending the requisite CHIPs. At the end of the lift each node's *balance*—the difference between the number of sent CHIPs and the number of received CHIPs—remains the same, but all the nodes have reduced their debt liability.

What makes the credit lift complex is that arbitrary nodes may lose connectivity or otherwise become *inactive* at any point in the protocol thus leaving lifts unfinished indefinitely. As such, each lift is given a time limit, represented by a timestamp at which time the lift can no longer be committed. Because we don't expect nodes on the network to have synchronized clocks, a *Referee* is appointed whose clock is considered authoritative and acts as a consensus object. When a node requests to commit before the timeout, the Referee provides a digital signature that is proof that the lift has been committed. If the lift is not committed before the timeout, the lift becomes *nullified* and the Referee digitally signs a statement to that effect.

1.1 Verified Properties and Assumptions

The key properties that underpin the security and functionality of the lift protocol, which we prove, are the following:

1. Lifts always eventually are committed or nullified for every active node.
2. At the final state of the lift, every active node agrees that the lift was committed or every active node agrees the lift was nullified.

3. The balance of every active node on the final state of the lift is equal to or greater than its initial balance.
4. Every active pair of nodes on the final state of the lift agree on their shared tally.

Our verification results are subject to the following assumptions that are explicit in the proofs:

– Only a single lift is in flight with every node in the cycle being distinct.
– Messages may experience arbitrarily long delays except that messages to and from the Referee are always eventually received.
– All nodes follow the protocol though they may do so with arbitrarily long delays between actions again with the exception that the Referee always eventually takes the next action

1.2 Tools and Approach

We use the Spin Model Checker to prove these properties hold in a system with one *Originator*—the node that initiates the lift algorithm—one Referee, and one *Relay* node—a node that participates in the lift that is not the Originator or the Referee. We then use Coq for our interactive theorem prover and produce an inductive proof that the temporal properties verified in Spin extend to an arbitrarily large system.

In Dill's work, he proves that if one subsystem conforms to another, then it can be safely substituted for the other. By safe substitution, we mean that any behavior of the new subsystem that could cause a failure is also a behavior of the original subsystem. Using the model checking results as a base case, we prove that the set of traces produced by a chain of $n + 1$ Relay nodes conforms to the set of traces produced by n Relay nodes and inductively argue that the properties verified by the model checker hold for any number of nodes.

We admit that using model checkers to verify properties of systems with arbitrarily large or even infinite state has been a common topic within the model-checking communities. To our knowledge, the various published methods reason inductively on the calculus used within the model checker either automatically [11,19], or manually [6]. Our proof approach here deviates from these works in two key ways:

1. We use conformance as the theoretical foundation for the inductive proof.
2. We use distinct and separate representations of the nodes for the model checking and the theorem proving, and we use a partial conformance equivalence to provide strong assurance that these representations are interchangeable.

Using this new method in our approach gives two key advantages:

1. New properties may be verified in the system through model checking without needing to change the inductive proof.

2. Using a different representation for the inductive proof allows for more flexibility in proof tactics and structure.

For this method to be sound and complete the representation used in the model checker and the representation used in the inductive proof must be interchangeable. Proving conformance equivalence between these two representations is sufficient to prove interchangeability. However, this work only succeeds in formally proving a partial conformance equivalence. We do however, provide evidence that gives strong assurance of conformance equivalence sufficient for our needs. We leave a method to formally prove conformance equivalence between these representations to future work.

A code archive with our Spin models and Coq proof script along with instructions to run the analysis is can be found at https://zenodo.org/records/13926804.

1.3 Organization

The rest of this paper is organized as follows: Sect. 2 defines the lift protocol. Section 3 is the inductive proof that generalizes the properties to any number of nodes. Section 4 summarizes the Spin verification of the base system. Section 5 describes the Coq proof for the induction. Section 6 gives strong assurance of conformance equivalence between the two representations used in Sect. 4 and Sect. 5. Section 7 discusses related work. And finally, Sect. 8 is the conclusion.

2 MyCHIPs Protocol

Figure 1 shows the Mealy machines that define the behavior for each type of node in the credit lift protocol: the Originator, the Referee, and any number of Relays [4]. Additional background for MyCHIPs can be found at the *gotchoices* website [3]. Each box in Fig. 1 represents the state of the lift from the perspective of the node type, and each arrow represents a change in state and an associated action that is taken during the transition from one state to the next. A '/' character delineates the condition that must hold for the transition to be taken from the action that is taken during the transition.

The state of each node type includes the network ID of its predecessor (Pred), its successor (Succ), and the Referee. A transition may be conditional on receiving a particular message from a particular network ID. This case is written as *sender* ? *message_type*. Additionally, actions may transmit a particular message to a particular network ID. This case is written as *recipient* ! *message_type*.

Each node in the lift begins in the *No Lift* state. The originator has an edge with *True* as its condition so at any time the originator can start the lift by sending a *Promise* to its predecessor in the cycle and transition to the *Pend* state. Each relay node when it receives this *Promise* message forwards the message to its predecessor and transitions to the *Pend* state.

The *Pend* state indicates that the node has not yet determined whether to commit or roll back the lift. At any time when a node is in the *Pend* state it can

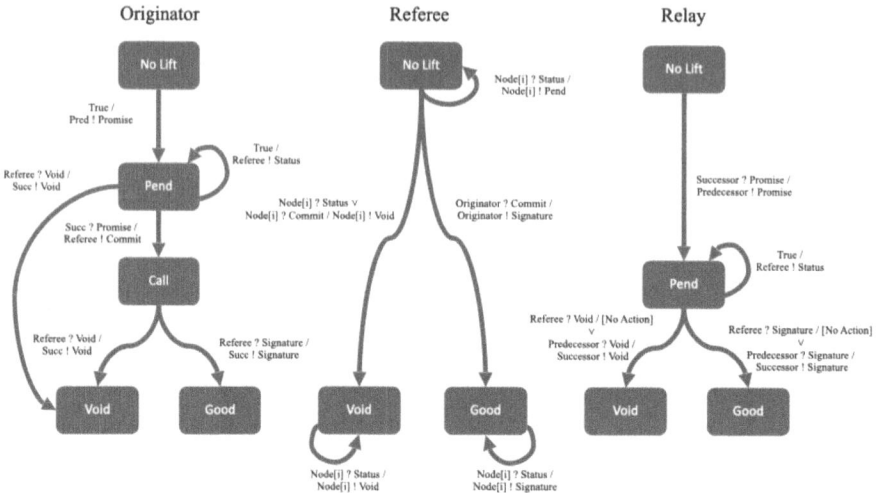

Fig. 1. Lift States Diagram

send a message to the Referee to ask if the lift has been completed. The referee will respond with one of three messages: *Void*, *Signature*, or *Pend*. The *Pend* message means no decision has been made, so this induces no state transition and no action. If the message is *Void* or *Signature*, this triggers the node to mark the lift as *Void* or *Good* respectively. This path through the states allows for lifts to be completed even when messages to and from peers in the cycle are unreliable. Note that transitions in this case have no action. So a node only propagates a *Void* or *Signature* message to its successor when it has received it from a peer, not from the referee.

If the messages to peers are successful, once the *Promise* message propagates around the cycle to the originator, this satisfies the transition condition so the originator can transition to the *Call* state. When it does so it sends a *Commit* message to the referee to request its signature. The referee can then return either a *Void* or *Signature* message. If *Void*, the originator transitions to the *Void* state. If *Signature*, the originator transitions to the *Good* state. In either case, it propagates the message to its successor. When these messages reach the relay nodes they similarly transition to *Void* or *Good* and forward the message to their successor. Eventually reaching the originator which ignores this message and the lift is complete.

3 Conformance Equivalence for Representation Switching

We would like to verify the stated properties of the lift protocol for a cycle of any number of nodes. To do this, we need to construct an inductive proof. The base case for the proof is the smallest complete system for which a lift can be performed: an Originator, a Referee, and a Relay. We generalize this step of the proof to an arbitrary set of properties and not just those stated in the introduction. For this generalization, we define a predicate $Prop(\mathit{Ref}, \mathit{Orig}, R_1, \cdots)$ that should only be true if the properties checked by the predicate hold in the given lift under all possible circumstances allowed by our assumptions. The lift itself is defined by the arguments to the predicate that name the Referee, Originator, and some sequence of one or more Relay nodes. The Originator is the head of the cycle for the lift and is followed by the sequence of relays.

Verifying properties in the base system is a natural fit for model checking. Spin is able to readily verify each of the properties stated at the outset of this paper by describing the Mealy machines directly in its input language Promela and expressing the desired properties in linear temporal logic. We state the result in the following lemma.

Lemma 1 (Base Case). *The base system consisting of one Referee, one Originator, and one Relay is correct.*

$$Prop(\mathrm{Ref}, \mathrm{Orig}, R_1) \text{ holds}$$

Proof. See Sect. 4.

The question now is how to prove out the inductive step that generalizes the model checking results to arbitrary lifts? For that, we turn to theorem proving. We tried defining the state machines as Automaton and composing them using Athalye's framework [2], but it didn't scale.

Instead, we used trace sets as our representation because our prior work has shown theorem provers tend to reason well over sets of traces [13,15]. We manually extract from the Mealy machines a set of constraints on sequential traces of events in the lift. These extracted constraints define what traces are allowed by the lift while the Mealy machines define how such traces are generated.

We compare this relationship between the two representations to black-box and white-box testing. The model checker is white-box meaning that it uses the internal state of each node to reason about all reachable message traces, while the theorem prover is black-box meaning that it uses constraints on, or properties of, traces to reason about all message traces. We differentiate these two models in our notation with R to denote a node as defined in the model checker and B to denote a node as defined in the theorem prover.

To prove that these two representations are equivalent, we utilize Dill's work on trace theory. In Dill's work, he defines a sufficient condition for one subsystem to conform to another. Dill assumes that if a subsystem experiences an internal failure the system as a whole will experience a failure. However, the goal of our analysis is to prove the properties of the system as a whole hold even when individual nodes in the system may fail. All of the behavior of each node, including the behavior of a node that experiences an internal error is considered in the possible behavior and not considered a failure. With this assumption, we can reduce the necessary condition for conformance to:

Definition 1 (Conformance).

$$Proj(P(R), R') \subseteq P(R') \longrightarrow R \preceq R'$$

where $P(R)$ and $P(R')$ are the sets of possible traces in R and R' respectively, and $Proj(T, R')$ projects the set of traces T to omit events that are not in the alphabet of R'.

Dill's work proves that conformance implies safe substitution so we admit the following theorem:

Theorem 1 (Conformance implies safe-substitution). *In the context of any system, if the component R conforms to the component R', then R' can be replaced by R, and all properties of the system are preserved.*

$$R \preceq R' \longrightarrow Prop(\cdots, R', \cdots) \longrightarrow Prop(\cdots, R, \cdots)$$

Conformance is a helpful tool in the inductive step as seen in Sect. 5, but it also proves useful to bridge the gap between our model checker and theorem prover representations. By showing *conformance equivalence* between each of these representations we can show that a node as defined in the model checker can safely be substituted with a node as defined in the theorem prover. Using this technique we can define Lemma 5 which allows for our inductive proof in Theorem 2 to be proven.

Theorem 2 (The properties hold on an arbitrarily large system). *The desired properties hold for an arbitrarily large number of nodes utilizing the MyCHIPs protocol.*

$$\forall n \in \mathbb{N}, n \geq 1 \longrightarrow Prop(\text{Ref}, \text{Orig}, R_1, \cdots, R_n)$$

Proof. By induction.
  Base Case: $Prop(\text{Ref}, \text{Orig}, R_1)$
  See Sect. 4.
  Inductive Step:

$$\forall n \in \mathbb{N}, Prop(\text{Ref}, \text{Orig}, R_1, \cdots, R_n) \longrightarrow$$
$$Prop(\text{Ref}, \text{Orig}, R_1, \cdots, R_n, R_{n+1})$$

See Lemma 5.

The proof for Lemma 5 relies on three Conformance properties defined in Lemma 2, Lemma 3 and Lemma 4.

Lemma 2 (Two nodes conform to one node). *In the context of a MyCHIPs system, a chain of two black-box Relay nodes conform to a single black-box Relay node.*

$$\forall n \in \mathbb{N}, (B_n, B_{n+1}) \preceq B_n$$

Proof. See Sect. 5.

Lemma 3 (Inductive node conforms to model checker node). *The representation of a node B as defined in the theorem prover conforms to the representation of a node R as defined in the model checker.*

$$B \preceq R$$

Proof. See Sect. 6.

Lemma 4 (Model checker node conforms to inductive node). *The representation of a node R as defined in the model checker conforms to the representation of a node B as defined in the theorem prover.*

$$R \preceq B$$

Unproven. See Sect. 6.

Using Theorem 1, Lemma 2, Lemma 3 and Lemma 4 we can show:

Lemma 5 (Inductive Step).

$$\forall n \in \mathbb{N}, Prop(\text{Ref}, \text{Orig}, R_1, \cdots, R_n) \longrightarrow Prop(\text{Ref}, \text{Orig}, R_1, \cdots, R_n, R_{n+1})$$

Proof.

$$Prop(\textit{Ref}, \textit{Orig}, R_1, \cdots, R_n) \land B \preceq R \longrightarrow \\ Prop(\textit{Ref}, \textit{Orig}, R_1, \cdots, R_{n-1}, B_n)$$

$$Prop(\textit{Ref}, \textit{Orig}, R_1, \cdots, R_{n-1}, B_n) \land (B_n, B_{n+1}) \preceq (B_n) \longrightarrow \\ Prop(\textit{Ref}, \textit{Orig}, R_1, \cdots, R_{n-1}, B_n, B_{n+1})$$

$$Prop(\textit{Ref}, \textit{Orig}, R_1, \cdots, R_{n-1}, B_n, B_{n+1}) \land R \preceq B \longrightarrow \\ Prop(\textit{Ref}, \textit{Orig}, R_1, \cdots, R_n, R_{n+1})$$

$$\therefore \forall n \in \mathbb{N}, Prop(\textit{Ref}, \textit{Orig}, R_1, \cdots, R_n) \longrightarrow \\ Prop(\textit{Ref}, \textit{Orig}, R_1, \cdots, R_n, R_{n+1})$$

Using Fig. 2 to help illustrate, the remainder of this section will describe the intuition of Lemma 5.

First, conformance requires that the alphabet of events of these subsystems be equivalent. Said differently, we define a standard interface that is sufficient for all the submodules we want to substitute. That interface includes 6 unique types of events: *Promise Send*, *Promise Receive*, *Commit Send*, *Commit Receive*, *Status Send*, and *Status Receive*. Figure 2a illustrates how these events map to messages passed in the system. The events are written from the perspective of

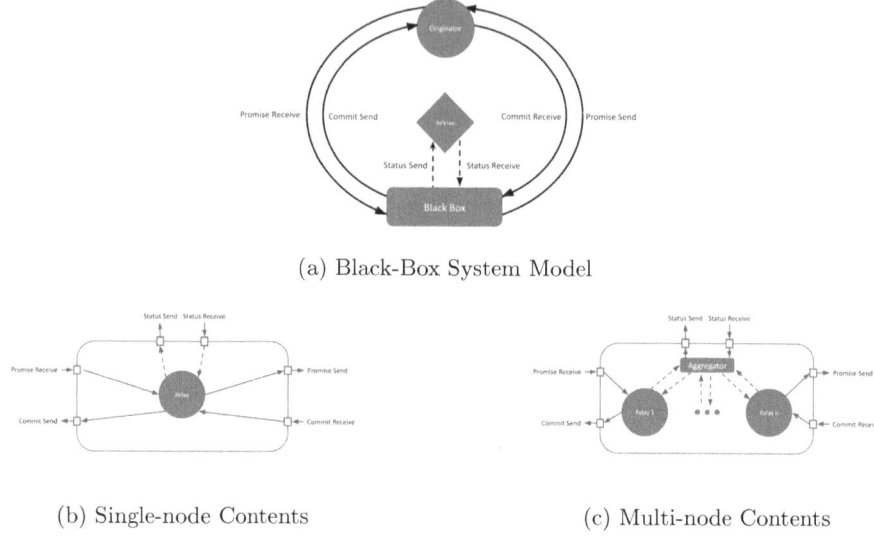

(a) Black-Box System Model

(b) Single-node Contents   (c) Multi-node Contents

Fig. 2. Utilizing Conformance

the black box at the bottom of the figure. When the black box emits a *Promise Send* event a *Promise* type message is sent to the originator as indicated by the arrow. Similarly when *Commit Receive* event is emitted by the black box this means that the originator has sent a *Commit* type message. Likewise, when a *Status Send* event is emitted a *Status* type message is sent to the referee. This is an optional event so the arrow is drawn with a dotted line.

Lemma 2 considers two subsystems where the first has exactly one Relay node and the second has a chain of two (or more) Relay nodes. These take the form of Fig. 2b and Fig. 2c, respectively. In the second subsystem, we introduce an *Aggregator* which connects requests made from any node inside the module to a single external channel that communicates with the Referee. We hide the details of the internal signals so that the alphabet of events matches the interface required for the black-box model. Lemma 2 states that the chain of 2 Relay nodes conforms to the single Relay node.

Lemma 3 and Lemma 4 consider two subsystems that each have exactly one Relay node. Both subsystems take the form visualized in Fig. 2b. The first module contains a Relay R using the model checking representation. The second module contains a Relay B that uses the theorem prover representation. Showing conformance equivalence between these two subsystems verifies that the two representations can be interchanged freely without changing the behavior of the system.

To understand the intuition of Lemma 5 first consider the system visualized in Fig. 2a with a Relay node R in place for the black box. This is equivalent to the system ($Ref, Orig, R_1$) of Lemma 1. Using Lemma 4 we substitute in a Relay node B. Using Lemma 2 we substitute a chain of two of these type of relay nodes

(B_n, B_{n+1}) in place of a single relay node B. This substitution is proven safe with a machine-checked proof described in Sect. 5. Finally, we consider each of those relay nodes B_n and B_{n+1} as individual black boxes and using Lemma 4 we switch back to the model checker representation to complete the proof that the properties hold for the system with one more relay node.

4 Model Checking

We would like to prove that the properties hold on a base system consisting of one Originator, one Relay node, and one Referee—$Prop(Ref, Orig, R_1)$. The temporal properties required for the MyCHIPs lift protocol can be proven by an exhaustive search of all possible states in this base system. The state required by the system can be split into two classes: the part of the state that is evaluated when checking the properties, and the state that is used only to determine which state transition to take. We provide a formal definition of the former and a description of the latter. The source code for the complete Spin model can be found in the code archive.

The system state is a combination of the individual states for each node. The state for the Originator and each Relay node can be represented as:

$$Node\ state \equiv \langle S, \Delta^s, \Delta^p \rangle; \quad S \in \{No\ Lift, Pend, Call, Void, Good\}; \quad \Delta^s, \Delta^p \in \mathbb{Z}$$

Here S represents the Mealy-machine state and Δ^s and Δ^p represent the change in balance with the node's successor or predecessor respectively. The state of the Referee is similar except that it omits the balances

$$Referee\ state \equiv \langle S \rangle$$

These are then composed together as a system state:

$$System\ state \equiv \langle S_O, S_1, S_{Ref}, \Delta^s_O, \Delta^p_O, \Delta^s_1, \Delta^p_1 \rangle$$

With the O subscript representing the *Originator*, 1 representing the *Relay* node, and *Ref* representing the *Referee*.

Each Mealy machine defined in Sect. 2 was implemented directly. This implementation consisted of three process types: Originator, Relay, and Referee. Each process has a unique set of edges marked as non-deterministic choices that are enabled only when the current state S matches the state of the edge in the diagram, and—if applicable—when the triggering message is present at the front of a message queue.

The message passing between nodes is modeled by creating three message channels for each node. One for messages received from the node's successor in the cycle, one for messages from the predecessor and one for messages from the Referee. The Referee also has a channel where it receives messages.

The details of the message are not important for the model, instead, each message will have a specific type. It is expected that nodes in a real system will verify that messages are valid as part of processing a received message. For the model, we will consider invalid messages as if they were never received. Once a valid message is received, the type of that message will inform the actions a node will take in response to the message.

When sending messages we model the possibility that a message might be lost in transit. One unique case is the possibility that all messages from a given node are dropped. This models the case where a node crashes or otherwise loses connection to the network.

The protocol requires that the Referee is consistently available so messages to and from the Referee are never dropped. Without this concession, the problem is isomorphic with the *two generals problem* [12], and it is impossible to guarantee eventual consensus on the lift status. However, the model will evaluate the possibilities that arise should those messages to and from the Referee take a very long time. Time is not modeled directly; rather, we allow all decisions that would be made by a node based on time to be a non-deterministic choice thus allowing for arbitrary, and even infinite, delays.

The Spin model verifies the following properties:

Definition 2 (Property 1)

$P_1 \equiv$ always eventually,
$$S_O \neq \text{Pend} \wedge S_1 \neq \text{Pend} \wedge S_{Ref} \neq \text{No Lift}$$

Definition 3 (Property 2)

$P_2 \equiv$ always eventually,
$$(S_O = \text{Good} \leftrightarrow S_1 = \text{Good} \leftrightarrow S_{Ref} = \text{Good})$$

Definition 4 (Property 3)

$P_3 \equiv$ always eventually,
$$(\Delta_O^s + \Delta_O^p \geq 0) \wedge (\Delta_1^s + \Delta_1^p \geq 0)$$

Definition 5 (Property 4)

$P_4 \equiv$ always eventually,
$$(\Delta_O^s = -\Delta_1^p) \wedge (\Delta_1^s = -\Delta_O^p)$$

To verify each of these properties, Spin evaluated 16K unique states with 173K edges between those states. Evaluating each property Spin found that there were no counterexamples. This result from Spin is the proof for Lemma 1 as described in Sect. 3.

5 Two Nodes Conform to One Node

We would like to prove that a chain of two Relay nodes conforms to a single Relay node in the context of a MyCHIPs system as implemented in the theorem prover. $\forall n \in \mathbb{N}, (B_n, B_{n+1}) \preceq (B_n)$. By Definition 1 this is can be reduced to: $Proj(P(B_n, B_{n+1}), B_n) \subseteq P(B_n) \longrightarrow (B_n, B_{n+1}) \preceq B_n$

The *partial order* defines the set possible traces of the model.

$$\forall T, \mathbb{V}(T, n) \leftrightarrow T \in P(B_n) \tag{1}$$

where $\mathbb{V}(T, n)$ is a predicate that is true when a trace T satisfies the *partial order* for a system of size n. When the predicate holds for a trace we say that trace is *valid* for the MyCHIPs protocol.

We use this predicate to state the conformance property necessary to prove safe substitution.

Theorem 3 (Valid projection implies conformance)

$$[\forall T \in P(B_n, B_{n+1}), \mathbb{V}(\mathbb{P}(T, n), n)] \longrightarrow (B_n, B_{n+1}) \preceq B_n$$

where $\mathbb{P}(T, n)$ *is a projection function that removes events from the trace that are not part of the alphabet of events in a system of size n. This is equivalent to a combination of* hide *and* rename *operations as described by Dill.*

Proof. Introduce:
$$\forall T \in P(B_n, B_{n+1}), \mathbb{V}(\mathbb{P}(T, n), n) \tag{2}$$

This is equivalent to:
$$\forall T \in Proj(P(B_n, B_{n+1}), B_n), \mathbb{V}(T, n) \tag{3}$$

Apply (1) and (3):
$$\forall T, T \in Proj(P(B_n, B_{n+1}), B_n) \longrightarrow T \in P(B_n) \tag{4}$$

By Definition 1 this is equivalent to:
$$Proj(P(B_n, B_{n+1}), B_n) \subseteq P(B_n) \tag{5}$$

$$\therefore (B_n, B_{n+1}) \preceq B_n \tag{6}$$

□

5.1 Machine-Checked Proof

We use Coq to prove that a chain of $n+1$ Relay nodes can be projected onto a chain of n nodes.

$$\forall T \in P(B_n, B_{n+1}), \mathbb{V}(T, n+1) \longrightarrow \mathbb{V}(\mathbb{P}(T, n), n) \tag{7}$$

This allows us to prove that the left-hand side of Theorem 3 holds for all valid systems, which in turn allows us to prove conformance.

The *partial order* and the *projection* are formally defined in Sects. 5.2 and 5.3. Section 5.4 describes the proof script briefly. For details of the proofs, we direct the reader to examine the machine-checked proof script which can be found in the code archive.

5.2 Partial Order Description

The Coq proof operates on the set of all traces of all events that might occur in the system. Each trace is represented by an ordered list of events.

The *partial order* is defined in *acts_valid*. There are eight rules that define the *partial order*:

1. *has_required_actions*,
2. *has_no_duplicate_receives*,
3. *all_receives_causal*,
4. *all_sends_triggered*,
5. *all_ids_in_range*,
6. *promise_forward_commit_backward*,
7. *phase_sequence_correct*,
8. *all_ref_receives_causal*.

All must hold for a trace of events to be considered valid. The rules were designed to match the possible traces that could be generated by a system of nodes for the given size. For brevity, only the most interesting rule *all_receives_causal* is defined here. For definitions of the other rules we direct the user to the machine-checked proof script in the code archive.

For the following definitions we use the notation $\mathbb{A}$ to represent the action type.

$$\mathbb{A} \equiv \{(Send, s, d, m), (Receive, d, m), (SendRef, s)(ReceiveRef, d)\}$$

where $s \in \mathbb{N}$ is the source identifier $0 \leq s < size$, $d \in \mathbb{Z}$ is the destination identifier $-1 \leq d \leq size$, and m is a message $m \in \{Promise, Commit\}$.

We also use the notation T where $T \equiv (a_1, a_2, ..., a_n)$, $a_i \in \mathbb{A}$, to represent a *trace* of events and denote the size of the system $S \in \mathbb{N}$ which is a count of the number of nodes in the cycle.

Definition 6 (all_receives_causal). *Every* Receive *action is preceded by a corresponding* Send *action. Given* $r = (Receive, d, m), r \in T \longrightarrow \exists a = (Send, s, d^*, m), a \in T \land d^* = d \mod S \land a \prec_T r$. *Where* $\prec_T$ *indicates that a occurs before r in the trace of events.*

Because we allow the destination of a *Send* to be an integer between -1 and the size *inclusive* this means that a node can send a message with a destination -1 and this always corresponds with the node with the maximum ID in the system. Additionally, a node might send a message with the destination n, but when we project to size n the node with ID n is removed from the system. This send—instead of being sent to a node that doesn't exist—now gets mapped to node 0 because the destination need only be equal modulo the size. This allows for the projection to work without the need to mutate events, which makes proving properties about the projected system much simpler.

5.3 Projection Definition

Projection is defined with the *projected* function. Given an action and the size of the system we would like to project onto the *projected* function determines if the given action is kept or omitted in the projected trace. It uses a special type $option\mathbb{A} \equiv \{None, Some\ a\}$ with $a \in \mathbb{A}$.

projected returns *None* if the given action, a, should be omitted and *Some a* if the action should be kept in the projected trace.

Definition 7 (projected).

$$a = (Send, s, d, m) \land s \geq S \longrightarrow projected(a, S) = None$$

$$a = (Send, s, d, m) \land d > S \longrightarrow projected(a, S) = None$$

$$a = (Receive, d, m) \land d \geq S \longrightarrow projected(a, S) = None$$

$$a = (SendRef, s) \land s \geq S \longrightarrow projected(a, S) = None$$

$$a = (ReceiveRef\ d) \land d \geq S \longrightarrow projected(a, S) = None$$

$$Otherwise \longrightarrow projected(a, S) = Some\ a$$

Definition 8 (project_to_size). *Given a size and a list of events, project_to_size, denoted* $\mathbb{P}$, *returns a new list of events that omits all events for which* projected *returns None.*

$$a \in T \land projected(a, S) = Some\ a \longrightarrow a \in \mathbb{P}(T, s)$$

$$projected(a, S) = None \longrightarrow a \notin \mathbb{P}(T, s)$$

5.4 Proofs

The machine-checked proof consists of a set of lemmas that progressively build from basic principles to Theorem 4. We support this claim with Lemmas 6–16.

Theorem 4. *larger_conforms_to_smaller*

$$\forall T \in [\mathbb{A}]\ \mathbb{V}(T, n+1) \land n > 1 \longrightarrow \mathbb{V}(\mathbb{P}(T, n), n).$$

Where $[\mathbb{A}]$ is an arbitrary sequence of events.

Lemmas 6 through 12 all have the form

Lemma 6 (through 12). *property_independent_of_proj*

$$\forall T \in [\mathbb{A}]\ n \in \mathbb{N}, n > 1 \land property(T, n+1) \longrightarrow property(\mathbb{P}(T, n), n)$$

For each of the properties: has_required_actions, has_no_duplicate_receives, all_ids_in_range, promise_forward_commit_backward, phase_sequence_correct, all_ref_receives_causal and all_sends_triggered.

Lemma 13. *valid_implies_all_receives_causal_in_proj*

$$\forall T \in [\mathbb{A}] \; n \in \mathbb{N}, n > 1 \wedge \mathbb{V}(T, n+1) \longrightarrow$$
$$\text{all\_receives\_causal}(\mathbb{P}(T,n), n)$$

Lemma 13 is the least trivial and the core problem that needed to be proven to show that the larger system conforms to the smaller system. The *Receive* causality property is difficult because a *Send* action that is associated with a given *Receive* in the original system may be removed in the projection. However, if the *Send* for a *Receive* is projected out there is an equivalent *Send* from that node's predecessor that now can be matched with that *Receive*.

This lemma is proven by careful case analysis. The proof itself spans nearly 500 lines and makes use of information known about the trace of events based on the *acts_valid* assumption.

We begin by introducing an action r which is the *Receive* in question. We assume it exists in the projected list and that it is indeed a *Receive* and use the sub-lemma *in_proj_in_orig* to show that the action must also appear in the original trace. With the knowledge that the action was in the original trace the assumption of the validity of the original trace allows us to show that there must be a corresponding *Send* for that *Receive* in the original trace. We now know that there is a *Receive* that is both in the original trace and in the projected trace and that there is a corresponding *Send* in the original trace. From here we examine the possible scenarios for what happens to the corresponding *Send* in the projection.

1. Both are unchanged. Neither the *Send* nor the *Receive* are removed and neither *wrap* around in the modulus.
2. The *Send* source alone is projected out. The source of the *Send* refers to a node that is no longer in the system.
3. The *Send* destination alone is projected out and the send now *wraps*.
4. Both the *Send* source and the send destination refer to a node that is removed.

Scenario 1 is pretty straightforward, the *Send* for the *Receive* is the same *Send* as in the original trace. We just show that the send remains in the projection.

Scenario 2 is tricky. The previously associated *Send* is now projected out. But the *Receive* is not projected out. To prove this scenario we use information about the required events in the original trace. After unpacking this information we can show that there must be a *Send* that would have gone to the node that was projected out. We show that this *Send wraps* due to the change in the size which means that that send is now associated with the *Receive* in question.

Scenario 3 is the dual of scenario 2. The same problem but from the other perspective. This is the case where a node that is not projected out sends to a node that is projected out. We show that that message *wraps* to connect with a new *Receive* which is our *Receive* in question.

Scenario 4 is an interesting case because it is only possible if we project out more than one node in a single step. The way the property is defined makes this case not possible because we have a system that is valid for size $n+1$ that is projected to size n. It takes some mathematics and a few properties about modular arithmetic to show that this case always leads to contradictions.

Once each of these cases were individually proven or shown to lead to contradictions the property proves out.

Lemma 14. *not_in_orig_not_in_proj*

$$\forall T \in [\mathbb{A}] \ a \in \mathbb{A} \ n \in \mathbb{N}, a \notin T \longrightarrow a \notin \mathbb{P}(T,n)$$

Lemma 15. *in_proj_in_orig*

$$\forall T \in [\mathbb{A}] \ a \in \mathbb{A} \ n \in \mathbb{N}, a \in \mathbb{P}(T,n) \longrightarrow a \in T$$

Lemma 16. *happens_before_independent_of_proj*

$$\forall T \in [\mathbb{A}] \ n \in \mathbb{N}, a \ b \in T \wedge a \ b \in \mathbb{P}(T,n) \wedge a \prec_T b \longrightarrow a \prec_{\mathbb{P}(T,n)} b$$

Lemmas 6–16 and Theorem 4 were certified correct by Coq. These results prove Lemma 2. Theorem 4 tells us that the set of possible traces of events is independent of the size of the system. Because this is true for a chain of n nodes we know that the possible traces of events in a chain of n nodes is equivalent to possible traces of events in a single node. With this and Theorem 3 we know that a chain of n Relay nodes conforms to a single node.

6 Conformance Equivalence

We would like to prove Lemma 3 and Lemma 4:

$$B \preceq R \ \wedge \ R \preceq B$$

By Definition 1 it is sufficient to prove:

$$Proj(P(B), R) \subseteq P(R) \wedge Proj(P(R), B) \subseteq P(B)$$

By design, when we construct traces for $P(R)$ we hide events that represent changes to the internal state of a node using the Hide operation described by Dill. Once this is done the alphabet of events in $P(R)$ will be equivalent to the alphabet of events in $P(B)$. This allows us to reduce the proof requirement to:

$$P(B) \subseteq P(R) \wedge P(R) \subseteq P(B)$$

We will consider the left side of the conjunction in Sect. 6.1 and the right side of the conjunction in Sect. 6.2.

6.1 Possible Traces of Model Checker Respect the *Partial Order*

If $P(R) \not\subseteq P(B)$ then there must exist a trace explored by Spin that does not respect the *partial order*. By encoding the *partial order* as properties in Spin, we can verify that the rules hold for all of the traces evaluated by Spin.

To implement the properties the Spin model needed to be extended to generate a data structure that recorded the trace of events. To simplify this structure we rely on the fact that messages received by nodes are *idempotent*. If a node receives a message more than once the state will be the same as if it received the message only once. Because of this we can use Boolean flags for each type of message at a given node, and make assertions about these Booleans to verify the *partial order*. The Spin model was adjusted to set these flags when a message was sent and when a message was received.

The constraint that certain events exist is parameterized on the number of nodes in the lift cycle. This parameter is referred to as the *size* of the system. The base system has the Originator and one Relay so the base system is of *size two*. Rather than parameterizing the properties we verify in Spin we write the properties exclusively for a system of *size two*. However, there are cases explored by Spin that when observing the trace of events appear to not include a Relay node. For example, consider the case where the Originator fails to send the first promise to the Relay. Eventually, the Originator requests the status from the referee and the referee will respond *Void* which rolls back the lift. However, this event trace does not contain any messages sent to or from the Relay. Because the Relay doesn't appear in the trace, this trace is indistinguishable from a system with no Relay, a system of *size one*.

In Theorem 2 we specify that the properties must hold for $n \geq 1$. The system of size one corresponds to $n = 0$ so this is a case we would like to ignore. But these cases where the system is effectively *size one* cause Spin to produce some counter-examples that violate the *has_required_actions* rule.

To exclude these counter-examples we introduce a new fairness property that eliminates cases that represent systems smaller than the base system. This property, *size_fair*, requires that each node eventually leave the *no_lift* state ensuring that each node in the system is represented at least once in the trace. We then write each of the properties in the form: $size\_fair \longrightarrow property$. After triggering on this fairness property, all of the *partial order* for all traces evaluated by Spin.

When adding a fairness property it is prudent to check that the fairness property can be met, otherwise, we risk the properties verifying because they are vacuously true. To ensure that the new *size_fair* property is achieved we write the property: $always\ !size\_fair$. We expect this property to fail which indicates that the fairness constraint does not make our properties vacuously true.

The implementation of these properties can be found in the code archive. Spin evaluated these properties in 11K unique states with 74K edges between those states and found no counterexamples. These results verify that $P(R) \subseteq P(B)$.

6.2 Possible Traces of the *Partial Order* are a Subset of the Definition

We would like to show: $P(B) \subseteq P(R)$. This work fails to produce a formal proof of this assertion. However, the following evidence provides high assurance that this assertion is true.

Consider the set of all the possible combinations of events for a single node. Because a node might emit a given event more than once, this set is infinitely large. Each of these events is associated with sending or receiving a message. Because messages are idempotent in MyCHIPs, the set of possible events we need to consider can be reduced into a small set of equivalence classes where each possible trace of events is represented by a canonical member that has only one of each type of event.

We can represent a canonical member of each of these equivalence classes as an ordered list of events. We use the following abbreviations for the alphabet of events: *Promise Send* $\equiv PS$, *Promise Receive* $\equiv PR$, *Commit Send* $\equiv CS$, *Commit Receive* $\equiv CR$, *Status Send* $\equiv SS$, and *Status Receive* $\equiv SR$.

Many of the possible orderings do not satisfy the *partial order*. After eliminating the impossible cases the remaining equivalence classes are:

1. $[PR, SS, SR]$
2. $[PR, PS, SS, SR]$
3. $[PR, PS, CR]$
4. $[PR, PS, CR, CS]$
5. $[PR, PS, SS, CR, SR]$
6. $[PR, PS, SS, SR, CR]$
7. $[PR, PS, SS, SR, CR, CS]$
8. $[PR, PS, SS, CR, SR, CS]$
9. $[PR, PS, SS, CR, CS, SR]$

We can evaluate if each of these equivalence classes is considered in the Spin model by generating a property that asserts that a trace of that class is not present. If this property fails we know that the trace is evaluated at least once in the Spin model.

The properties used to generate a witness trace for each equivalence class are defined as an automaton. The implementation of these automata can be found in the code archive. After executing the model checker to search for each of these new properties, each equivalence class was observed in the Spin model. Spin evaluated relatively few system states before finding each witness trace. The fastest search only evaluated 36 states and the longest evaluated 8K states. This allows us to conclude with high assurance that $P(B) \subseteq P(R)$.

This together with $P(R) \subseteq P(B)$ as proven in Sect. 6.1 allows us to conclude:

$$B \preceq R \land R \preceq B$$

7 Related Work

Numerous bodies of work solve similar problems or utilize similar constructs. The following list is particularly relevant.

Dill's logic has been extended to reason about software and distributed message-passing systems [9,17]. Rajamani and Rehof's work uses conformance to show the substitutability of a model implementation and a definition implementation similar to this work's method in Sect. 6. Their method uses the CCS and π-calculus along with custom notation for non-deterministic choices. This work uses a partial order which at first brush seemed easier to define and prove in the Coq theorem prover. It is possible that Rajamani and Rehof's method may be equally easy, but an exploration of that is left as future work.

Clarke, Grümberg, and Browne describe a method for verifying properties on systems with many identical subprocesses. This method relies on a manual proof of bisimulation which is difficult to prove if these subprocesses can make non-deterministic choices as the nodes in the MyCHIPs system can. The method described in this work uses some similar techniques to the method described by Clarke *et al.* but presents it in terms of conformance and operates on an alternative representation of the system for the inductive proof.

German and Sistla build on the work of Clarke, Grümberg, and Browne, but present a fully automatic method that can reason about multiple identical processes without any manual proofs of those processes.

German and Sistla's work is often referenced when describing methods for model-checking infinite state spaces [1,5,14]. The methods adjust the state matching to weaken the equivalence relation. Two states may be unique due to some infinite variable (a natural number) being different but they are equivalent in all ways we care about for verifying a property. This allows model-checking to operate on an infinite state space and is somewhat akin to abstract interpretation.

Fischer, Lynch, and Paterson [10], prove the impossibility of consensus on even a Boolean, with even one faulty (or malicious) process. However, this is only true if the processes don't have synchronized clocks. This proof shows the necessity of the Referee with strong reachability requirements for the lift algorithm to always eventually reach consensus on the lift's final status.

Schneider summarizes and frames many fault-tolerant distributed algorithms in the framework of state machines [18]. He shows how many common algorithms are isomorphic to, and can be derived using, the state machine approach. It is a helpful method to characterize and compare different approaches and this work is inspired by these methods.

Delzanno, Tatarek, and Traverso, model check a common consensus algorithm called Paxos in Spin. Their Spin constructs provide helpful examples of how distributed algorithms are efficiently modeled [7].

Ogles, Aldous and Mercer prove that *doesn't commute*, a weakened version of the happens-before relation, is sound for certain common classes of task parallel programs. They present a machine-checked proof that proves properties for all traces constrained by a partial order. The methods demonstrated by Ogles,

Aldous and Mercer were used as inspiration for some of this work's machine-checked proofs [15].

8 Conclusion

Conformance can be an effective tool to verify that model-checking results extend to arbitrarily large systems. It provides a framework for showing equivalence between a representation of a system suitable for model checking and a representation of a system suitable for inductive proofs.

For the MyCHIPs system, this work proved that the representation used in the model checker can be safely substituted into the theorem prover. However, it failed to produce a formal proof that the representation used in the theorem prover can be safely substituted into the model checker. However, we provide evidence that gives high assurance of the conformance equivalence between the two representations.

Because the inductive proof only relies on the *partial order*, new properties of the MyCHIPs lift protocol can be verified through model checking without making adjustments to the machine-checked proof.

For some systems, the high assurance this work provides is insufficient. Future work will need to prove conformance equivalence between the two representations. We foresee three plausible paths forward:

1. Add a function in the Coq proof script to produce a complete set of traces for the base system. Then, use a similar method used in Sect. 6.1 to produce a witness for each of these traces.
2. Define the representation of the system used for the model checker directly in Coq. While this representation is not conducive to inductive reasoning, proving Lemma 4 may be possible.
3. Prove that a weaker equivalence than a conformance equivalence is sufficient to prove that the properties extend to a system of arbitrary size.

While the method applied in this work requires considerable manual effort, a similar method could be used to create automated approaches that combine many verification techniques.

References

1. Abdulla, P., Haziza, F., Holík, L.: Parameterized verification through view abstraction. Int. J. Softw. Tools Technol. Transf. **18**(5), 495–516 (2016). https://doi.org/10.1007/s10009-015-0406-x
2. Athalye, A.: CoqIOA : a formalization of IO automata in the Coq proof assistant. Master's thesis, Massachusetts Institute of Technology (2017)
3. Bateman, K.: MyCHIPs digital money (2023). http://gotchoices.org/mychips/intro.html
4. Bateman, K.: MyCHIPs protocol description 1.3. MyCHIPs Protocol Description 1.3 (2023). https://github.com/gotchoices/MyCHIPs/blob/master/doc/learn-protocol.md

5. Bozga, M., Iosif, R., Sifakis, J.: Verification of component-based systems with recursive architectures. Theoret. Comput. Sci. **940**, 146–175 (2023). https://doi.org/10.1016/j.tcs.2022.10.022. https://www.sciencedirect.com/science/article/pii/S0304397522006181
6. Clarke, E.M., Grumberg, O., Browne, M.C.: Reasoning about networks with many identical finite-state processes. In: Proceedings of the Fifth Annual ACM Symposium on Principles of Distributed Computing, PODC 1986, pp. 240–248. Association for Computing Machinery, New York (1986). https://doi.org/10.1145/10590.10611
7. Delzanno, G., Tatarek, M., Traverso, R.: Model checking Paxos in Spin. Electron. Proc. Theor. Comput. Sci. **161**, 131–146 (2014). https://doi.org/10.4204/eptcs.161.13
8. Dill, D.L.: Trace Theory for Automatic Hierarchical Verification of Speed-Independent Circuits, vol. 24. MIT Press, Cambridge (1989)
9. Driscoll, E., Burton, A., Reps, T.: Checking conformance of a producer and a consumer. In: Proceedings of the 19th ACM SIGSOFT Symposium and the 13th European Conference on Foundations of Software Engineering, ESEC/FSE 2011, pp. 113–123. Association for Computing Machinery, New York (2011). https://doi.org/10.1145/2025113.2025132
10. Fischer, M.J., Lynch, N.A., Paterson, M.S.: Impossibility of distributed consensus with one faulty process. In: Proceedings of the 2nd ACM SIGACT-SIGMOD Symposium on Principles of Database Systems, PODS 1983, pp. 1–7. Association for Computing Machinery, New York (1983). https://doi.org/10.1145/588058.588060
11. German, S.M., Sistla, A.P.: Reasoning about systems with many processes. J. ACM **39**(3), 675–735 (1992). https://doi.org/10.1145/146637.146681
12. Gray, J.N.: Notes on data base operating systems. In: Bayer, R., Graham, R.M., Seegmüller, G. (eds.) Operating Systems. LNCS, vol. 60, pp. 393–481. Springer, Heidelberg (1978). https://doi.org/10.1007/3-540-08755-9_9
13. Huang, Y., Ogles, B., Mercer, E.: A predictive analysis for detecting deadlock in MPI programs. In: 2020 35th IEEE/ACM International Conference on Automated Software Engineering (ASE), pp. 18–28 (2020)
14. Lowe, G.: Parameterized verification of systems with component identities using view abstraction. Int. J. Softw. Tools Technol. Transf. **24**(2), 287–324 (2022). https://doi.org/10.1007/s10009-022-00648-0
15. Ogles, B., Aldous, P., Mercer, E.: Proving data race freedom in task parallel programs using a weaker partial order. In: 2019 Formal Methods in Computer Aided Design (FMCAD), pp. 55–63 (2019). https://doi.org/10.23919/FMCAD.2019.8894270
16. Ozanne, L.: Learning to exchange time: Benefits and obstacles to time banking. Int. J. Community Currency Res. **14** (2010)
17. Rajamani, S.K., Rehof, J.: Conformance checking for models of asynchronous message passing software. In: Brinksma, E., Larsen, K.G. (eds.) CAV 2002. LNCS, vol. 2404, pp. 166–179. Springer, Heidelberg (2002). https://doi.org/10.1007/3-540-45657-0_13
18. Schneider, F.B.: Implementing fault-tolerant services using the state machine approach: a tutorial. ACM Comput. Surv. **22**(4), 299–319 (1990). https://doi.org/10.1145/98163.98167

19. Schuppan, V., Biere, A.: Liveness checking as safety checking for infinite state spaces. Electron. Notes Theor. Comput. Sci. **149**(1), 79–96 (2006). https://doi.org/10.1016/j.entcs.2005.11.018. https://www.sciencedirect.com/science/article/pii/S1571066106000557. Proceedings of the 7th International Workshop on Verification of Infinite-State Systems (INFINITY 2005)
20. Valek, L.: The time bank implementation and governance: Is PRINCE 2 suitable? Procedia Technol. **16**, 950–956 (2014). https://doi.org/10.1016/j.protcy.2014.10.048. https://www.sciencedirect.com/science/article/pii/S2212017314002758

Correctness Witnesses for Concurrent Programs: Bridging the Semantic Divide with Ghosts

Julian Erhard[1,5](✉), Manuel Bentele[2,6], Matthias Heizmann[4], Dominik Klumpp[2], Simmo Saan[3], Frank Schüssele[2], Michael Schwarz[1], Helmut Seidl[1], Sarah Tilscher[1,5], and Vesal Vojdani[3]

[1] Technical University of Munich, Garching, Germany
julian.erhard@tum.de
[2] University of Freiburg, Freiburg, Germany
[3] University of Tartu, Tartu, Estonia
[4] University of Stuttgart, Stuttgart, Germany
[5] Ludwig-Maximilians-Universität München, Munich, Germany
[6] Hahn-Schickard, Villingen-Schwenningen, Germany

Abstract. Static analyzers are typically complex tools and thus prone to contain bugs themselves. To increase the trust in the verdict of such tools, *witnesses* encode key reasoning steps underlying the verdict in an exchangeable format, enabling independent validation of the reasoning by other tools. For the correctness of concurrent programs, no agreed-upon witness format exists—in no small part due to the divide between the semantics considered by analyzers, ranging from interleaving to thread-modular approaches, making it challenging to exchange information. We propose a format that leverages the well-known notion of *ghosts* to embed the claims a tool makes about a program into a modified program with ghosts, such that the validity of a witness can be decided by analyzing this program. Thus, the validity of witnesses with respect to the interleaving and the thread-modular semantics coincides. Further, thread-modular invariants computed by an abstract interpreter can naturally be expressed in the new format using ghost statements. We evaluate the approach by generating such ghost witnesses for a subset of concurrent programs from the SV-COMP benchmark suite, and pass them to a model checker. It can confirm 75% of these witnesses—indicating that ghost witnesses can bridge the semantic divide between interleaving and thread-modular approaches.

Keywords: Software Verification · Correctness Witnesses · Concurrency · Ghost Variables · Abstract Interpretation · Model Checking

1 Introduction

While static analysis tools can help developers write bug-free programs, these tools sometimes return incorrect verdicts due to bugs in the tools themselves.

To increase the trust in the verdicts of static analysis tools, *witnesses* have been proposed as artifacts that contain further information about static analysis results [8,9]. For indicating that a program property does not hold, a witness may exhibit a program execution that violates the property (*violation witness*). For substantiating that a property holds throughout all possible executions, a suitable set of program invariants may be provided that guides the static analysis tool towards proving the property of interest (*correctness witness*)—and may expose errors in reasoning when invariants can be shown to be violated. *Validators* are static analysis tools that consume witnesses and report whether they can re-establish verdicts. They are, e.g., used in SV-COMP [5]. Here, we are interested in a witness format suitable for certifying the correctness of concurrent programs. We extend the format for correctness witnesses by Ayaziová et al. [3], which allows expressing invariants per location, with *ghost variables*. Ghost variables have not only been proposed for the verification of sequential programs [71] but have also been employed for concurrent programs [2,42,66]. Ghost variables—sometimes referred to as auxiliary variables—are additional program variables introduced to ease the specification and verification of intricate program properties. These variables allow encoding the progress of other threads, making it possible to relate observations that the current thread may make to this progress. A witness then specifies a set of ghost variables and how they evolve via *ghost updates* inserted at existing program locations. Additionally, it contains *invariants*, which can refer both to program and ghost variables and, thus, to properties that may be difficult to express without ghosts.

```
1   unsigned int used = 0;              1   unsigned int used = 0;

2   main:                                2   main:
3     create (t₁);                       3     create (t₁);
                                                 atomic { assert (🔒 == 0 ⟹ used == 0); }
4     lock (m);                          4     atomic { lock (m); 🔒 = 1; }
5     assert (used == 0);                5     assert (used == 0);
6     unlock (m);                        6     atomic { unlock (m); 🔒 = 0; }

7   t₁:                                  7   t₁:
8     lock (m);                          8     atomic { lock (m); 🔒 = 1; }
9     used = 47;                         9     used = 47;
10    used = 0;                          10    used = 0;
11    unlock (m);                        11    atomic { unlock (m); 🔒 = 0; }
```

(a) Concurrent program. (b) Concurrent program with ghosts.

List.1. Example program without(a) and with ghost statements (b). Parts highlighted in blue show ghost statements from a witness.

Example 1. The program from Listing 1a is an example of a *resource invariant*: the variable used is always 0 except when thread t_1 holds the mutex m. We cannot state this property with C assertions. Instead, in Listing 1b we add a ghost variable 🔒 and maintain it such that it indicates whether the mutex m is currently held by any thread. The assertion in *main* states that, when no thread is in a critical section, the value of the global variable `used` is 0.

A core requirement for a witness format is that it facilitates the exchange of information between tools. Different tools employ different semantics to formalize the behavior of (sequentially consistent) concurrent programs. Some may use an interleaving semantics [49], while others turn to a thread-modular semantics [48,60,72]. After showing that safety of programs coincides for the interleaving and a thread-modular semantics (Sect. 2), we introduce a witness format that allows instrumenting a program with ghost statements (Sect. 3) that are executed atomically with existing statements. We show how, by this construction, the validity of witnesses with respect to the interleaving and the thread-modular semantics also coincides (Sect. 4). We exemplify how ghost witnesses are naturally suited to express information obtained for concurrent programs by a thread-modular abstract interpreter (Sect. 5). Our format for ghost witness (Sect. 6) extends the existing SV-COMP witness format, easing adoption by other software verifiers. For the experimental evaluation, we automatically generate witnesses using GOBLINT [68,81], and validate them using the model checker ULTIMATE GEMCUTTER [29,44] (Sect. 7).

2 Two Views on Concurrent Programs

We first introduce the notion of programs that we will consider and present an interleaving and a thread-modular semantics for those programs. To simplify the presentation, we model a core subset of the C language, which is later extended by allowing the insertion of ghost statements. The language supports dynamic thread creation, locking and unlocking of mutexes, and reading and writing of global variables. Function calls, pointers and heap memory are skipped to not overcomplicate the exposition.

2.1 Programs

A program is given by a set of global variables $\mathcal{G}$, a set of mutexes $\mathcal{M}$, a set of local states $\mathcal{L}$ and a finite number of named control-flow graphs T_{templ}, called *thread templates*, one of which is `main`. We demand that $\mathcal{G} \cap \mathcal{M} = \emptyset$, and further, that there is a set of global declarations $\mathcal{D}$ that provides types and initial values for all global variables. Local states are a type-correct mapping from local variables $\mathcal{X}$ to values $\mathcal{V}$. For simplicity, we assume that all threads use the same set of local variables, and we omit procedures. Each control-flow graph consists of a finite set of nodes $\mathcal{N}$ and labeled edges $\mathcal{E}$. The sets of program points are disjoint between the different control-flow graphs. An edge $(u, a, v) \in \mathcal{E}$ consists of a source node u, an action a, and a sink node v. We demand that for a given u, v, there may be at most one a, such that $(u, a, v) \in \mathcal{E}$. Each control-flow graph has a dedicated initial node with no incoming edges, at which execution is meant to start. The following statements are supported:

Correctness Witnesses for Concurrent Programs   77

Locking/Unlocking mutexes. The actions lock(m) and unlock(m), where m is some mutex.

Thread creation. The action create(t), where t is the name of some thread template.

Local Update. The action l := f l, where f is a pure function taking the local state l, yielding the new local state of the thread.

Global Read. The action l := f l g, where f is a pure function taking the local state l and the value of some global variable g, yielding the new local state of the thread.

Global Write. The action g := f l, where f is some pure function taking the local state l, updating the global variable.

Assertion. The action assert(p l), where p is some pure function mapping the local state l to a boolean value.

Guard. The actions Pos(c l) and Neg(c l), realizing branching on some condition c over the local state l, where c is a pure function yielding a boolean.

We refer to the language of programs with these actions as Lang.

2.2 Interleaving Semantics

The interleaving semantics is a concrete semantics of concurrent programs where all actions in an execution are totally ordered. To distinguish threads, they are identified with a thread id computed from their creation history. The set of all possible thread ids for a program, in the following often simply called threads, is referred to as T. The thread id of the initial thread is the empty sequence, while for a created thread, the thread id is obtained by concatenating the thread id of the parent thread with the number of threads the parent thread has created before. A program configuration (or state) $(L, M, G) \in \mathcal{S} : (T \rightharpoonup (\mathcal{N} \times \mathcal{L})) \times (\mathcal{M} \rightharpoonup T) \times (\mathcal{G} \to \mathcal{V})$ is a triple, where L is a partial mapping of threads T to program nodes $\mathcal{N}$ and local states $\mathcal{L}$, M is a partial mapping from mutexes to the threads that hold them, and G is a mapping of global variables to values. Each thread has a local variable self that contains its thread id that is only set when the thread is created. We write $M\, m \uparrow$, if the value of M is not defined for m, i.e., the mutex m is not held by any thread in M. An execution of a program is given by a sequence of program configurations $(s_i)_{1 \leq i \leq k}$, with $s_i \in \mathcal{S}$, for some natural number $k \geq 1$, interleaved with a sequence $((e_i, t_i))_{1 \leq i \leq k-1}$, with $e_i \in \mathcal{E}, t_i \in T$, of edges annotated with the thread taking them. For notational convenience, let us denote an execution step via an edge (u, a, v) from state s_i to state s_{i+1} taken be thread t, as follows:

$$s_i \xrightarrow{(u,a,v)}_t s_{i+1}$$

We demand that the interleavings are *consistent*. The set of consistent interleavings is defined inductively:

I1 The sequence consisting of some initial state $s_0 = (\{t_0 \mapsto (\mathsf{st}_{\mathsf{main}}, l_0, 0)\}, \emptyset, g_0)$ where l_0 is the initial local state, i.e., maps self to the initial thread id t_0

and all other local variables to some initial value 0, and g_0 maps all globals to their initial value according to the global declarations $\mathcal{D}$, is consistent.

I2 Let i be a consistent interleaving ending in state $s = (L, M, G)$. Then i can be extended to a consistent interleaving with $s \xrightarrow{(u,a,v)}_t s' = (L', M', G')$ if
  (a) the edge (u, a, v) is in the control-flow graph for the prototype of t
  (b) t is at program node u in s
  (c) a is an *admissible* action for thread t, i.e.,
    – if $a \equiv$ unlock(m) it must hold that M m $= t$,
    – if $a \equiv$ lock(m) it must hold that M m $\uparrow$,
    – if $a \equiv$ Pos(c l) or $a \equiv$ Neg(c l), c l must evaluate to true, and false respectively, on the current local state l for t,
  (d) s' *reflects* the effect of action a of thread t (for a detailed description see the extended version of the paper [25]) and stores the new node v for t.

The set of all such consistent sequences of a program P forms the set of interleaving executions $[\![P]\!]_\mathcal{I}$. We denote the set of all consistent interleavings by $\mathcal{I}$. An interleaving of Listing 1a is as follows:

```
   create(t₁)    lock(m)   used = 47  used = 0   unlock(m)   lock(m)   tmp = used   assert(...)
○────────────○────────────○──────────○──────────○───────────○─────────○────────────○───────────○
    main         t₁          t₁         t₁         t₁         main       main          main
```

Here, the read of the global variable is extracted from the assertion, to follow the restrictions of Lang. We say that an assertion is *violated in an interleaving* if and only if it evaluates to false in the interleaving. We say a program is *safe* w.r.t. the interleaving semantics if and only if there are no interleavings of the program in which any of its assertions is violated.

2.3 Thread-Modular Semantics

The interleaving semantics assumes the existence of a global observer that can order all events totally. In contrast to that, the local and global trace semantics [72,73] order the events of a program execution in a *partial order* of local configurations of threads. A local configuration (n, u, l) consists of a local state l, a number $n \in \mathbb{N}$ of steps performed by the thread, and the program node u the thread is at. The local state contains the values of local variables, including the id of the thread in the variable self. This semantics requires programs from a language Lang$_{\mathcal{M}_\mathcal{G}}$, where there is a dedicated mutex for each global, $\mathcal{M}_\mathcal{G} = \{m_g \mid g \in \mathcal{G}\} \subseteq \mathcal{M}$, and at each access to a global variable g by a thread t, the thread holds the mutex $m_g \in \mathcal{M}_\mathcal{G}$. This can be achieved by e.g. inserting actions lock(m_g) and unlock(m_g) before and after every access to g. In the partial order of a global trace, events within the same thread are ordered according to the program order. Additionally, there exist special sets of *observable* events and *observer* events, which contain those events where information from one thread is published to be observed by another thread. In our setting, unlock(m) is an observable event with the corresponding observer being lock(m). A global trace orders every observing event after the event it observes.

We refer to the reflexive transitive closure of the union of the program order, the synchronizing actions, and the create order as the *causality order*. A *global trace* may contain multiple maximal elements. We call global traces that have a unique maximal element *local traces*. The thread to which this maximal element belongs, is called the *ego-thread*. For a given ego-thread reaching some configuration, the local trace records all other local configurations that may have transitively influenced the ego-thread.

Values of global variables are not contained in a local program state; instead, the value of a global g is determined by looking at the last write that was performed to g in the local trace. Global traces can be viewed as acyclic graphs; for the program in Listing 1, the following shows an example local trace, where mutex m_{used} is abbreviated with m_u:

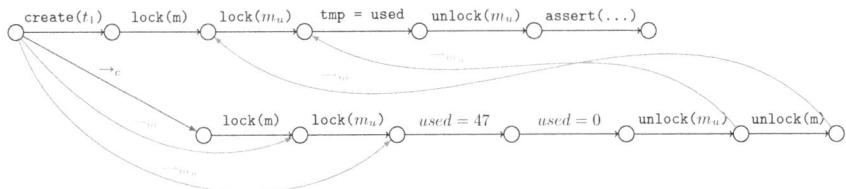

Following the formal description by Schwarz et al. [72], requirements for a consistent global trace in particular are:

Causality Order The partial order has a unique least element.
Create Order Each thread except the initial thread is created by exactly one create action, and each create action creates at most one thread.
Lock Order For a given mutex m, each lock operation is preceded by exactly one unlock operation, or it is the first lock operation of the mutex m. Each unlock operation for m is followed by at most one lock operation of m.
Reads of Globals The value read from a global variable must agree with the last write performed to that variable in the global trace or the initial value.

Schwarz et al. [72] give a fixpoint formulation of the set of local traces of a program and thread-modular analyses that compute abstractions of these. We denote the set of global traces and the set of local traces of a program $P \in \text{Lang}_{\mathcal{M}_G}$ by $[\![P]\!]_{\mathcal{GT}}$ and $[\![P]\!]_{\mathcal{LT}}$, respectively. We say an assertion is *violated in a local trace* if it evaluates to false in that local trace. We say a program is *safe w.r.t. the local trace semantics* if and only if there are no local traces of the program that violate any of its assertions.

2.4 Equivalence of Interleaving Semantics and Local Trace Semantics w.r.t. Safety

We show that the interleaving and the local trace semantics agree on which programs are safe. This later allows to show that, when a ghost witness is encoded as a program, these semantics agree on the validity of witnesses. To formalize the correspondence between interleavings and global traces, we introduce some definitions. The function $\mathsf{threads}_\mathcal{I}$ returns the set of threads appearing in the L component of the last state of an interleaving. Similarly, we define $\mathsf{threads}_{\mathcal{GT}}$ that returns the set of threads whose local configurations appear in a global trace.

Definition 1 (Coincidence). *Consider a program P from $\mathsf{Lang}_{\mathcal{M}_\mathcal{G}}$, one of its global traces $gt \in [\![P]\!]_{\mathcal{GT}}$ and one of its interleavings $i \in [\![P]\!]_\mathcal{I}$. Let $t \in \mathsf{threads}_\mathcal{I}(i) \cap \mathsf{threads}_{\mathcal{GT}}(gt)$. Let us consider the sequence A_{gt} of local configurations and steps of the thread t in gt and denote it by:*

$$\sigma_0 \xrightarrow{(u_0,a_0,v_0)}_t \sigma_1 \ldots \sigma_{k-1} \xrightarrow{(u_{k-1},a_{k-1},v_{k-1})}_t \sigma_k$$

Similarly, consider the subsequence A_i of i for steps taken by t and their start and target states:

$$s_0 \xrightarrow{(u_0',a_0',v_0')}_t s_1 \quad s_1' \xrightarrow{(u_1',a_1',v_1')}_t s_2 \ldots \xrightarrow{(u_{l-1}',a_{l-1}',v_{l-1}')}_t s_l$$

By construction, for any two consecutive states s_j, s_j' appearing in A_i, only threads different from t may have taken steps, and thus $L_j(t) = L_j'(t)$, where L_j and L_j' are the first components of s_j and s_j', respectively. By projecting the states to the local variables of t and its current location, adding the number of steps performed by t, and fusing the consecutive states with no steps in between, we obtain a sequence A_i' of the same type as A_{gt}. We say that the interleaving i and the global trace gt coincide w.r.t the thread t, in case that A_{gt} is equal to A_i'. If the same threads appear in i and gt and they coincide w.r.t. all of these, we say they coincide.

Definition 2 (Create-complete Global Trace). *We say that a global trace gt is* create-complete, *if for each created thread appearing in $\mathsf{threads}_{\mathcal{GT}}\ gt$, the* create *action that created the thread is part of gt and for each* create *action, a configuration of the created thread appears.*

Lemma 1. *Let $\mathcal{I} = [\![P_1]\!]_\mathcal{I}$ be the set of all interleavings for some program $P_1 \in \mathsf{Lang}_{\mathcal{M}_\mathcal{G}}$, and let $\mathcal{GT} = [\![P_1]\!]_{\mathcal{GT}}$.*

1. *For any interleaving $i \in \mathcal{I}$, there exists a create-complete global trace $gt \in \mathcal{GT}$ such that i and gt coincide.*
2. *For any create-complete global trace $gt \in \mathcal{GT}$, there exists an interleaving $i \in \mathcal{I}$ such that i and gt coincide.*

Proof. A proof can be found in the extended version of the paper [25].

Theorem 1. *A program $P \in \mathsf{Lang}_{\mathcal{M}_\mathcal{G}}$ is safe w.r.t. interleaving semantics if and only if it is safe w.r.t. the local trace semantics.*

Proof. We consider the following two directions:

1. For any local trace $lt \in [\![P]\!]_{\mathcal{LT}}$ violating an assertion, there is an interleaving $i \in [\![P]\!]_{\mathcal{I}}$ violating the same assertion.
2. For any interleaving $i \in [\![P]\!]_{\mathcal{I}}$ violating an assertion, there is a local trace $lt \in [\![P]\!]_{\mathcal{LT}}$ violating the same assertion.

For claim 1, we extend the local trace lt to a create-complete global trace by inserting **create** actions and configurations for created threads where they are still missing. With Lemma 1, it follows that there is a coinciding interleaving i for this global trace. If the assertion S at step k of thread t in lt is violated, the k-th step of t in the interleaving violates the assertion S as well, as the local state of the thread t is the same in the interleaving and the global trace at the preceding state.

For claim 2, for the interleaving i, we obtain a create-complete global trace gt. Assume the assertion S at the k-th step of thread t in i is violated. From gt, we extract the local trace lt ends with the last action of t in gt. At the k-th step of thread t in lt, the assertion is violated, as the local state of the thread t there is the same as in the interleaving before the k-th step of thread t.

3 Ghost Witnesses

To show the correctness of a program, we express invariants that hold at a specific location in the program, in a common format for *witnesses*. For sequential programs, this can be achieved with a mapping from locations to invariants that hold at specified location. However, for concurrent programs, the program state may depend on the interleavings. Therefore, in this setting, invariants should be able to reason about different interleavings. To this end, we allow additional *ghost variables* in the witness that can be also used in invariants, and *ghost updates* of the corresponding variables. Ghost updates may not modify local or global variables of the *original program* P, but may only modify ghost local and global variables. Thus, ghost updates may be global or local writes on ghost variables, or global reads. First, we define witnesses:

Definition 3 (Ghost Witness). *A* ghost witness *for a program P is a tuple $(\mathcal{D}_\mathcal{A}, \mathcal{X}_\mathcal{A}, \mathbb{U}, \mathbb{I})$. There, $\mathcal{D}_\mathcal{A}$ is a set of ghost global declarations, i.e., triples of the form (name, type, value), for an identifier name, its type type and a value value describing the ghost variable's initial value, where newly introduced globals are disjoint from the ones existing in P. $\mathcal{X}_\mathcal{A}$ is a set of* ghost local variables *(with their types) disjoint from $\mathcal{X}$. The* ghost updates *are given by a partial function $\mathbb{U}$ from edges to (non-empty) sequences of ghost statements. The* location invariants *are given by a partial function $\mathbb{I}$ from nodes to boolean expression over ghost variables, global program variables, and local variables.*

In the following, we refer to ghost witnesses also simply as *witnesses*. We define the semantics of a witness W for a program P via an instrumented *ghost program* $P_\mathcal{A}^W$ that can be obtained from the witness. This instrumentation adds the declarations from $\mathcal{D}_\mathcal{A}$ and $\mathcal{X}_\mathcal{A}$ and inserts the ghost updates $\mathbb{U}$ and the invariants $\mathbb{I}$ into the control flow graph. However, for the instrumentation of the ghost updates, we combine them with statements from the original program into *atomic blocks*, i.e., a sequence of multiple statements that can be only executed together, as a single atomic action. Therefore, we first extend our programming language with atomic blocks, before we continue with the semantics of a witness.

3.1 Atomic Blocks

Given an arbitrary statement a_1 and statements that are always admissible (i.e., local update, global read, global write) $a_2, \ldots, a_n$, we define an atomic block atomic$\{a_1; \ldots; a_n\}$. The interleaving semantics in Sect. 2.2 is extended such that consistent steps may involve atomic blocks. Such an atomic block may only be executed if the first statement a_1 is admissible. The subsequent state obtained after the atomic action needs to reflect the sequential composition of the effects of the individual steps $a_1; \ldots; a_n$. Thus, an atomic block may only be executed as a whole. In case that a_k, with $1 \leq k \leq n$, is an assertion, we say that a_k is violated if and only if the assertion evaluates to false after executing $a_1; \ldots; a_{k-1}$ on the preceding state of the interleaving. We call the extension of the language Lang with programs containing atomic blocks Lang$_{Atomic}$.

3.2 Witness-Instrumented Programs

We define how, given a program P and a witness $W = (\mathcal{D}_{\mathbb{A}}, \mathcal{X}_{\mathbb{A}}, \mathbb{U}, \mathbb{I})$, a program can be constructed from P containing the instrumentation from the witness. For a given P and W, we define a helper function Ψ_W that yields an instrumentation for asserting the invariants supplied via $\mathbb{I}$, and a function δ_W that replaces actions on edges with atomic blocks consisting of the original action and the ghost update, if any. More precisely,

- The partial map $\Psi_W : \mathcal{N} \to \mathcal{E}$ yields for a node u, where $\mathbb{I}(u) = i$ and $(u, a', v) \in \mathcal{E}$, for some a', the edge $(u, \text{atomic}\{a_1; \ldots; a_k\}, (u,v))$, where $a_1, \ldots, a_k$ is a sequence of actions to evaluate and then assert the invariant i. There, (u, v) is a new node. If for u the mapping $\mathbb{I}(u)$ is undefined, $\Psi_W u$ is undefined as well.
- The map $\delta_W : \mathcal{E} \to \mathcal{E}$ transforms each edge (u, a, v) of the original program into an edge (n, a', v). In case that $\mathbb{U}((u, a, v))$ is specified, we set $a' = \text{atomic}\{a; a_1; \ldots; a_k\}$ where $a_1; \ldots; a_k$ is the sequence of updates provided by $\mathbb{U}((u, a, v))$. In case $\mathbb{U}((u, a, v))$ is undefined, $a' = a$. In case $\Psi_W u$ is defined, we set $n = (u, v)$, and otherwise $n = u$.

Definition 4 (Witness-instrumented Program).
Given a witness $W = (\mathcal{D}_{\mathbb{A}}, \mathcal{X}_{\mathbb{A}}, \mathbb{U}, \mathbb{I})$ for a program P, the witness-instrumented program $P_{\mathbb{A}}^W$ is obtained by adding $\mathcal{D}_{\mathbb{A}}$ to the set of global declarations and adding $\mathcal{X}_{\mathbb{A}}$ to the set of local variables of P. The set of nodes in the resulting program is obtained by combining the existing set of nodes from P with the new nodes added by Ψ_W. The set of edges is obtained by applying Ψ_W to the nodes in P and applying δ_W to the edges of P.

Now, validity of a witness w.r.t. the interleaving semantics can be defined:

Definition 5 (Valid Witness w.r.t. Interleaving Semantics). *A witness W for a program P is valid w.r.t. the interleaving semantics if and only if the witness-instrumented program $P_🝆^W$ is safe w.r.t. the interleaving semantics.*

3.3 Preservation of Safety Properties in Ghost Programs

Here, we go on to show that if the original program is unsafe, then a program obtained by instrumentation with a witness will be unsafe as well. Additionally, if an assertion is safe in the original program, the same assertion will be safe in any program obtained by instrumentation with a witness.

Assume that a verifier produced the verdict that a program P is safe, and produces a witness W that yields the witness-instrumented program $P_🝆^W$. By checking that the witness is valid, it should follow that P is safe. In other words, violations of safety of the original program should be preserved in the witness-instrumented program. To obtain from an interleaving of P an interleaving of $P_🝆^W$, the main idea is to extend the states to ghost variables, add edges introduced by Ψ_W and to replace steps taken on an edge e with a step on the edge introduced by $\delta_W\, e$. This notion is formalized by the definition of $\pi_🝆$:

Definition 6. *Given a program P, a witness W, and an interleaving $i \in [\![P]\!]_\mathcal{I}$, the function $\pi_🝆$ is defined inductively and*

G1 extends the initial state with local and global ghost variables with their initial values,
G2 for each subsequent step (u, a, v) by a thread t in the interleaving, adds
  – two steps taken by thread t corresponding to edges e and e', if $\Psi_W\, u = e$ and $\delta_W\, (u, a, v) = e'$.
  – one step taken by thread t corresponding to edges e, if $\Psi_W\, u$ is undefined and $\delta_W\, (u, a, v) = e$.

As by requirement the ghost statements do not differ with regard to their treatment of non-ghost variables, the local and global states for these variables evolve in the resulting interleaving in the same manner as in the input interleaving. We use the function π_P that takes a state (L, M, G), and yields a state where all local and global variables that are not defined by P are removed. How one obtains a consistent interleaving of $P_🝆^W$ from a consistent interleaving of P is formalized in Lemma 2.

Lemma 2. *Let $P \in$ Lang be a program and W a witness for P. Given an interleaving $i \in [\![P]\!]_\mathcal{I}$, then $\pi_🝆 i \in [\![P_🝆^W]\!]_\mathcal{I}$, i.e., is a consistent interleaving of $P_🝆^W$. For the last state s of i and the last state s' of $\pi_🝆 i$ it holds that $\pi_P\, s' = s$.*

The proof is by induction over the length of the interleaving i. A proof can be found in the extended version [25]. Building on Lemma 2, we can now say that violations in the original program are preserved in the witness-instrumented program.

Theorem 2. *Let $P \in \mathsf{Lang}$ be a program and $W = (\mathcal{D}_{\mathbb{A}}, \mathcal{X}_{\mathbb{A}}, \mathbb{U}, \mathbb{I})$ a witness for P. If there is an interleaving $i \in [\![P]\!]_{\mathcal{I}}$ that violates an assertion a at edge (u, a, v) taken by thread t, the interleaving $\pi_{\mathbb{A}} i \in [\![P_{\mathbb{A}}^W]\!]_{\mathcal{I}}$ violates the assertion a at an edge $\delta_W(u, a, v)$ taken by thread t.*

Proof. Let i' be the prefix of i that ends at the state before the assertion is violated. From Lemma 2, it follows that $\pi_{\mathbb{A}} i' \in [\![P_{\mathbb{A}}]\!]_{\mathcal{I}}$, and the state of variables, owners of mutexes and locations of threads in the last state of $\pi_{\mathbb{A}} i'$ agree with those in the last state of i'. In case that $\Psi_W(u, a, v)$ is undefined, the interleaving can be prolonged by a step $(u, a', v) = \delta_W(u, a, v)$ taken by thread t, where a' contains the assertion a that will fail. If $\Psi_W(u, a, v)$ is defined, the interleaving can be prolonged by two steps $(u, a'', (u, v))$ and $((u, v), a', v)$ taken by thread t, where $(u, a'', (u, v)) = \Psi_W u$ and $((u, v), a', v) = \delta_W(u, a, v)$. Then, the assertion a is contained in a' and will be violated.

Corollary 1 (Preservation of Correctness). *Let P be a program and W be a witness for P. If W is valid, then P is safe.*

A further property of interest is that the witness format ensures that assertions that cannot be violated in the original program are unaffected by the instrumentation.

Theorem 3. *Let W be a witness for a program $P \in \mathsf{Lang}$. If there is an interleaving $i \in [\![P_{\mathbb{A}}^W]\!]_{\mathcal{I}}$ that violates an assertion a at an edge $\delta_W(u, a, v)$, then there is an interleaving $i' \in [\![P]\!]_{\mathcal{I}}$ that violates the assertion a at an edge (u, a, v).*

A proof sketch constructing the interleaving i' can be found in the extended version of the paper [25].

Corollary 2 (Preservation of Violation). *Let P be a program and W a witness for P. If W is not valid because an assertion from P can be violated in $P_{\mathbb{A}}^W$, then P is unsafe.*

4 Valid Witnesses in the Local Trace Semantics

For witness-instrumented programs, we have introduced the notion of atomic blocks to the language. While this notion is convenient for the interleaving semantics, it is less clear for the thread-modular local trace semantics, where communication between threads is assumed to happen via observable and observing events. Thus, we discuss how atomic blocks can be encoded via lock and unlock actions, i.e., via critical sections. Then we show the equivalence of the validity of witnesses w.r.t. the interleaving and the local traces semantics by considering programs where atomic blocks are represented in this manner.

Critical sections allow encoding atomic blocks in a way that preserves safety, as has been noted in the literature [51] and used for practical implementations of atomic blocks [54]. A witness validator therefore may choose to take either view: analyzing the ghost program w.r.t. the semantics that considers atomic blocks or w.r.t. a semantics that encodes atomic blocks as critical sections.

We consider the procedure split, that takes programs $P \in \mathsf{Lang}_{Atomic}$ with atomic blocks and yields programs in $\mathsf{Lang}_{\mathcal{M}_\mathcal{G}}$ where these are encoded as critical sections. In the resulting program, all statements accessing a global variable g have to be embedded into critical sections protected by the mutex m_g. Additionally, each atomic block is encoded as a critical section protected by all mutexes of the form $m_g \in \mathcal{M}_\mathcal{G}$, if g is read or written in the atomic block. These lock-operations on mutexes inserted there are performed following some total order on $\mathcal{M}_\mathcal{G}$. In Listing 1, one of the atomic blocks of the introductory example from Listing 2 is encoded via critical sections.

```
lock(m_A);
  lock(m);
  A = 0;
unlock(m_A);
```

List. 2. Example for the encoding of an atomic block as a critical section.

Theorem 4. *Let $P \in \mathsf{Lang}_{Atomic}$ be a program. The program P is safe with respect to the interleaving semantics if and only if $\mathsf{split}\, P \in \mathsf{Lang}_{\mathcal{M}_\mathcal{G}}$ is safe with respect to the interleaving semantics.*

A proof-sketch for the direction that if $\mathsf{split}\, P$ is safe w.r.t. the interleaving semantics, also P is safe w.r.t. the interleaving semantics, can be found in the extended version [25]. We remark that such an encoding of atomic blocks via mutexes may introduce deadlocks which do not unduly restrict the set of reachable states. Using this encoding, we can define the validity of a witness with respect to the local trace semantics.

Definition 7 (Valid Witness w.r.t. Local Trace Semantics). *A witness W for a program $P \in \mathsf{Lang}$, resulting in the witness-instrumented program P_A^W, is valid w.r.t. the local trace semantics if and only if the program $\mathsf{split}\, P_A^W$ is safe w.r.t. the local trace semantics.*

Finally, we can relate this notion back to the validity of witnesses in the interleaving semantics. With the definition of a valid witness w.r.t. the local trace semantics referring back to safety and by encoding atomicity in witness-instrumented programs via critical sections, we can lift the agreement of the two semantics on safety to an agreement on validity of ghost witnesses.

Theorem 5. *Let W be a witness for a program $P \in \mathsf{Lang}$. The witness is valid with respect to the interleaving semantics if and only if it is valid with respect to the local trace semantics.*

Proof. By Definition 5, the witness W is valid with respect to the interleaving semantics if and only if the witness-instrumented program P_A^W is safe. According to Theorem 4, the program P_A^W is safe with respect to the interleaving semantics if and only if the program $\mathsf{split}\, P_A^W$ is safe with respect to the interleaving semantics. As $\mathsf{split}\, P_A^W \in \mathsf{Lang}_{\mathcal{M}_\mathcal{G}}$, according to Theorem 1, the program $\mathsf{split}\, P_A^W$ is safe with respect to the interleaving semantics if and only if it is safe with respect to the local trace semantics. With Definition 7, the statement follows.

5 Witness Generation with Thread-Modular Abstract Interpretation

Here, we outline how the analysis results of two thread-modular analyses based on abstract interpretation can naturally be expressed using ghosts. While both analyses [72,73] are implemented in GOBLINT and proven correct relative to the local trace semantics, their abstractions differ: one computes relational invariants per mutex, while the other computes non-relational invariants per global.

The relational *mutex-meet* analysis by Schwarz et al. [73] assumes that each global variable g is (write-)protected by a set of mutexes $\mathcal{M}[g]$ which are held at every write to g. For every mutex m, the analysis computes a *mutex invariant* $[m]$ containing relational information between those global variables $\mathcal{G}[m] = \{g \mid m \in \mathcal{M}[g]\}$ only written when m is held. The computed invariant holds whenever no thread holds the mutex m, i.e., it may be violated while some thread has exclusive access, but holds again once m is released. To express mutex invariants, we introduce for each mutex m a boolean ghost variable m_{locked}, indicating whether any thread has locked m. The variable is initialized to false and corresponding ghost variable updates are added to the witness: every lock of m is instrumented with $m_{\text{locked}} = \text{true}$ and every unlock with $m_{\text{locked}} = \text{false}$. This allows every mutex invariant $[m]$ to be added to the witness as $\neg m_{\text{locked}} \implies [m]$. Listing 1b is an example obtained by such an instrumentation.

GOBLINT additionally distinguishes two phases of the program: when it is single-threaded and when it has become multithreaded. Thread-modular analysis are only used for the latter, while, in the former phase, global variables are analyzed flow-sensitively to, e.g., retain precision during the initialization phase of the program. We introduce another boolean ghost variable *multithreaded*, indicating whether the program has become multithreaded. It is initialized to false and all thread creation actions that may potentially create the first additional thread are instrumented with *multithreaded* = true. Thus, every invariant $[m]$ is actually added as $(\textit{multithreaded} \land \neg m_{\text{locked}}) \implies [m]$.

The non-relational *protection* analysis [72] computes for each global variable g a *protected invariant* $[g]$ which describes values of g when no thread holds a mutex from $\mathcal{M}[g]$. Every protected invariant $[g]$ is added to the witness as $(\textit{multithreaded} \land \bigwedge_{m \in \mathcal{M}[g]} \neg m_{\text{locked}}) \implies [g]$.

The analyses and their computed invariants are flow-insensitive on global variables, so the invariants should be valid at all program locations that fulfill certain conditions. Repeating the flow-insensitive invariants at every location is costly for validators; instead, we add flow-insensitive invariants after each thread create in the program.

6 Ghost Witnesses for C Programs

Here, we propose a uniform format for correctness witnesses with ghost variables (based on Definition 3) for concurrent C programs. The format enables witness exchange between verifiers and witness validators. Our format is compatible with an existing format [3] for correctness witnesses of sequential programs utilized by the community around the International Competition on Software Verification (SV-COMP) [6] and is realized in YAML [27].

A witness in this format contains invariants and specifies at which locations of the C program these invariants hold. Rather than referring to a specific control-flow graph, locations are expressed by a line number and a column number in the source C file. Corresponding to the elements of the tuple $(\mathcal{D}_\triangle, \mathcal{X}_\triangle, \mathbb{U}, \mathbb{I})$ from Definition 3, the format defines the following types of entries in the witness:

ghost_variables A ghost global variable declaration consists of a primitive C type, a unique identifier, and an initial value given as a (side effect-free) C expression over global variables of the program.

ghost_updates A ghost update consists of an identifier (referring to a ghost variable declaration), a location at which the update is performed, and a (side effect-free) C expression over program variables (which must be in scope at the location) and ghost variables.

location_invariant A location invariant consists of a location and an invariant represented as a (side effect-free) C expression over program variables (which must be in scope at the location) and ghost variables.

A more detailed description of these entries can be found in the extended version of the paper [25]. We omit the definition of local ghost variables, as in C the reading from and writing to global variables does not require local variables as in our formalization. A full YAML schema for witnesses is available [24] and documented [26]. Listings 3 and 4 show two excerpts of a correctness witness for the program in Example 1, the full witness can also be found in the extended version [25].

Before defining *validity* for these witnesses, we briefly discuss the semantics of concurrent C. Notably, the semantics of concurrent C programs is not based on the interleaving of well-defined atomic steps as in our formal model; the granularity of steps is implementation-defined. For instance, depending on compiler and target platform, a write of a 64 bit shared variable may be performed atomically, or it may be split into two write accesses of 32 bit, or even more granularly. Since actions of other threads may interleave arbitrarily with these accesses, leading to unexpected results, C forbids (non-atomic) concurrent accesses to shared variable, and declares such *data races* as undefined behaviour. As we are concerned with proving correctness (wrt. reachability properties) on C programs, we only consider well-behaved C programs without such data races. Moreover, C allows atomic operations on shared data to specify different *memory orders*, allowing for certain weak memory models to be used. We do not support these memory orders, and always assume *sequential consistency* [49].

```
- entry_type: invariant_set
  metadata: ...
  content:
    - invariant:
        type: location_invariant
        location: { line: 4, ... }
        value: ⌂ == 0 ⟹ used == 0
        format: c_expression
```

The validity of our witnesses for C programs is based on Definition 5. The witness format for sequential programs by Ayaziová et al. [3] specifies that each `location_invariant` must hold immediately before executing the statement or declaration it is attached to, i.e., the statement or declaration at the specified location. In concurrent programs we must additionally define which global states the evaluation of an invariant may observe. In particular, as the granularity of write accesses by other threads to shared variables is unclear, an invariant could potentially observe arbitrary intermediate states of such a write. This is problematic for witnesses, as they cannot express useful information over the global state in that case. To address this issue, we use the notion of *sequence points* in C, a set of well-defined points in an execution where all side effects of the previous actions are guaranteed to have been performed and no side effects of subsequent actions have taken place [39]. For validity of a witness in our format, it suffices that every location invariant in the witness holds whenever one thread is at the specified location and every other thread is at a sequence point. Additionally, each location invariant is evaluated atomically, i.e., no other thread can interleave while the location invariant is evaluated.

```
- entry_type: ghost_instrumentation
  metadata: ...
  content:
    ghost_variables:
      - name: 🔒
        type: int
        scope: global
        initial:
          value: 0
          format: c_expression
    ghost_updates:
      - location: { line: 4, ... }
        updates:
          - variable: 🔒
            value: 1
            format: c_expression
```

List. 4. Ghost declaration and update.

The effect of the declaration of a ghost variable g with type T and initial value v corresponds to inserting a global declaration `T g = v;` in the C program. The declaration is executed after the global variables of the program are declared. As a result, the evaluation of v can access the initial value of the program's global variables. The update of a ghost variable g with the value v has the same semantics as the C statement `g = v;`. As described in Definition 4, the update is executed atomically with the action at the given witness location. Unlike the witnesses in the control flow graph as in Definition 3, we cannot in general attach the ghost update to an arbitrary statement, as we cannot assume that this statement is executed atomically (e.g., it may contain a nested function call or might execute other side-effects). Therefore, we allow ghost updates only at locations that point to the beginning of one of the following statements.

- If the statement is a call to a function from the pthread library (e.g. for thread-creation or locking of a mutex; the full list can be found in the extended version of the paper [25]), then the ghost update is performed at the sequence point after evaluating the arguments of the function and together with the action leading to the next sequence point (e.g., locking of the mutex for `pthread_mutex_lock`, thread creation for `pthread_create`).

– If the statement is an assignment, the ghost update is performed after the assignment, and it happens atomically with the actual write (i.e., only the write itself is performed atomically, not the evaluation of the expressions). Performing the write atomically may reduce the possible interleavings, but for programs without data races the behaviour is equivalent.

In the benchmarks used by SV-COMP, there are two additional functions that allow one to define an atomic block in C [7]. The verifiers are instructed to assume that the code between __VERIFIER_atomic_begin and __VERIFIER_atomic_end is executed in one atomic step. As we aim for our witnesses to be used by the SV-COMP community, we also allow ghost updates at the call of these two functions. For a call to __VERIFIER_atomic_begin (resp. __VERIFIER_atomic_end), the update is performed after (resp. before) the call to this additional function, i.e., the update is performed atomically with the code between these calls.

7 Validation with Software Model Checking

Witnesses expressed in the format described in the previous section can be exchanged with, and validated by, other verification tools, to confirm the results of the verification using tools with a significant technological and/or conceptual difference, and thereby increase trust in the verification tool or potentially uncover bugs in the verification. To demonstrate this, we implemented witness validation in the software model checking tool ULTIMATE GEMCUTTER [29,44], which is developed as part of the ULTIMATE program analysis framework [78].

ULTIMATE GEMCUTTER is based on an interleaving semantics (Sect. 2.2). It uses a *commutativity relation* between statements to identify pairs of statements (of different threads) whose relative order does not affect program correctness. In its analysis, GEMCUTTER groups interleavings into equivalence classes based on the induced *Mazurkiewicz equivalence* [53], i.e., it considers interleavings as equivalent if they only differ in the order of commuting statements. GEMCUTTER proves correctness for one representative per equivalence class, using the *trace abstraction* algorithm [37], a counterexample-guided abstraction refinement scheme. If the proof for the representatives succeeds, GEMCUTTER soundly concludes correctness of the entire program. To validate witnesses, GEMCUTTER instruments the given program with the ghost code and invariants from the witness (similar to Definition 4), and applies its usual verification algorithm to the instrumented program.

7.1 Experimental Setup

Complementary to the theoretical equivalence results of Theorem 5, we empirically evaluate the impact of witness validation regarding the following questions:

Q1 Can witnesses generated by a verifier be validated by a different tool based on different semantics, analysis approaches, and technological foundations?

Table 1. Number of *confirmed* witnesses. In *witness confirmation* mode, GEMCUTTER checks validity of witness invariants while ignoring program correctness.

witnesses	protection-🅖	mutexmeet-🅖	protection	mutexmeet
witnesses for correct programs				
total	181	217	159	230
confirmed	171	193	122	167
rejected	0	0	0	0
out of resources	5	19	34	56
witnesses for incorrect programs				
total	282	288	295	300
confirmed	192	164	224	130
rejected	0	0	0	0
out of resources	85	117	68	163

Q2 What is the overhead of witness validation? Can witnesses help to accelerate the analysis, compared to verification without a witness?

Q3 Can witness exchange and validation with tools based on a different approach help to find bugs in verification tools?

For our evaluation, we analysed the programs in the *ConcurrencySafety-Main* category of SV-COMP [6] with GOBLINT and generated correctness witnesses. For each of the two different analyses (*mutex-meet* and *protection*) described in Sect. 5, we generated witnesses with and without ghost variables, yielding four witness benchmark sets. The witnesses without ghosts simply contain location invariants within critical sections, immediately after lock operations. The non-ghost invariants only refer to variables protected by mutexes held at their locations, implicitly using locations to encode concurrency information. Thus, the number of invariants, their locations and the invariant expressions themselves are incomparable between the two kinds of witnesses. We excluded witnesses that do not contain any (non-trivial) invariants. Note that, as an abstract interpreter, GOBLINT cannot prove incorrectness of programs, but is still able to generate valid invariants for incorrect programs (though they are, naturally, insufficient to prove correctness). Hence, the generated witness benchmark sets include witnesses for incorrect programs, where the invariants reflect information that GOBLINT has derived about the program, and can be checked by our *witness confirmation* analysis.

We validated these witnesses using GEMCUTTER. The validation was executed on an AMD Ryzen Threadripper 3970X at 3.7 GHz with a time limit of 900 s, a memory limit of 16 GB and a CPU core limit of 2. For reliable measurements, the experiments were carried out using the BENCHEXEC framework [12]. An artifact allowing reproduction of our experiments is available [23].

Table 2. Validated witnesses and required time. The second line indicates verification results on those benchmarks where a witness in the corresponding witness set existed.

witnesses	protection-🔒		mutexmeet-🔒		protection		mutexmeet	
	#	time h:m:s	#	time h:m:s	#	time h:m:s	#	time h:m:s
validation	119	1:31:33	123	1:29:27	100	2:01:29	110	2:37:12
verification	120	1:07:27	131	1:18:29	101	44:34	118	1:07:43

7.2 Results

Table 1 shows the number of witnesses in each witness benchmark set, and how many of them are *confirmed* resp. rejected by GEMCUTTER. In the *witness confirmation* mode used here, GEMCUTTER only checks if the asserts of the instrumented program that correspond to witness invariants can be violated, while ignoring the asserts already present in the original program.

Despite the significant differences between GOBLINT and GEMCUTTER, the communication via the witness format is successful (**Q1**). The analysed witnesses contain non-trivial information: on average, witnesses in the set "protection-🔒" (resp. "mutexmeet-🔒") contain 44.1 invariant entries (resp. 43.9) and 3.8 ghost updates (resp. 3.6), whereas the witnesses in "protection" (resp. "mutexmeet") contain on average 179 (resp. 749.9) invariant entries. For witnesses that are not confirmed, GEMCUTTER either runs out of time or memory (see Table 1), or crashes due to ghost updates at unsupported locations (12 cases) or unsupported C features in the program or witness (58 cases).

Table 2 shows the results of running GEMCUTTER in full *witness validation* mode (the mode used in SV-COMP), where the witness invariants must be confirmed and correctness of the program must be proven. As we are working with *correctness* witnesses, we only consider correct programs for the validation. For GOBLINT witnesses validated by GEMCUTTER, witness validation generally incurs an overhead rather than accelerating the verification (**Q2**). However, for almost all programs verified by GEMCUTTER, the corresponding witnesses can also be validated with a moderate slowdown. In two cases, validation of a witness from the "mutexmeet-🔒" resp. "protection-🔒" set succeeds (in less than 250 s) while the verification of the corresponding program times out.

Over the duration of this work, its implementation and (re-)evaluation revealed 29 bugs in GOBLINT and GEMCUTTER and lead to their fixing in most cases (**Q3**). These bugs already existed prior to and independently of this work, and crucially, they were revealed on a subset of the very same SV-COMP benchmarks that have been used over the years to test and evaluate these tools. A full overview of these bugs is provided in the extended version of the paper [25]. In particular, four soundness-critical bugs were discovered in the verification performed by GOBLINT, and 13 bugs in GOBLINT's witness production (including missing or incorrect invariants). In ULTIMATE, five issues concerned the support for ACSL specifications (which is reused as part of the witness instrumentation), two bugs concerned the translation of C code to ULTIMATE's internal representation, one bug concerned the internal representation, and one bug was specific to witness validation. Notably, the bugs regarding ACSL support and translation of C code are soundness-critical, and cannot be caught by correctness checks inside ULTIMATE that operate on the internal representation. Thus, the witness exchange and validation significantly improved both tools.

Threats to Validity. We conducted experiments with GOBLINT and GEMCUTTER, producing ghost witnesses in GOBLINT (albeit for two different analysis approaches) and validating them in GEMCUTTER. While further experiments with more analyzers would offer additional evidence of the suitability of the format, GOBLINT and GEMCUTTER employ radically different approaches and technologies; thus, we consider our experiments strong evidence that the format succeeds in exchanging information across such divides. We do not study the exchange of ghost witnesses in the opposite direction here. Lastly, the evaluation was performed on tasks from the ConcurrencySafety-Main category of SV-COMP, implying the usual concerns about generalizability to real-world programs of any experiment performed on SV-COMP benchmarks also apply here.

8 Related Work

The exchange of information between analyzers through an analysis-agnostic exchange format is actively researched and has led to the development of cooperative verification [11,36]. Witnesses are central to SV-COMP with many tools specifically developing witness generation and validation approaches [10,56,69,77]. The first generation relied on automata exported as GraphML [8–10], which has been deprecated in favor of a YAML-based format [3]. There are other means of exchanging analysis information that focus on external results, such as SARIF (Static Analysis Results Interchange Format) [65], or conversely, work within a given approach, such as abstract interpretation [1] or model checking [35,55]. The goal of our witnesses, however, is to expose reasoning about concurrent programs that can be transferred across the semantic divide.

Witness exchange aims to increase trust in the analysis results, and can effectively reveal latent bugs in the analyzers. Making program verification tools more dependable is an active area of research. Novel testing methods have been developed specifically for testing analyzers [17,32,41,43]. Ideally, we would like to formally prove analyzers correct, and there has been significant progress in this direction [4,16,19,33,40,76,80]; however, none of these target the full range of advanced techniques needed for efficient analysis of real-world concurrent programs, which are still under active research and development.

Ghost variables are crucial for the (relative) completeness of the proof systems of Owicki and Gries [66] and Lamport [47]. The usefulness of these and similar approaches was shown in case studies [22,34,52,74]. Nipkow and Nieto [64] formalize the Owicki-Gries method in Isabelle/HOL and extend the method to parametric programs [62,63]. Recent work extend the Owicki-Gries method to weak memory [20,21,46,82] and persistent memory [13,67].

We showed that our ghost instrumentation preserve the safety of the original program. Filliâtre et al. [31] present an ML-style programming language and ensure non-interference via the type system, allowing a bisimulation proof of ghost code erasure. For concurrent code, Zhang et al. [83] present a structural approach to establish erasure and Schmaltz [70] gives bisimulation proofs for a significant portion of low-level C with ghost updates.

Our approach to generating ghost witnesses is reminiscent of a form of resource invariant synthesis [18]. There are other approaches worth exploring, such as generating ghosts for counting proofs Farzan et al. [28] or inferring conditional history variables from counter-examples [79]. Hoenicke et al. [38] explore how thread-modular reasoning can avoid ghosts by introducing proof systems that consider interleavings of up to k threads. Farzan et al. [30] present an approach to simplifying proofs, reducing the complexity of the ghost state.

Thread-local concrete semantics have been used to justify thread-modular data-flow analysis [60] and concurrent separation logic [14,15,45,50,61,75]. For abstract interpretation, Miné [57] presents a thread-modular abstract interpretation framework as layers of abstractions of the interleaving semantics, and in subsequent work [58,59], the interleaving semantics is encoded into a local semantics by using auxiliary variables to track the control location of other threads. In contrast, we have shown equivalence for a purely thread-local trace semantics where threads only learn of other thread actions through synchronizing events.

9 Conclusion

We have introduced a witness format for communicating information about concurrent programs between static analysis tools—bridging the divide between interleaving and thread-modular views of the concrete semantics. We build on the notion of ghost variables, which is well-established for arguing about the correctness of concurrent programs. The ghost witness can be easily encoded via a program transformation. This way, the notion of a *valid* witness is independent of the semantic view, which is crucial for an exchange format. We extend an existing and established format to minimize the technical burden for other tools to support ghost witnesses. For two types of thread-modular analyses implemented in the GOBLINT abstract interpreter, we have shown how the invariants computed can naturally be expressed in the new format, highlighting that its expressiveness is useful for existing analyses. A validation of the witnesses that GOBLINT generated for a set of concurrent SV-COMP tasks showed that most could be validated by the model checker GEMCUTTER building on an interleaving view of concurrency. Altogether, the notion of ghost witnesses can be leveraged as a building block for exchanging more expressive invariants of concurrent programs between static analysis tools. Future work may study how the results of other static analyzers for concurrent programs are best encoded via ghost code and how to cleverly exploit this information in thread-modular validators. Additionally, it would be interesting to explore the applicability of ghosts in violation witnesses for concurrent programs.

Acknowledgments. This work was supported in part by the Deutsche Forschungsgemeinschaft (DFG) under project numbers 503812980 and 378803395/2428 ConVeY, the Shota Rustaveli National Science Foundation of Georgia (FR-21-7973), and the European Union and the Estonian Research Council via project TEM-TA119.

References

1. Albert, E., Puebla, G., Hermenegildo, M.: Abstraction-carrying code. In: Baader, F., Voronkov, A. (eds.) LPAR 2005. LNCS (LNAI), vol. 3452, pp. 380–397. Springer, Heidelberg (2005). https://doi.org/10.1007/978-3-540-32275-7_25
2. Apt, K.R., de Boer, F.S., Olderog, E.: Verification of Sequential and Concurrent Programs. Texts in Computer Science. Springer (2009). ISBN 978-1-84882-744-8. https://doi.org/10.1007/978-1-84882-745-5
3. Ayaziová, P., Beyer, D., Lingsch-Rosenfeld, M., Spiessl, M., Strejček, J.: Software verification witnesses 2.0. In: Model Checking Software. Lecture Notes in Computer Science, vol. 14624, pp. 184–203. Springer, Cham (2024). https://doi.org/10.1007/978-3-031-66149-5_11
4. Becker, B., Marché, C.: Ghost code in action: automated verification of a symbolic interpreter. In: Chakraborty, S., Navas, J.A. (eds.) VSTTE 2019. LNCS, vol. 12031, pp. 107–123. Springer, Cham (2020). https://doi.org/10.1007/978-3-030-41600-3_8
5. Beyer, D.: Competition on software verification and witness validation: SV-COMP 2023. In: Tools and Algorithms for the Construction and Analysis of Systems, pp. 495–522, Springer, Cham (2023). https://doi.org/10.1007/978-3-031-30820-8_29
6. Beyer, D.: State of the art in software verification and witness validation: SV-COMP 2024. In: Finkbeiner, B., Kovács, L. (eds.) Tools and Algorithms for the Construction and Analysis of Systems - 30th International Conference, TACAS 2024, Luxembourg City, Luxembourg, 6–11 April 2024, Proceedings, Part III. Lecture Notes in Computer Science, vol. 14572, pp. 299–329. Springer, Cham (2024). https://doi.org/10.1007/978-3-031-57256-2_15
7. Beyer, D.: SV-COMP 2024 - 13th competition on software verification (2024). https://sv-comp.sosy-lab.org/2024/rules.php. Accessed 29 Sept 2024
8. Beyer, D., Dangl, M., Dietsch, D., Heizmann, M.: Correctness witnesses: exchanging verification results between verifiers. In: Zimmermann, T., Cleland-Huang, J., Su, Z. (eds.) Proceedings of the 24th ACM SIGSOFT International Symposium on Foundations of Software Engineering, FSE 2016, Seattle, WA, USA, 13–18 November 2016, pp. 326–337. ACM (2016). https://doi.org/10.1145/2950290.2950351
9. Beyer, D., Dangl, M., Dietsch, D., Heizmann, M., Stahlbauer, A.: Witness validation and stepwise testification across software verifiers. In: Nitto, E.D., Harman, M., Heymans, P. (eds.) Proceedings of the 2015 10th Joint Meeting on Foundations of Software Engineering, ESEC/FSE 2015, Bergamo, Italy, 30 August–4 September 2015, pp. 721–733. ACM (2015). https://doi.org/10.1145/2786805.2786867
10. Beyer, D., Friedberger, K.: Violation witnesses and result validation for multi-threaded programs. In: Margaria, T., Steffen, B. (eds.) ISoLA 2020. LNCS, vol. 12476, pp. 449–470. Springer, Cham (2020). https://doi.org/10.1007/978-3-030-61362-4_26

11. Beyer, D., Kanav, S.: CoVeriTeam: on-demand composition of cooperative verification systems. In: TACAS 2022. LNCS, vol. 13243, pp. 561–579. Springer, Cham (2022). https://doi.org/10.1007/978-3-030-99524-9_31
12. Beyer, D., Löwe, S., Wendler, P.: Reliable benchmarking: requirements and solutions. Int. J. Softw. Tools Technol. Transfer **21**(1), 1–29 (2017). https://doi.org/10.1007/s10009-017-0469-y
13. Bila, E.V., Dongol, B., Lahav, O., Raad, A., Wickerson, J.: View-based Owicki–Gries reasoning for persistent x86-TSO. In: ESOP 2022. LNCS, vol. 13240, pp. 234–261. Springer, Cham (2022). https://doi.org/10.1007/978-3-030-99336-8_9
14. Brookes, S.: A semantics for concurrent separation logic. Theor. Comput. Sci. **375**(1–3), 227–270 (2007). https://doi.org/10.1016/J.TCS.2006.12.034
15. Brookes, S.: On grainless footprint semantics for shared-memory programs. In: Jacobs, B., Silva, A., Staton, S. (eds.) Proceedings of the 30th Conference on the Mathematical Foundations of Programming Semantics, MFPS 2014, Ithaca, NY, USA, 12–15 June 2014. Electronic Notes in Theoretical Computer Science, vol. 308, pp. 65–86. Elsevier (2014). https://doi.org/10.1016/J.ENTCS.2014.10.005
16. Cachera, D., Pichardie, D.: A certified denotational abstract interpreter. In: Kaufmann, M., Paulson, L.C. (eds.) ITP 2010. LNCS, vol. 6172, pp. 9–24. Springer, Heidelberg (2010). https://doi.org/10.1007/978-3-642-14052-5_3
17. Casso, I., Morales, J.F., López-García, P., Hermenegildo, M.V.: Testing your (static analysis) truths. In: LOPSTR 2020. LNCS, vol. 12561, pp. 271–292. Springer, Cham (2021). https://doi.org/10.1007/978-3-030-68446-4_14
18. Clarke, E.M.: Synthesis of resource invariants for concurrent programs. In: Aho, A.V., Zilles, S.N., Rosen, B.K. (eds.) Conference Record of the Sixth Annual ACM Symposium on Principles of Programming Languages, San Antonio, Texas, USA, January 1979, pp. 211–221. ACM Press (1979). https://doi.org/10.1145/567752.567772
19. Correnson, A., Steinhöfel, D.: Engineering a formally verified automated bug finder. In: Chandra, S., Blincoe, K., Tonella, P. (eds.) Proceedings of the 31st ACM Joint European Software Engineering Conference and Symposium on the Foundations of Software Engineering, ESEC/FSE 2023, San Francisco, CA, USA, 3–9 December 2023, pp. 1165–1176. ACM (2023). https://doi.org/10.1145/3611643.3616290
20. Dalvandi, S., Doherty, S., Dongol, B., Wehrheim, H.: Owicki-Gries reasoning for C11 RAR. In: Hirschfeld, R., Pape, T. (eds.) 34th European Conference on Object Oriented Programming, ECOOP 2020, 15–17 November 2020, Berlin, Germany (Virtual Conference). LIPIcs, vol. 166, pp. 11:1–11:26, Schloss Dagstuhl - Leibniz-Zentrum für Informatik (2020). https://doi.org/10.4230/LIPICS.ECOOP.2020.11
21. Dalvandi, S., Dongol, B., Doherty, S., Wehrheim, H.: Integrating Owicki-Gries for C11-style memory models into Isabelle/HOL. J. Autom. Reason. **66**(1), 141–171 (2022). https://doi.org/10.1007/S10817-021-09610-2
22. Dijkstra, E.W.: Finding the correctness proof of a concurrent program. In: Bauer, F.L., Broy, M. (eds.) Program Construction, International Summer School, July 26 - August 6, 1978, Marktoberdorf, Germany, Lecture Notes in Computer Science, vol. 69, pp. 24–34, Springer (1978). https://doi.org/10.1007/BFB0014652
23. Erhard, J., et al.: Artifact for the VMCAI'2025 paper "Correctness Witnesses for Concurrent Programs: Bridging the Semantic Divide with Ghosts" (2024). https://doi.org/10.5281/zenodo.13863579
24. Erhard, J., et al.: Correctness witness schema (2024). https://ultimate-pa.github.io/concurrency-witnesses/correctness-witness-schema.yml. Accessed 29 Sept 2024

25. Erhard, J., et al.: Correctness witnesses for concurrent programs: bridging the semantic divide with ghosts (extended version) (2024). https://doi.org/10.48550/arXiv.2411.16612
26. Erhard, J., et al.: Format for correctness witnesses, version 2.1 (2024). ultimate-pa.github.io/concurrency-witnesses/index.html. Accessed 29 Sept 2024
27. Evans, C., Ben-Kiki, O., döt Net, I., Müller, T., Antoniou, P., Aro, E., Smith, T.: YAML Ain't Markup Language (YAML^{TM}) Version 1.2 (2021). https://yaml.org/spec/1.2.2/
28. Farzan, A., Kincaid, Z., Podelski, A.: Proofs that count. In: Jagannathan, S., Sewell, P. (eds.) The 41st Annual ACM SIGPLAN-SIGACT Symposium on Principles of Programming Languages, POPL 2014, San Diego, CA, USA, 20–21 January 2014, pp. 151–164. ACM (2014). https://doi.org/10.1145/2535838.2535885
29. Farzan, A., Klumpp, D., Podelski, A.: Sound sequentialization for concurrent program verification. In: Jhala, R., Dillig, I. (eds.) PLDI 2022: 43rd ACM SIGPLAN International Conference on Programming Language Design and Implementation, San Diego, CA, USA, 13–17 June 2022, pp. 506–521. ACM (2022). https://doi.org/10.1145/3519939.3523727
30. Farzan, A., Klumpp, D., Podelski, A.: Commutativity simplifies proofs of parameterized programs. Proc. ACM Program. Lang. **8**(POPL), 2485–2513 (2024). https://doi.org/10.1145/3632925
31. Filliâtre, J., Gondelman, L., Paskevich, A.: The spirit of ghost code. Formal Methods Syst. Des. **48**(3), 152–174 (2016). https://doi.org/10.1007/S10703-016-0243-X
32. Fleischmann, M., Kaindlstorfer, D., Isychev, A., Wüstholz, V., Christakis, M.: Constraint-based test oracles for program analyzers. In: Proceedings of the 39th International Conference on Automated Software Engineering (ASE 2024). ACM (2024). https://doi.org/10.1145/3691620.3695035
33. Franceschino, L., Pichardie, D., Talpin, J.-P.: Verified functional programming of an abstract interpreter. In: Drăgoi, C., Mukherjee, S., Namjoshi, K. (eds.) SAS 2021. LNCS, vol. 12913, pp. 124–143. Springer, Cham (2021). https://doi.org/10.1007/978-3-030-88806-0_6
34. Gries, D.: An exercise in proving parallel programs correct. Commun. ACM **20**(12), 921–930 (1977). https://doi.org/10.1145/359897.359903
35. Gurfinkel, A., Chechik, M.: Proof-like counter-examples. In: Garavel, H., Hatcliff, J. (eds.) TACAS 2003. LNCS, vol. 2619, pp. 160–175. Springer, Heidelberg (2003). https://doi.org/10.1007/3-540-36577-X_12
36. Haltermann, J., Wehrheim, H.: Information exchange between over- and under-approximating software analyses. In: Schlingloff, B.H., Chai, M. (eds.) Software Engineering and Formal Methods. Lecture Notes in Computer Science, pp. 37–54. Springer, Cham (2022). ISBN 978-3-031-17108-6. https://doi.org/10.1007/978-3-031-17108-6_3
37. Heizmann, M., Hoenicke, J., Podelski, A.: Refinement of trace abstraction. In: Palsberg, J., Su, Z. (eds.) SAS 2009. LNCS, vol. 5673, pp. 69–85. Springer, Heidelberg (2009). https://doi.org/10.1007/978-3-642-03237-0_7
38. Hoenicke, J., Majumdar, R., Podelski, A.: Thread modularity at many levels: a pearl in compositional verification. In: Castagna, G., Gordon, A.D. (eds.) Proceedings of the 44th ACM SIGPLAN Symposium on Principles of Programming Languages, POPL 2017, Paris, France, 18–20 January 2017, pp. 473–485. ACM (2017). https://doi.org/10.1145/3009837.3009893

39. International Organization for Standardization and International Electrotechnical Commission: ISO/IEC 9899:1999 - Programming Languages - C. Technical report, ISO/IEC JTC 1/SC 22 (1999). http://www.open-std.org/jtc1/sc22/wg14/www/docs/n1124.pdf. ISO/IEC 9899:1999 (C99) draft
40. Jourdan, J., Laporte, V., Blazy, S., Leroy, X., Pichardie, D.: A formally-verified C static analyzer. In: Rajamani, S.K., Walker, D. (eds.) Proceedings of the 42nd Annual ACM SIGPLAN-SIGACT Symposium on Principles of Programming Languages, POPL 2015, Mumbai, India, 15–17 January 2015, pp. 247–259. ACM (2015). https://doi.org/10.1145/2676726.2676966
41. Kaindlstorfer, D., Isychev, A., Wüstholz, V., Christakis, M.: Interrogation testing of program analyzers for soundness and precision issues. In: Proceedings of the 39th International Conference on Automated Software Engineering (ASE 2024). ACM (2024). https://doi.org/10.1145/3691620.3695034
42. Keller, R.M.: Formal verification of parallel programs. Commun. ACM **19**(7), 371–384 (1976). https://doi.org/10.1145/360248.360251
43. Klinger, C., Christakis, M., Wüstholz, V.: Differentially testing soundness and precision of program analyzers. In: Zhang, D., Møller, A. (eds.) Proceedings of the 28th ACM SIGSOFT International Symposium on Software Testing and Analysis, ISSTA 2019, Beijing, China, 15–19 July 2019, pp. 239–250. ACM (2019). https://doi.org/10.1145/3293882.3330553
44. Klumpp, D., Dietsch, D., Heizmann, M., Schüssele, F., Ebbinghaus, M., Farzan, A., Podelski, A.: Ultimate GemCutter and the axes of generalization. In: Fisman, D., Rosu, G. (eds.) TACAS 2022. LNCS, vol. 13244, pp. 479–483. Springer, Cham (2022). https://doi.org/10.1007/978-3-030-99527-0_35
45. Krebbers, R., Jung, R., Bizjak, A., Jourdan, J.-H., Dreyer, D., Birkedal, L.: The essence of higher-order concurrent separation logic. In: Yang, H. (ed.) ESOP 2017. LNCS, vol. 10201, pp. 696–723. Springer, Heidelberg (2017). https://doi.org/10.1007/978-3-662-54434-1_26
46. Lahav, O., Vafeiadis, V.: Owicki-Gries reasoning for weak memory models. In: Halldórsson, M.M., Iwama, K., Kobayashi, N., Speckmann, B. (eds.) ICALP 2015. LNCS, vol. 9135, pp. 311–323. Springer, Heidelberg (2015). https://doi.org/10.1007/978-3-662-47666-6_25
47. Lamport, L.: Proving the correctness of multiprocess programs. IEEE Trans. Software Eng. **3**(2), 125–143 (1977). https://doi.org/10.1109/TSE.1977.229904
48. Lamport, L.: Time, clocks, and the ordering of events in a distributed system. Commun. ACM **21**(7), 558–565 (1978). https://doi.org/10.1145/359545.359563
49. Lamport, L.: How to make a multiprocessor computer that correctly executes multiprocess programs. IEEE Trans. Computers **28**(9), 690–691 (1979). https://doi.org/10.1109/TC.1979.1675439
50. Ley-Wild, R., Nanevski, A.: Subjective auxiliary state for coarse-grained concurrency. In: Giacobazzi, R., Cousot, R. (eds.) The 40th Annual ACM SIGPLAN-SIGACT Symposium on Principles of Programming Languages, POPL 2013, Rome, Italy - 23–25 January 2013, pp. 561–574. ACM (2013). https://doi.org/10.1145/2429069.2429134
51. Lomet, D.B.: Process structuring, synchronization, and recovery using atomic actions. In: Wortman, D.B. (ed.) Proceedings of an ACM Conference on Language Design for Reliable Software (LDRS), Raleigh, North Carolina, USA, 28–30 March 1977, pp. 128–137. ACM (1977). https://doi.org/10.1145/800022.808319
52. Lubachevsky, B.D.: An approach to automating the verification of compact parallel coordination programs I. Acta Informatica **21**, 125–169 (1984). https://doi.org/10.1007/BF00289237

53. Mazurkiewicz, A.: Trace theory. In: Brauer, W., Reisig, W., Rozenberg, G. (eds.) ACPN 1986. LNCS, vol. 255, pp. 278–324. Springer, Heidelberg (1987). https://doi.org/10.1007/3-540-17906-2_30
54. McCloskey, B., Zhou, F., Gay, D., Brewer, E.A.: Autolocker: synchronization inference for atomic sections. In: Morrisett, J.G., Jones, S.L.P. (eds.) Proceedings of the 33rd ACM SIGPLAN-SIGACT Symposium on Principles of Programming Languages, POPL 2006, Charleston, South Carolina, USA, 11–13 January 2006, pp. 346–358. ACM (2006). https://doi.org/10.1145/1111037.1111068
55. Meolic, R., Fantechi, A., Gnesi, S.: Witness and counterexample automata for ACTL. In: de Frutos-Escrig, D., Núñez, M. (eds.) FORTE 2004. LNCS, vol. 3235, pp. 259–275. Springer, Heidelberg (2004). https://doi.org/10.1007/978-3-540-30232-2_17
56. Milanese, M., Miné, A.: Generation of violation witnesses by under-approximating abstract interpretation. In: Dimitrova, R., Lahav, O., Wolff, S. (eds.) Verification, Model Checking, and Abstract Interpretation - 25th International Conference, VMCAI 2024, London, United Kingdom, 15–16 January 2024, Proceedings, Part I. Lecture Notes in Computer Science, vol. 14499, pp. 50–73. Springer, Cham (2024). https://doi.org/10.1007/978-3-031-50524-9_3
57. Miné, A.: Static analysis of run-time errors in embedded real-time parallel C programs. Log. Methods Comput. Sci. **8**(1) (2012). https://doi.org/10.2168/LMCS-8(1:26)2012
58. Miné, A.: Relational thread-modular static value analysis by abstract interpretation. In: McMillan, K.L., Rival, X. (eds.) VMCAI 2014. LNCS, vol. 8318, pp. 39–58. Springer, Heidelberg (2014). https://doi.org/10.1007/978-3-642-54013-4_3
59. Monat, R., Miné, A.: Precise thread-modular abstract interpretation of concurrent programs using relational interference abstractions. In: Bouajjani, A., Monniaux, D. (eds.) VMCAI 2017. LNCS, vol. 10145, pp. 386–404. Springer, Cham (2017). https://doi.org/10.1007/978-3-319-52234-0_21
60. Mukherjee, S., Padon, O., Shoham, S., D'Souza, D., Rinetzky, N.: Thread-local semantics and its efficient sequential abstractions for race-free programs. In: Ranzato, F. (ed.) SAS 2017. LNCS, vol. 10422, pp. 253–276. Springer, Cham (2017). https://doi.org/10.1007/978-3-319-66706-5_13
61. Nanevski, A., Ley-Wild, R., Sergey, I., Delbianco, G.A.: Communicating state transition systems for fine-grained concurrent resources. In: Shao, Z. (ed.) ESOP 2014. LNCS, vol. 8410, pp. 290–310. Springer, Heidelberg (2014). https://doi.org/10.1007/978-3-642-54833-8_16
62. Nieto, L.P.: Completeness of the Owicki-Gries system for parameterized parallel programs. In: Proceedings of the 15th International Parallel & Distributed Processing Symposium (IPDPS-01), San Francisco, CA, USA, 23–27 April 2001, p. 150. IEEE Computer Society (2001). https://doi.org/10.1109/IPDPS.2001.925138
63. Nieto, L.P.: Verification of parallel programs with the Owicki-Gries and Rely-Guarantee methods in Isabelle, HOL. Ph.D. thesis, Technical University Munich, Germany (2002). https://mediatum.ub.tum.de/?id=601717
64. Nipkow, T., Nieto, L.P.: Owicki/Gries in Isabelle/HOL. In: Finance, J.-P. (ed.) FASE 1999. LNCS, vol. 1577, pp. 188–203. Springer, Heidelberg (1999). https://doi.org/10.1007/978-3-540-49020-3_13
65. OASIS SARIF Technical Committee: Static analysis results interchange format (SARIF) version 2.1.0. OASIS standard, Organization for the Advancement of Structured Information Standards (OASIS) (2020). https://docs.oasis-open.org/sarif/sarif/v2.1.0/os/sarif-v2.1.0-os.pdf

66. Owicki, S.S., Gries, D.: Verifying properties of parallel programs: an axiomatic approach. Commun. ACM **19**(5), 279–285 (1976). https://doi.org/10.1145/360051.360224
67. Raad, A., Lahav, O., Vafeiadis, V.: Persistent Owicki-Gries reasoning: a program logic for reasoning about persistent programs on Intel-x86. Proc. ACM Program. Lang. **4**(OOPSLA), 151:1–151:28 (2020). https://doi.org/10.1145/3428219
68. Saan, S., et al.: Goblint: Autotuning thread-modular abstract interpretation. In: Tools and Algorithms for the Construction and Analysis of Systems, pp. 547–552. Springer, Cham (2023). https://doi.org/10.1007/978-3-031-30820-8_34
69. Saan, S., Schwarz, M., Erhard, J., Seidl, H., Tilscher, S., Vojdani, V.: Correctness witness validation by abstract interpretation. In: Dimitrova, R., Lahav, O., Wolff, S. (eds.) Verification, Model Checking, and Abstract Interpretation - 25th International Conference, VMCAI 2024, London, United Kingdom, 15–16 January 2024, Proceedings, Part I. Lecture Notes in Computer Science, vol. 14499, pp. 74–97. Springer, Cham (2024). https://doi.org/10.1007/978-3-031-50524-9_4
70. Schmaltz, S.B.: Towards the pervasive formal verification of multi-core operating systems and hypervisors implemented in C. Ph.D. thesis, Universität des Saarlandes (2012). https://doi.org/10.22028/D291-26525
71. Schreiber, T.: Auxiliary variables and recursive procedures. In: Bidoit, M., Dauchet, M. (eds.) TAPSOFT'97: Theory and Practice of Software Development, 7th International Joint Conference CAAP/FASE, Lille, France, 14–18 April 1997, Proceedings. Lecture Notes in Computer Science, vol. 1214, pp. 697–711. Springer (1997). https://doi.org/10.1007/BFB0030635
72. Schwarz, M., Saan, S., Seidl, H., Apinis, K., Erhard, J., Vojdani, V.: Improving thread-modular abstract interpretation. In: Drăgoi, C., Mukherjee, S., Namjoshi, K. (eds.) SAS 2021. LNCS, vol. 12913, pp. 359–383. Springer, Cham (2021). https://doi.org/10.1007/978-3-030-88806-0_18
73. Schwarz, M., Saan, S., Seidl, H., Erhard, J., Vojdani, V.: Clustered relational thread-modular abstract interpretation with local traces. In: Wies, T. (ed.) Programming Languages and Systems - 32nd European Symposium on Programming, ESOP 2023, Paris, France, 22–27 April 2023, Proceedings. Lecture Notes in Computer Science, vol. 13990, pp. 28–58. Springer, Cham (2023). https://doi.org/10.1007/978-3-031-30044-8_2
74. Semenyuk, M., Dongol, B.: Ownership-based Owicki-Gries reasoning. In: Hong, J., Lanperne, M., Park, J.W., Cerný, T., Shahriar, H. (eds.) Proceedings of the 38th ACM/SIGAPP Symposium on Applied Computing, SAC 2023, Tallinn, Estonia, 27–31 March 2023, pp. 1685–1694. ACM (2023). https://doi.org/10.1145/3555776.3577636
75. Sergey, I., Nanevski, A., Banerjee, A.: Mechanized verification of fine-grained concurrent programs. In: Grove, D., Blackburn, S.M. (eds.) Proceedings of the 36th ACM SIGPLAN Conference on Programming Language Design and Implementation, Portland, OR, USA, 15–17 June 2015, pp. 77–87. ACM (2015). https://doi.org/10.1145/2737924.2737964
76. Stade, Y., Tilscher, S., Seidl, H.: The top-down solver verified: building confidence in static analyzers. In: Gurfinkel, A., Ganesh, V. (eds.) Computer Aided Verification - 36th International Conference, CAV 2024, Montreal, QC, Canada, 24–27 July 2024, Proceedings, Part I. Lecture Notes in Computer Science, vol. 14681, pp. 303–324. Springer, Cham (2024). https://doi.org/10.1007/978-3-031-65627-9_15
77. Švejda, J., Berger, P., Katoen, J.-P.: Interpretation-based violation witness validation for C: NITWIT. In: TACAS 2020. LNCS, vol. 12078, pp. 40–57. Springer, Cham (2020). https://doi.org/10.1007/978-3-030-45190-5_3

78. Ultimate developers: Ultimate program analysis framework (2024). https://ultimate-pa.org/. Accessed 29 Sept 2024
79. Vick, C., McMillan, K.L.: Synthesizing history and prophecy variables for symbolic model checking. In: Dragoi, C., Emmi, M., Wang, J. (eds.) Verification, Model Checking, and Abstract Interpretation - 24th International Conference, VMCAI 2023, Boston, MA, USA, 16–17 January 2023, Proceedings. Lecture Notes in Computer Science, vol. 13881, pp. 320–340. Springer, Cham (2023). https://doi.org/10.1007/978-3-031-24950-1_15
80. de Vilhena, P.E., Pottier, F., Jourdan, J.: Spy game: verifying a local generic solver in Iris. Proc. ACM Program. Lang. **4**(POPL), 33:1–33:28 (2020). https://doi.org/10.1145/3371101
81. Vojdani, V., Apinis, K., Rõtov, V., Seidl, H., Vene, V., Vogler, R.: Static race detection for device drivers: the Goblint approach. In: Proceedings of the 31st IEEE/ACM International Conference on Automated Software Engineering. ACM (2016). https://doi.org/10.1145/2970276.2970337
82. Wright, D., Dalvandi, S., Batty, M., Dongol, B.: Mechanised operational reasoning for C11 programs with relaxed dependencies. Formal Aspects Comput. **35**(2), 10:1–10:27 (2023). https://doi.org/10.1145/3580285
83. Zhang, Z., Feng, X., Fu, M., Shao, Z., Li, Y.: A structural approach to prophecy variables. In: Agrawal, M., Cooper, S.B., Li, A. (eds.) TAMC 2012. LNCS, vol. 7287, pp. 61–71. Springer, Heidelberg (2012). https://doi.org/10.1007/978-3-642-29952-0_12

Parameterized Verification of Systems with Precise (0,1)-Counter Abstraction

Paul Eichler[1](✉)[iD], Swen Jacobs[1][iD], and Chana Weil-Kennedy[2][iD]

[1] CISPA Helmholtz Center for Information Security, Saarbrücken, Germany
{paul.eichler,jacobs}@cispa.de
[2] IMDEA Software Institute, Madrid, Spain
chana.weilkennedy@imdea.org

Abstract. We introduce a new framework for verifying systems with a parametric number of concurrently running processes. The systems we consider are well-structured with respect to a specific well-quasi order. This allows us to decide a wide range of verification problems, including control-state reachability, coverability, and target, in a fixed finite abstraction of the infinite state-space, called a 01-counter system. We show that several systems from the parameterized verification literature fall into this class, including reconfigurable broadcast networks (or systems with lossy broadcast), disjunctive systems, synchronizations and systems with a fixed number of shared finite-domain variables. Our framework provides a simple and unified explanation for the properties of these systems, which have so far been investigated separately. Additionally, it extends and improves on a range of the existing results, and gives rise to other systems with similar properties.

Keywords: Parameterized Verification · Finite Abstraction

1 Introduction

Concurrent systems often consist of an arbitrary number of uniform user processes running in parallel, possibly with a distinguished controller process. Given a description of the user and controller protocols and a desired property, the *parameterized model checking problem* (PMCP) is to decide whether the property holds in the system, regardless of the number of user processes. The PMCP is well-known to be undecidable in general [6], even when the property is control-state reachability and all processes are finite-state [39]. However, a long line of research has valiantly strived for the identification of decidable fragments that support interesting models and properties [1,3,14,21,23,25,26,31].

One of the most prominent techniques for the identification of fragments with decidable PMCP are well-structured transition systems (WSTS) [1,2,29,30]. The WSTS framework puts a number of restrictions on the system, most importantly the compatibility of its transition relation with a well-quasi order (wqo) on its (infinite) set of states, which in turn allows to decide some PMCP problems, including coverability. However, while many of the works on parameterized verification share certain techniques, systems with different communication primitives have usually been studied separately, and it is hard to keep an overview of which problems are decidable for which class of systems, and why.

In this paper, we show that a range of systems, previously studied using different techniques, can be unified in a single framework. Our framework gives a surprisingly simple explanation of existing decidability results for these systems, extends both the class of systems and the types of properties that can be verified, and allows us to prove previously unknown complexity bounds for some of these problems. While the main condition of our framework resembles that of WSTS, i.e., compatibility of transitions with a wqo, we do not make use of any WSTS techniques. Instead, we show that $(0,1)$-counter abstraction (or simply 01-abstraction), i.e., a binary abstraction that does not count the number of processes in a given state, but only distinguishes whether it is occupied or not, is precise for all systems satisfying the condition. This abstraction is not only fixed for the whole class of systems, but may also be much more concise than the abstraction obtained by using WSTS techniques. The wqo $\preceq_0$ we consider is an extension of the "standard" wqo for component-based systems, in which two configurations of the system are only comparable if they agree on which local states currently are occupied (by at least one process), and which are not occupied by any process.

Parameterized Systems and Related Work. The systems we consider are based on one control process and an arbitrary number of identical user processes. Processes change state synchronously according to a step relation, usually based on local transitions that may be synchronized based on transition labels. In particular, our framework supports the following communication (or synchronization) primitives from the parameterized verification literature:

- *Lossy broadcast* [19], where processes can send broadcast messages that may or may not be received by the other processes. This model is equivalent to the widely studied system model of *reconfigurable broadcast networks* (RBN) [8,9,13,17], where processes communicate via broadcast to their neighbors in the underlying communication topology, which can reconfigure at any time. Here, we frame them as lossy broadcast in a clique topology, since we also assume all other systems to be arranged in a clique topology.
- *Disjunctive guards* [21], where transitions of a process depend on the existence of another process that is in a certain local state. Systems with disjunctive guards (or: *disjunctive systems*) have been studied extensively in the literature [3,7,21,22,32,33]. We note that this model is equivalent to immediate observation (IO) protocols [28], a subclass of population protocols [5], where a process observes the state of another process and changes its own state accordingly. IO protocols are also known to be equivalent to the restriction of RBN in which broadcast transitions must be self-loops [10].
- *Synchronization*, where transitions are labeled with actions and in every step of the system all processes synchronize on the same action. This model is studied for example in the context of controller synthesis [12,16]. There, the

goal is to decide if, for a given protocol followed by a parametric number of processes, a controller strategy exists that eventually puts all processes in the final state f. In [16], the problem is posed in a stochastic setting. Synchronization protocols may be seen as a restriction of (non-lossy) broadcast protocols [24,26].
- *Asynchronous shared memory* (ASM) [27] allows processes to communicate through finite-domain shared variables, but without locks and non-trivial read-modify-write operations, i.e., a transition cannot read and write a variable simultaneously. ASM systems (also called register protocols [15]) are known to be equivalent to RBN with regard to reachability properties [10]. In [27] the authors go beyond what we consider in this work, as they consider that processes can also be pushdown machines or even Turing machines, and show that decidability can be preserved under certain restrictions.

Considering related verification techniques, close to ours in spirit is $(0, 1, \infty)$-counter abstraction [37], with the crucial difference that their technique is approximative, while ours is precise for the systems we consider. Additionally, 01-counter abstraction has already been used for parameterized verification and repair in previous work [11,33], but for more restricted classes of systems and, again, with correctness arguments specific to these classes. In contrast, we provide a general criterion for correctness of the abstraction for a much broader class of systems and properties.

In addition, there has been a lot of work on the parameterized verification of systems with more powerful communication primitives, such as pairwise rendezvous [3,31] or (non-lossy) broadcast [26]. While these also fall into the class of WSTS, they are not compatible with the wqo $\preceq_0$, and 01-abstraction is not precise for them. Accordingly, the complexity of parameterized verification problems is in general much higher for these systems.

Contributions. We introduce a common framework for the verification of parameterized systems that are well-structured with respect to the wqo $\preceq_0$.
- We prove that for all such systems, 01-abstraction is sound and complete for safety properties, and that lossy broadcast protocols, disjunctive systems, synchronization protocols, and ASM fall into this class, as well as systems based on a novel *guarded* synchronization primitive, and systems with combinations of these primitives (Sect. 3).
- We show that a cardinality reachability (CRP) problem, which subsumes classical parameterized problems like coverability and target, is PSPACE-complete for our class of systems (Sect. 4).
- We show how the 01-abstraction can be leveraged to decide finite trace properties of a fixed number of processes in the parameterized system, and slightly improve known results on properties over infinite traces for disjunctive systems (Sect. 5).
- We show that under modest additional assumptions on the systems, the complexity of the CRP is significantly lower (Sect. 6).

Note that full proofs and additional details can be found in the extended version of this paper [20].

2 Preliminaries

Multisets. A *multiset* on a finite set E is a mapping $C \colon E \to \mathbb{N}$, i.e. for any $e \in E$, $C(e)$ denotes the number of occurrences of element e in C. We sometimes consider C as a vector of length the cardinality of E, and denote it as $\mathbf{c} \in \mathbb{N}^E$. Given $e \in E$, we denote by $\mathbf{e}$ the multiset consisting of one occurrence of element e. Operations on $\mathbb{N}$ like addition or comparison are extended to multisets by defining them component-wise on each element of E. Subtraction is allowed in the following way: if $\mathbf{c}, \mathbf{d}$ are multisets on set E then for all $e \in E$, $(\mathbf{c} - \mathbf{d})(e) = \max(\mathbf{c}(e) - \mathbf{d}(e), 0)$. We call $|\mathbf{c}| = \sum_{e \in E} \mathbf{c}(e)$ the *size* of $\mathbf{c}$. The support $[\![\mathbf{c}]\!]$ of $\mathbf{c}$ is the set of elements $e \in E$ such that $\mathbf{c}(e) \geq 1$.

Counter System. Intuitively, a counter system explicitly keeps track of the state of the controller process, and for user processes keeps track of *how many* user processes are in which local state. Let us formalize this idea.

Definition 1. *A counter system (CS) is a triple $\mathcal{C} = (C, Q, \mathcal{T})$ where C is the finite set of states of the controller, Q is the finite set of states of the users and $\mathcal{T}$ is the step relation such that $\mathcal{T} \subseteq (C \times \mathbb{N}^Q) \times (C \times \mathbb{N}^Q)$, where $|\mathbf{v}| = |\mathbf{v}'|$ whenever $((c, \mathbf{v}), (c', \mathbf{v}')) \in \mathcal{T}$, i.e., steps are size-preserving. A configuration of $\mathcal{C}$ is a pair $(c, \mathbf{v}) \in C \times \mathbb{N}^Q$. We may fix initial states $c_0 \in C$ and $Q_0 \subseteq Q$; an initial configuration is then any $(c_0, \mathbf{v}_0)$ such that $\mathbf{v}_0(q) = 0$ for all q not in Q_0. The size of a configuration is $|(c, \mathbf{v})| = 1 + |\mathbf{v}|$.*

If $((c, \mathbf{v}), (c', \mathbf{v}')) \in \mathcal{T}$ then we say there is a *step* from $(c, \mathbf{v})$ to $(c', \mathbf{v}')$, also denoted $(c, \mathbf{v}) \to (c', \mathbf{v}')$. We denote by $\xrightarrow{*}$ the reflexive and transitive closure of the step relation. A sequence of steps is called a *path* of $\mathcal{C}$. A path is a *run* if it starts in an initial configuration. A configuration $(c, \mathbf{v})$ is *reachable* if there is a run that ends in $(c, \mathbf{v})$.

Remark 1. In contrast to some of the results in this area, our model supports an additional distinguished controller process, which may execute a different protocol than the user processes. It is known that in some settings the model with a controller is strictly more expressive than the model without [3].

Moreover, since our model also supports multiple initial states for the user processes, our results extend to the case of any fixed number of distinguished processes, and any fixed number of different types of user processes.[1] To keep notation simple, we will use a single controller and a single type of user process throughout the paper.

Well-Quasi Order. Let S be the (infinite) set of configurations of a CS. A *well-quasi order (wqo)* on S is a relation $\preceq \subseteq S \times S$ that is reflexive and transitive,

---

[1] To see this, note first that if a system has multiple controllers, we can encode all of them as a single controller by simply considering their (finite-state) product. To support k different types of user processes with state sets $Q_1, \ldots, Q_k$ such that $Q_i \cap Q_j = \emptyset$ for all $i \neq j$, we simply construct one big user process with state set $Q_1 \cup \cdots \cup Q_k$, and similarly let the union of all individual initial states be the initial states of the constructed system.

and is such that every infinite sequence $s_0, s_1, \ldots$ of elements from S contains an increasing pair $s_i \preceq s_j$ with $i < j$.

A wqo commonly used on configurations of a CS is defined as follows:

$$(c, \mathbf{v}) \preceq (d, \mathbf{w}) \Leftrightarrow (c = d \wedge \forall q \in Q : \mathbf{v}(q) \leq \mathbf{w}(q))$$

We define our wqo $\preceq_0$ as the following refinement of $\preceq$:

$$(c, \mathbf{v}) \preceq_0 (d, \mathbf{w}) \Leftrightarrow ((c, \mathbf{v}) \preceq (d, \mathbf{w}) \wedge \forall q \in Q : (\mathbf{v}(q) = 0 \Leftrightarrow \mathbf{w}(q) = 0))$$

Compatibility. We say that a CS $\mathcal{C}$ is *forward $\preceq$-compatible*[2] for a wqo $\preceq$ if whenever there is a step $(c, \mathbf{v}) \to (c', \mathbf{v}')$ and $(c, \mathbf{v}) \preceq (d, \mathbf{w})$, then there exists a step $(d, \mathbf{w}) \to (d', \mathbf{w}')$ with $(c', \mathbf{v}') \preceq (d', \mathbf{w}')$. We say $\mathcal{C}$ is *backward $\preceq$-compatible* if whenever there is a step $(c, \mathbf{v}) \to (c', \mathbf{v}')$ and $(c', \mathbf{v}') \preceq (d', \mathbf{w}')$, then there exists a step $(d, \mathbf{w}) \to (d', \mathbf{w}')$ with $(c, \mathbf{v}) \preceq (d, \mathbf{w})$. $\mathcal{C}$ is *fully $\preceq$-compatible* if it is forward and backward $\preceq$-compatible.

01-Counter System. The idea of the $(0, 1)$-counter abstraction is that we only distinguish whether a given local state is occupied or not. This is formalized through an abstraction function $\alpha : C \times \mathbb{N}^Q \to C \times \{0, 1\}^Q$ such that $\alpha(c, \mathbf{v}) = (c, \mathbf{v}^\alpha)$, where $\mathbf{v}^\alpha(q) = 1$ if $\mathbf{v}(q) \geq 1$ and $\mathbf{v}^\alpha(q) = 0$ if $\mathbf{v}(q) = 0$. We define the abstraction of a given CS via α.

Definition 2. *The* 01-counter system *(01-CS) of* $\mathcal{C} = (C, Q, \mathcal{T})$ *is the tuple* $\mathcal{C}_\alpha = (C \times \{0, 1\}^Q, \mathcal{T}_\alpha)$*, where* $\mathcal{T}_\alpha \subseteq (C \times \{0, 1\}^Q) \times (C \times \{0, 1\}^Q)$ *is such that* $(c, \mathbf{v}^\alpha) \to (c', \mathbf{v}'^\alpha) \in \mathcal{T}_\alpha$ *if there exists a concrete step* $(c, \mathbf{v}) \to (c', \mathbf{v}') \in \mathcal{T}$ *with* $\alpha(c, \mathbf{v}) = (c, \mathbf{v}^\alpha)$ *and* $\alpha(c', \mathbf{v}') = (c', \mathbf{v}'^\alpha)$. *Given initial states* $c_0 \in C$ *and* $Q_0 \subseteq Q$ *of* $\mathcal{C}$, *an initial configuration of* $\mathcal{C}_\alpha$ *is any* $(c_0, \mathbf{v}^\alpha)$ *such that* $\mathbf{v}^\alpha(q) = 0$ *for all q not in Q_0.*

Remark 2. Unlike for CSs, in a 01-CS it is not the case that steps occur only between configurations of the same size. For example, we may have $(c, \mathbf{v}) \to (c, \mathbf{v}')$ in a CS $\mathcal{C}$ where $\mathbf{v}(q) = 1$ for all states $q \in Q$, and the step sends all the user processes to a state p in $\mathbf{v}'$. Then the corresponding 01-CS $\mathcal{C}_\alpha$ has a step $(c, \mathbf{v}^\alpha) \to (c, \mathbf{v}'^\alpha)$ such that $\mathbf{v} = \mathbf{v}^\alpha$ and $\mathbf{v}'^\alpha(q) = 1$ if $q = p$ and 0 elsewhere, i.e., $|\mathbf{v}^\alpha| = |Q| + 1$ while $|\mathbf{v}'^\alpha| = 1$.

The following link between the wqo $\preceq_0$ and 01-CSs follows directly from the definition.

Lemma 1. *Let* $\mathcal{C} = (C, Q, \mathcal{T})$ *a CS, and* $(c, \mathbf{v}^\alpha)$ *a configuration in* $C \times \{0, 1\}^Q$. *Let* $(d, \mathbf{w})$ *a configuration in* $C \times \mathbb{N}^Q$. *Then* $(c, \mathbf{v}^\alpha) \preceq_0 (d, \mathbf{w})$ *if and only if* $\alpha(d, \mathbf{w}) = (c, \mathbf{v}^\alpha)$.

---

[2] This is sometimes called *strong compatibility* in the literature.

2.1 Types of Steps

We formally define the communication primitives mentioned in Sect. 1. Steps between configurations are defined as multisets of (local) transitions that are taken by different processes at the same time, i.e., every process takes at most one transition in a step. We say a process in configuration $(c, \mathbf{v})$ *takes a transition* $p \to q$ if the process moves from state p to q, resulting in a new configuration $(c', \mathbf{v}')$ equal to $(c, \mathbf{v} - \mathbf{p} + \mathbf{q})$ if $p, q \in Q$ and $\mathbf{v}(p) \geq 1$, and equal to $(q, \mathbf{v})$ if $p, q \in C$ and $c = p$. The conditions $\mathbf{v}(p) \geq 1$ and $c = p$ ensure that there is a process in $(c, \mathbf{v})$ that can take the transition.

The following definitions of transition types and the steps they induce come from the literature on systems in which all steps are of one type, e.g., RBN [17,19] in which all steps are lossy broadcasts. Note that we allow the same transition to be taken by more than one process in a single step, even if the classical definitions would consider this several consecutive steps. We discuss the resulting differences after formally defining our steps.

Internal. *Internal transitions* are of the form (p, q) with $p, q \in C$ or $p, q \in Q$, also denoted $p \to q$. These induce *internal steps* where one or more processes take the same transition $p \to q$, i.e., $(c, \mathbf{v}) \to (c, \mathbf{v} - i \cdot \mathbf{p} + i \cdot \mathbf{q})$ if $p, q \in Q$ and $\mathbf{v}(p) \geq i$, and $(p, \mathbf{v}) \to (q, \mathbf{v})$ if $p, q \in C$.

Lossy Broadcast [19]. Let Σ be a finite alphabet. *Lossy broadcast transitions* are of the form (p, l, q) with $l \in \{!a, ?a \mid a \in \Sigma\}$ and $p, q \in C$ or $p, q \in Q$. We sometimes denote a transition (p, l, q) by $p \xrightarrow{l} q$. Transitions with l of the form $!a$ are *broadcast transitions*, and transitions with l of the form $?a$ are *receive transitions*. A *lossy broadcast step* from a configuration $(c, \mathbf{v})$ is made up of one or more processes taking the same broadcast transition $p \xrightarrow{!a} p'$, and an arbitrary number $k \geq 0$ of processes taking receive transitions $p_1 \xrightarrow{?a} p'_1, \ldots, p_k \xrightarrow{?a} p'_k$. If $(c', \mathbf{v}')$ is the resulting configuration, we denote the step by $(c, \mathbf{v}) \to (c', \mathbf{v}')$.

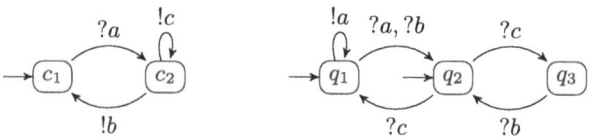

Fig. 1. A lossy broadcast protocol with two controller states and three user states.

Example 1. Figure 1 depicts a lossy broadcast protocol. The initial configurations are the $(c_1, \mathbf{v}_0)$ such that $[\![\mathbf{v}_0]\!] \subseteq \{q_1, q_2\}$. From configuration $(c_1, \mathbf{v})$ with $\mathbf{v} = (2, 1, 0)$, there is a step to $(c_2, \mathbf{v}')$ with $\mathbf{v}' = (1, 2, 0)$: a user process takes broadcast $q_1 \xrightarrow{!a} q_1$, the controller takes receive $c_1 \xrightarrow{?a} c_2$ and the other user takes receive $q_1 \xrightarrow{?a} q_2$. Depending on which processes receive the a broadcast, there is also a step on a from $(c_1, \mathbf{v})$ to $(c_1, \mathbf{v})$, $(c_2, \mathbf{v})$ and $(c_1, \mathbf{v}')$.

Disjunctive Guard [21]. *Disjunctive guard transitions* are of the form $(p, G_\exists, q)$ where $p, q \in C$ or $p, q \in Q$, and $G_\exists \subseteq C \cup Q$. We denote the transition by $p \xrightarrow{G_\exists} q$. A configuration $(c, \mathbf{v})$ *satisfies* $G_\exists$ if $(\mathbf{v})(r) \geq 1$ for some $r \in G_\exists$ or if $c \in G_\exists$. A *disjunctive guard step* on transition $p \xrightarrow{a} q$ is only enabled from configurations $(c, \mathbf{v})$ that satisfy $G_\exists$. Then, it consists of one or more processes taking the transition $p \xrightarrow{G_\exists} q$ (like in internal steps), such that the resulting configuration $(c', \mathbf{v}')$ also satisfies $G_\exists$, i.e., a moving process cannot be the one that satisfies the guard. We denote the step by $(c, \mathbf{v}) \to (c', \mathbf{v}')$.

Synchronization [12]. Let Σ be a finite set of labels. *Synchronization transitions* are of the form (p, a, q) with $a \in \Sigma$ and $p, q \in C$ or $p, q \in Q$, also denoted $p \xrightarrow{a} q$. In a *synchronization step on a* from a configuration $(c, \mathbf{v})$, all processes take a transition with label a, if such a transition is available in their current state (otherwise they stay in their current state). If $(c', \mathbf{v}')$ is the resulting configuration, then we denote the step by $(c, \mathbf{v}) \to (c', \mathbf{v}')$.

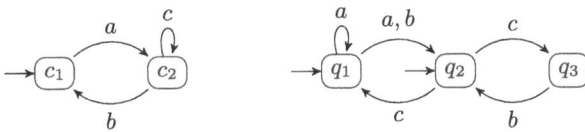

Fig. 2. A synchronization protocol with two controller states and three user states.

Example 2. Consider the synchronization protocol depicted in Fig. 2. From the configuration $(c_1, \mathbf{v})$ with $\mathbf{v} = (2, 1, 0)$, there is a step on letter a to $(c_2, (2, 1, 0))$, $(c_2, (1, 2, 0))$, or to $(c_2, (0, 3, 0))$. The user process initially in q_2 does not move because there is no a-transition from q_2.

Asynchronous Shared Memory (ASM) [27]. *ASM transitions* are of the form (p, l, q) with $l \in \{w(a), r(a) \mid a \in C\}$ and $p, q \in Q$. In systems with ASM transitions, we assume that a transition (a, b) is available for every $a, b \in C$. We also denote a transition (p, l, q) by $p \xrightarrow{l} q$, and (a, b) by $a \to b$. Transitions with l of the form $w(a)$ are *write transitions*, and transitions with l of the form $r(a)$ are *read transitions*. Intuitively, the controller keeps track of the value of the shared variable, and the user processes can read that value or give an instruction to update the value[3]. An *ASM step* from a configuration $(c, \mathbf{v})$ is either a write step or a read step: A *write step* is made up of one or more user processes taking a transition $p \xrightarrow{w(a)} q$ and the controller taking transition $c \to a$. A *read step* is made up of one or more user processes taking a transition $p \xrightarrow{r(a)} q$ and the controller taking transition $a \to a$, ensuring that $a = c$.

---

[3] The restriction to a single variable is for simplicity, our results extend to multiple finite-domain variables.

Examples for systems with disjunctive guards and ASM can be found in the extended version [20]. Note that all of these types of steps are between configurations of the same size. Any finite sets C, Q and any combination of transitions of the types described above define a set of steps $\mathcal{T}$ such that $\mathcal{C} = (C, Q, \mathcal{T})$ is a CS. As mentioned above, one of our steps sometimes corresponds to several consecutive steps in the classical definitions of the literature. More precisely, the same broadcast transition, disjunctive transition, read transition or write transition can be taken by an arbitrary number of distinct processes in the same step. Note that these "accelerated" steps do not change the reachability properties of our systems: if $(c, \mathbf{v}) \to (c', \mathbf{v}')$ is a step in our definition, then $(c, \mathbf{v}) \xrightarrow{*} (c', \mathbf{v}')$ is a sequence of steps in the classic definition.

3 Reduction of Parameterized Safety Verification to the 01-CS

We first prove that for $\preceq_0$-compatible systems, parameterized safety verification, i.e., regarding properties of finite runs, can be reduced to their 01-abstraction. Then we show that all system types introduced in Sect. 2.1 are $\preceq_0$-compatible, as well as some new system types.

Lemma 2. *If a given CS $\mathcal{C}$ is fully $\preceq_0$-compatible, then there exists a run $(c_0, \mathbf{v}_0) \to (c_1, \mathbf{v}_1) \to \ldots \to (c_n, \mathbf{v}_n)$ in $\mathcal{C}$ if and only if there exists a run $(c_0, \mathbf{v}_0^\alpha) \to (c_1, \mathbf{v}_1^\alpha) \to \ldots \to (c_n, \mathbf{v}_n^\alpha)$ in the corresponding 01-CS $\mathcal{C}_\alpha$ such that $\alpha(c_i, \mathbf{v}_i) = (c_i, \mathbf{v}_i^\alpha)$ for each i.*

Proof. Let $\mathcal{C} = (C, Q, \mathcal{T})$ be a $\preceq_0$-compatible CS, and let $\mathcal{C}_\alpha$ be its 01-CS. Assume there exists a run $(c_0, \mathbf{v}_0) \to \ldots \to (c_n, \mathbf{v}_n) \to (c, \mathbf{v})$ in $\mathcal{C}$. Then by definition of $\mathcal{C}_\alpha$, for each step $(c_i, \mathbf{v}_i) \to (c_{i+1}, \mathbf{v}_{i+1})$ of the sequence there exists an abstract step $(c_i, \mathbf{v}_i^\alpha) \to (c_{i+1}, \mathbf{v}_{i+1}^\alpha)$ in $\mathcal{C}_\alpha$, where $\alpha(c_i, \mathbf{v}_i) = (c_i, \mathbf{v}_i^\alpha)$.

In the other direction, assume that there exists a run $(c_0, \mathbf{v}_0^\alpha) \to \ldots \to (c_n, \mathbf{v}_n^\alpha) \to (c, \mathbf{v}^\alpha)$ in $\mathcal{C}_\alpha$. We prove by induction on n that there exists a run in $\mathcal{C}$ with the desired properties.

Base case: $n = 0$. If $(c_0, \mathbf{v}_0^\alpha) \to (c, \mathbf{v}^\alpha)$ is a step in $\mathcal{C}_\alpha$, then by definition there exist $\mathbf{v}_0, \mathbf{v}$ with $\alpha(c_0, \mathbf{v}_0) = (c_0, \mathbf{v}_0^\alpha)$ and $\alpha(c, \mathbf{v}) = (c, \mathbf{v}^\alpha)$ and $(c_0, \mathbf{v}_0) \to (c, \mathbf{v})$.

Induction step: $n \to n+1$. Let $(c_0, \mathbf{v}_0^\alpha) \xrightarrow{*} (c_n, \mathbf{v}_n^\alpha)$ be a run of $\mathcal{C}_\alpha$ and $(c_n, \mathbf{v}_n^\alpha) \to (c, \mathbf{v}^\alpha)$ a step in $\mathcal{C}_\alpha$. By induction hypothesis, there exists a run $(c_0, \mathbf{v}_0) \xrightarrow{*} (c_n, \mathbf{v}_n)$ of n steps in $\mathcal{C}$ with $\alpha(c_0, \mathbf{v}_0) = (c_0, \mathbf{v}_0^\alpha)$ and $\alpha(c_n, \mathbf{v}_n) = (c_n, \mathbf{v}_n^\alpha)$. By definition of the 01-CS, there exist $\mathbf{w}_n, \mathbf{w}$ with $\alpha(c_n, \mathbf{w}_n) = (c_n, \mathbf{v}_n^\alpha)$, $\alpha(c, \mathbf{w}) = (c, \mathbf{v}^\alpha)$ and step $(c_n, \mathbf{w}_n) \to (c, \mathbf{w})$ in $\mathcal{C}$. Configurations $\mathbf{w}_n, \mathbf{v}_n$ (and $\mathbf{v}_n^\alpha$) are equal to 0 on the same states, so there exists $\mathbf{x}_n$ such that $(c, \mathbf{w}_n) \preceq_0 (c, \mathbf{x}_n)$ and $(c, \mathbf{v}_n) \preceq_0 (c, \mathbf{x}_n)$. And therefore by Lemma 1 also $\alpha(c_n, \mathbf{x}_n) = (c_n, \mathbf{v}_n^\alpha)$. Then, by backward $\preceq_0$-compatibility of $\mathcal{C}$ we get that there is a run that reaches $(c_n, \mathbf{x}_n)$ in n steps, and by forward $\preceq_0$-compatibility we get that there exists a step $(c_n, \mathbf{x}_n) \to (c, \mathbf{x})$ for some $(c, \mathbf{x})$ such that $\alpha(c, \mathbf{x}) = (c, \mathbf{v}^\alpha)$. □

Example 3. Consider the lossy broadcast step in the system of Fig. 1 from $(c_2, (1, 0, 2))$ to $(c_1, (0, 2, 1))$ where the controller broadcasts b, one of the two processes in q_3 takes $q_3 \xrightarrow{?b} q_2$ and the process in q_1 takes $q_1 \xrightarrow{?b} q_2$. Configuration $(c_2, (2, 0, 3))$ is such that $(c_2, (1, 0, 2)) \preceq_0 (c_2, (2, 0, 3))$. We describe a step from this configuration to a configuration $(c, \mathbf{v})$ such that $(c_1, (0, 2, 1)) \preceq_0 (c, \mathbf{v})$: the controller broadcasts b, two processes take $q_3 \xrightarrow{?b} q_2$ and two processes take $q_1 \xrightarrow{?b} q_2$.

We can prove full $\preceq_0$-compatibility for the types of steps introduced in Sect. 2.1.

Lemma 3. *CSs induced by one of the following types of steps are fully $\preceq_0$-compatible: lossy broadcast, disjunctive guard, synchronization, or ASM.[4]*

We give the proof for CSs induced by lossy broadcasts steps, and relegate the other (similar) proofs to the extended version [20].

Proof (Partial). Let $\mathcal{C} = (C, Q, \mathcal{T})$ be a CS with only lossy broadcast steps. To prove forward $\preceq_0$-compatibility, assume there is a step $(c, \mathbf{v}) \rightarrow (c', \mathbf{v}')$ and $(c, \mathbf{v}) \preceq_0 (d, \mathbf{w})$. This step is made up of $j \geq 1$ processes taking a broadcast transition $p_0 \xrightarrow{!a} p'_0$ and k processes taking receive transitions $p_1 \xrightarrow{?a} p'_1, \ldots, p_k \xrightarrow{?a} p'_k$ for some $k \geq 0$. Since $(c, \mathbf{v}) \preceq_0 (d, \mathbf{w})$, for every state $q \in Q$, $\mathbf{v}(q) \leq \mathbf{w}(q)$ and $c = d$. Therefore a step with the same $j + k$ transitions can be taken from $(d, \mathbf{w})$. Call $(d'', \mathbf{w}'')$ the resulting configuration. We want to check $(c', \mathbf{v}') \preceq_0 (d'', \mathbf{w}'')$ or modify the step from $(d, \mathbf{w})$ to make it true. This means we have to satisfy the following three conditions:

1. $d'' = c'$: either $c = c'$, in which case no transition is taken by the controller in either step and $d = d'' = c = c'$, or $c \neq c'$, in which case $c = p_i, c' = p'_i$ for some $i \in \{0, ..., k\}$ and this transition is taken in both steps, so $d'' = p'_i = c'$.
2. $\mathbf{w}''(q) \geq \mathbf{v}'(q)$ for all $q \in Q$: the same transitions are taken from $\mathbf{w} \geq \mathbf{v}$.
3. $\mathbf{w}''(q) = 0$ if and only if $\mathbf{v}'(q) = 0$ for all $q \in Q$: if there are no such states then we are done; otherwise suppose $\mathbf{v}'(q) = 0$ and $\mathbf{w}''(q) > 0$. This entails $\mathbf{w}(q) > 0$ and thus also $\mathbf{v}(q) > 0$ by definition of $\preceq_0$. Informally, this means state q was emptied in the step $(c, \mathbf{v}) \rightarrow (c', \mathbf{v}')$; one of the transitions taken is of the form $q = p_i \xrightarrow{a\star} p'_i$ with $\star \in \{!, ?\}$. We modify the step from $(d, \mathbf{w})$ by adding $\mathbf{w}(q) - \mathbf{v}(q)$ iterations of $p_i \xrightarrow{\star a} p'_i$, i.e., enough to empty q. The resulting configuration $(d', \mathbf{w}')$ is such that $(c', \mathbf{v}') \preceq_0 (d', \mathbf{w}')$.

Backward $\preceq_0$-compatibility can be proven in a similar way. □

As a new communication primitive, we can extend synchronization transitions (as introduced in Sect. 2.1) to *guarded synchronizations*, which are additionally labeled with a pair $(G_\exists, G_\forall)$ with $G_\exists, G_\forall \subseteq (C \cup Q)$, and then denoted $p \xrightarrow{a, (G_\exists, G_\forall)} q$. The step is defined as for synchronization steps, except that a

---
[4] Internal steps can be seen as a special case of lossy broadcast, disjunctive guard, or ASM steps.

synchronization step guarded by $(G_\exists, G_\forall)$ is only enabled from a configuration $(c, \mathbf{v})$ with $S = [\![\mathbf{v}]\!] \cup \{c\}$ if $S \cap G_\exists \neq \emptyset$ (there exists a $g \in S$ with $g \in G_\exists$), and $S \subseteq G_\forall$ (for all $g \in S$ we have $g \in G_\forall$). That is, they are interpreted as a disjunctive and a conjunctive guard, respectively, and we can mix both types of guards, even in the same transition.

Example 4 (Guarded Synchronization). Consider the synchronization protocol from Example 2, assuming that the action on a is guarded by $G_\exists = \{c_1, q_1\}$ and $G_\forall = Q \setminus \{q_3\}$. Then it is enabled from $(c_1, \mathbf{v})$ with $\mathbf{v} = (2, 1, 0)$ (as there is a process in $c_1 \in G_\exists$, and no processes are in q_3). It is however not enabled from $(c_2, \mathbf{w})$ with $\mathbf{w} = (0, 2, 0)$ (as there is no process in a state from $G_\exists$), and also not from $(c_1, \mathbf{w}')$ with $\mathbf{w}' = (0, 2, 1)$ (as one process is in $q_3 \notin G_\forall$).

CSs with guarded synchronizations are also fully $\preceq_0$-compatible.

Lemma 4. *CSs induced by guarded synchronization steps are fully $\preceq_0$-compatible.*

Proof. To prove forward $\preceq_0$-compatibility, assume there is a step $(c, \mathbf{v}) \xrightarrow{a,(G_\exists, G_\forall)} (c', \mathbf{v}')$ and $(c, \mathbf{v}) \preceq_0 (d, \mathbf{w})$. Note that, as the step is enabled from $(c, \mathbf{v})$, there exists a state $q_\exists \in G_\exists$ that is also in $[\![\mathbf{v}]\!] \cup \{c\}$, and every state in $[\![\mathbf{v}]\!] \cup \{c\}$ is also in $G_\forall$. By definition of $\preceq_0$ it follows that for $(d, \mathbf{w})$ there is at least one process in $q_\exists$ and all states in $(C \cup Q) \setminus G_\forall$ remain empty. Consequently, the synchronization on a is also enabled in $(d, \mathbf{w})$ and forward compatibility follows by forward compatibility of synchronization actions.

To prove backward $\preceq_0$-compatibility, assume there is a step $(c, \mathbf{v}) \xrightarrow{a,(G_\exists, G_\forall)} (c', \mathbf{v}')$ and $(c', \mathbf{v}') \preceq_0 (d', \mathbf{w}')$. By backward compatibility of synchronization steps, we know that there must exist a configuration $(d, \mathbf{w})$ such that $(c, \mathbf{v}) \preceq_0 (d, \mathbf{w})$. By the same reasoning as above the configuration does also satisfy both guards $G_\exists$ and $G_\forall$. □

This result can be considered surprising, as the combination of disjunctive and conjunctive guards for internal transitions leads to undecidability [22]. It is key that we use synchronizations and not internal transitions here.

However, note that in each of the compatibility proofs, it is enough to prove $\preceq_0$-compatibility for a single (arbitrary) step of the system. Therefore, we can also mix different types of steps in the same CS.

Theorem 1. *A CS is fully $\preceq_0$-compatible if its steps can be partitioned into sets such that $\preceq_0$-compatibility holds for each set. In particular, a CS is fully $\preceq_0$-compatible if each of its steps is induced by one of the following transition types: internal, disjunctive guard, lossy broadcast, (guarded) synchronization, or ASM.*

Remark 3. Note that Theorem 1 does not make a statement about transitions that combine the characteristics of different types of transitions. Nonetheless, compatibility with $\preceq_0$ holds for many extensions of the types of steps we consider.

In particular, all of them can be extended with disjunctive guards, and even with conjunctions of disjunctive guards, i.e., requiring multiple disjunctive guards to be satisfied at the same time (as in [32]). Moreover, shared finite-domain steps can have several shared variables encoded into the controller states.

Compatibility with $\preceq_0$ is related to what is sometimes called the "copycat property". Informally, this property holds if, whenever a process can move from p to p' in a step $(c, \mathbf{v}) \to (c', \mathbf{v}')$, then any additional processes that are in p in a configuration $(c, \mathbf{v} + i \cdot \mathbf{p})$ can also move to p' in a sequence of steps $(c, \mathbf{v} + i \cdot \mathbf{p}) \xrightarrow{*} (c, \mathbf{v}' + i \cdot \mathbf{p}')$, "copying" the movement of the first process. We use this property implicitly to prove $\preceq_0$-compatibility, and prove or reprove it for all the systems considered here.

4 Parameterized Reachability Problems

We define the type of parameterized problems we consider and show that we can solve them in polynomial space using the 01-CS. Then, we introduce a class of $\preceq_0$-compatible CSs that have not been considered in the literature before, and use them to prove PSPACE-hardness of any of these problems.

4.1 The Cardinality Reachability Problem

Inspired by Delzanno et al. [17], we define a *cardinality constraint* φ as a formula in the following grammar, where $c \in C$, $a \in \mathbb{N}$, and $q \in Q$:

$$\varphi ::= \mathsf{ctrl} = c \mid \mathsf{ctrl} \neq c \mid \#q \geq a \mid \#q = 0 \mid \varphi \wedge \varphi \mid \varphi \vee \varphi$$

The satisfaction of cardinality constraints is defined in the natural way, e.g., $(c, \mathbf{v}) \models \mathsf{ctrl} = c'$ if $c = c'$, and $(c, \mathbf{v}) \models \#q \geq a$ if $\mathbf{v}(q) \geq a$. In [17], the

mic propositions $\mathsf{ctrl} = c$ nor $\mathsf{ctrl} \neq c$ (since they do not have a controller process), but there is $\#q \leq b$ for any $b \in \mathbb{N}$ (which is not supported in the 01-abstraction, except for the special case $\#q = 0$).

Given a CS $\mathcal{C}$ and a cardinality constraint φ, the *cardinality reachability problem (CRP)* asks whether a configuration $(c, \mathbf{v})$ with $(c, \mathbf{v}) \models \varphi$ is reachable in $\mathcal{C}$.

- Let $CC[\geq a]$ be the class of cardinality constraints in which atomic propositions are only of the form $\#q \geq a$ for any $a \in \mathbb{N}$.
- Let $CC[\geq a, = 0]$ be the class of cardinality constraints in which atomic propositions are only of the form $\#q = 0$ or $\#q \geq a$ for any $a \in \mathbb{N}$.
- Let $CC[\mathsf{ctrl}, \geq a, = 0]$ be the class of cardinality constraints in which atomic propositions are of the form $\mathsf{ctrl} = c, \mathsf{ctrl} \neq c, \#q = 0$ or $\#q \geq a$ for any $a \in \mathbb{N}$, i.e., the maximal class.

For a given $\varphi \in CC[\mathsf{ctrl}, \geq a, = 0]$, let $\varphi_\alpha = \varphi[\#q \geq a \mapsto \#q \geq 1]_{a \in \mathbb{N}^+}$, i.e., the result of replacing every atomic proposition of the form $\#q \geq a$ with the proposition $\#q \geq 1$ if $a \in \mathbb{N}^+$. We write *CRP for S* to denote that we consider the CRP problem for a cardinality constraint in $S \in \{CC[\geq a], CC[\geq a, = 0], CC[\mathsf{ctrl}, \geq a, = 0]\}$.

Many parameterized reachability problems can be expressed in CRP format, e.g., coverability, control-state reachability, or the target problem [18].

4.2 Deciding the CRP for $\preceq_0$-compatible Counter Systems

We show that CRP is PSPACE-complete for CSs given a light restriction. We start by showing that checking CRP in a fully $\preceq_0$-compatible CS can be reduced to checking the 01-CS.

Lemma 5. *Let $\mathcal{C}$ be a fully $\preceq_0$-compatible CS, $\mathcal{C}_\alpha$ its 01-CS and let $\varphi \in CC[\mathsf{ctrl}, \geq a, = 0]$.*

1. *If a configuration $(c, \mathbf{v})$ that satisfies φ is reachable in $\mathcal{C}$, then $(c, \mathbf{v}^\alpha) = \alpha(c, \mathbf{v})$ satisfies φ_α and is reachable in $\mathcal{C}_\alpha$.*
2. *If an abstract configuration $(c, \mathbf{v}^\alpha)$ that satisfies φ_α is reachable in $\mathcal{C}_\alpha$, then there exists $(c, \mathbf{v})$ that satisfies φ, is reachable in $\mathcal{C}$, and with $\alpha(c, \mathbf{v}) = (c, \mathbf{v}^\alpha)$.*

Proof. To prove (1), assume there is a configuration $(c, \mathbf{v}) \models \varphi$ that is reachable in $\mathcal{C}$. Then $(c, \mathbf{v}^\alpha) = \alpha(c, \mathbf{v})$ is reachable in $\mathcal{C}_\alpha$ by Lemma 2, and the fact that α of an initial configuration in $\mathcal{C}$ is an initial configuration in $\mathcal{C}_\alpha$. It is easy to see that $(c, \mathbf{v}) \models \varphi$ entails $(c, \mathbf{v}^\alpha) \models \varphi_\alpha$ by definition. To prove (2), assume there is a configuration $(c, \mathbf{v}^\alpha) \models \varphi_\alpha$ reachable in $\mathcal{C}_\alpha$. By Lemma 1, any $(c, \mathbf{v}) \succeq_0 (c, \mathbf{v}^\alpha)$ maps to $(c, \mathbf{v}^\alpha)$ by α. Choose $\mathbf{v}$ such that $\mathbf{v}(q) = 0$ if $\mathbf{v}^\alpha(q) = 0$ and $\mathbf{v}(q) = A$ otherwise, where A is the highest lower bound in φ, that is $A \geq a$ for all $\#q \geq a$ appearing in φ. Then $(c, \mathbf{v}) \models \varphi$, and by Lemma 2, $(c, \mathbf{v})$ is reachable in $\mathcal{C}$. □

Let $\mathcal{C} = (C, Q, \mathcal{T})$ be a $\preceq_0$-compatible CS. A product of wqos is a wqo [35], so $\preceq_0 \times \preceq_0$ is a wqo on $(C \times \mathbb{N}^Q)^2$. Given a wqo $\preceq$ on a set S, it is the case that for every subset $X \subseteq S$ there exists a finite subset $Y \subseteq X$ of minimal elements such that for every $x \in X$ there exists $y \in Y$ with $y \preceq x$ [36, Thm. 1.1]. This subset is called the *finite basis* of X, and it is unique if the wqo is antisymmetric (our $\preceq_0$ is antisymmetric). Applying this to the step relation $\mathcal{T}$ of $\mathcal{C}$ and the wqo $\preceq_0 \times \preceq_0$ implies the existence of a finite basis Y of $\mathcal{T}$, since $\mathcal{T} \subseteq (C \times \mathbb{N}^Q)^2$. Then define

$$B_\mathcal{C} = max(\mathbf{v}(q) \mid ((c, \mathbf{v}), (c', \mathbf{v}')) \in Y, q \in Q),$$

i.e., the maximal number of user processes per state in any step in the basis Y.

Remark 4. The constant $B_\mathcal{C}$ is usually small in counter systems. For example, for $\mathcal{C}$ a counter system with only lossy broadcast steps, $B_\mathcal{C}$ is bounded by $|Q|$: a step depends on one broadcast transition and an arbitrary number of receive transitions. In the worst case, a minimal step is such that, for a given state

p, the broadcast is $p \xrightarrow{!a} p'$ and there are receive transitions $p \xrightarrow{?a} q$ for every $q \in Q \setminus \{p'\}$.

For disjunctive guards, $B_\mathcal{C} \leq 2$; for synchronizations, $B_\mathcal{C} \leq |Q|$; and for ASM, $B_\mathcal{C} \leq 1$. For a CS with several types of these steps, $B_\mathcal{C}$ is bounded by the maximum of the $B_\mathcal{C}$ given here.

We say that a fully $\preceq_0$-compatible CS $\mathcal{C} = (C, Q, \mathcal{T})$ is *polynomially abstractable* if $B_\mathcal{C}$ is polynomial in $|C|$ and $|Q|$, and membership in $\mathcal{T}$ can be checked in PTIME. All the types of systems that we have considered so far are polynomially abstractable.

Theorem 2. *Let $\mathcal{C}$ be a polynomially abstractable CS for which $B_\mathcal{C}$ is known. Then the CRP for $\mathcal{C}$ and $\varphi \in CC[\mathsf{ctrl}, \geq a, = 0]$ is in PSPACE.*

Proof. Let $\mathcal{C} = (C, Q, \mathcal{T})$ be a CS that is $\preceq_0$-compatible and polynomially abstractable for a known $B_\mathcal{C}$, let $\mathcal{C}_\alpha$ be its 01-CS and let $\varphi \in CC[\mathsf{ctrl}, \geq a, = 0]$. By Lemma 5, it suffices to check whether there exists an abstract configuration $(c, \mathbf{v}^\alpha)$ that satisfies φ_α and that is reachable in $\mathcal{C}_\alpha$. There are at most $|C| \cdot 2^{|Q|}$ abstract configurations. We explore the abstract system $\mathcal{C}_\alpha$ non-deterministically, guessing an initial configuration, then a path from this configuration. At each step, we check if the current configuration $(c, \mathbf{v}^\alpha)$ satisfies φ_α (this can be done in polynomial time in the number of states). If it does not, we guess a step $(c, \mathbf{v}^\alpha) \to (c', \mathbf{v}'^\alpha)$ in $\mathcal{C}_\alpha$. To do this, we guess a configuration $(c, \mathbf{v})$ of $\mathcal{C}$ with $1 \leq \mathbf{v}(q) \leq B_\mathcal{C}$ for all q such that $\mathbf{v}^\alpha(q) = 1$, and with $\mathbf{v}(q) = 0$ elsewhere. We guess a configuration $(c', \mathbf{v}')$ of the same size as $(c, \mathbf{v})$. We check if $(c, \mathbf{v}) \to (c', \mathbf{v}')$ is a step of $\mathcal{T}$ (which by assumption can be done in polynomial time). If it is, then $\alpha(c, \mathbf{v}) \to \alpha(c', \mathbf{v}')$ is a step in $\mathcal{C}_\alpha$.

Let Y be the finite basis of $\mathcal{T}$. It is enough to check whether there exists a step with only counters under $B_\mathcal{C}$ because $(c, \mathbf{v}^\alpha) \to (c', \mathbf{v}'^\alpha)$ is a step in $\mathcal{C}_\alpha$ if and only if there exists a step $(c, \mathbf{v}) \to (c', \mathbf{v}') \in Y$ with $\alpha(c, \mathbf{v}) = (c, \mathbf{v}^\alpha)$ and $\alpha(c', \mathbf{v}') = (c', \mathbf{v}'^\alpha)$. Indeed, $(c, \mathbf{v}) \to (c', \mathbf{v}') \in Y$ implies an abstract step because $Y \subseteq \mathcal{T}$. In the other direction, if $(c, \mathbf{v}^\alpha) \to (c', \mathbf{v}'^\alpha)$ is a step in $\mathcal{C}_\alpha$ then there exists $(d, \mathbf{w}) \to (d', \mathbf{w}') \in \mathcal{T}$ with $\alpha(d, \mathbf{w}) = (c, \mathbf{v}^\alpha)$ and $\alpha(d', \mathbf{w}') = (c', \mathbf{v}'^\alpha)$. By definition of Y there exists $(c, \mathbf{v}) \to (c', \mathbf{v}') \in Y$ with $(c, \mathbf{v}) \preceq_0 (d, \mathbf{w})$ and $(c', \mathbf{v}') \preceq_0 (d', \mathbf{w}')$. By Lemma 1, this entails $\alpha(c, \mathbf{v}) = (c, \mathbf{v}^\alpha)$ and $\alpha(c', \mathbf{v}') = (c', \mathbf{v}'^\alpha)$. This procedure is in polynomial space in the number of states and $B_\mathcal{C}$ because each configuration can be written in polynomial space, all checks can be performed in polynomial time, and by Savitch's Theorem PSPACE = NPSPACE so we can give a non-deterministic algorithm. □

PSPACE-Hardness of the CRP. Our upper bound on the complexity of CRP for $\preceq_0$-compatible CSs is higher than some of the existing complexity results for systems that fall into this class[5]. We show that this complexity is unavoidable,

---

[5] E.g., for RBN without a controller, CRP for $CC[\geq 1]$ is in PTIME, and for $CC[\geq 1, = 0]$ it is in NP [19].

implying that the class of fully $\preceq_0$-compatible systems is more expressive than its instances considered in the literature.

We prove PSPACE-hardness by a reduction of the intersection non-emptiness problem for deterministic finite automata (DFA) [34] to the CRP. The detailed construction can be found in the extended version [20]. The idea is to view the DFA as systems communicating via synchronization transitions, where the set of actions is the input alphabet. The intersection of the languages accepted by the automata is then non-empty iff some configuration is reachable such that in each automaton there is at least one accepting state covered by a process. This constraint can be encoded into a constraint $\varphi \in CC[\geq 1]$ and the construction does not use a controller, therefore deciding the CRP even in this restricted setting is PSPACE-hard. As a consequence, we get PSPACE-completeness for the CRP of $\preceq_0$-compatible systems.

Theorem 3. *Let $\mathcal{C}$ be a polynomially abstractable CS for which $B_{\mathcal{C}}$ is known. Then the CRP for $\mathcal{C}$ and $\varphi \in CC[\mathsf{ctrl}, \geq a, = 0]$ is* PSPACE-*complete.*

5 Parameterized Model Checking of Trace Properties

A large part of the parameterized verification literature has focused on model checking of stutter-insensitive trace properties of a single process, or a fixed number k of processes [3,7,21,22,32]. We sketch how our framework improves existing results in this area, including for liveness properties.

Trace Properties. Given a $CS\ \mathcal{C} = (C, Q, \mathcal{T})$, a *trace* of the controller is a finite word $w \in C^*$ obtained from a run ρ of $\mathcal{C}$ by projection on the first element of each configuration, and removing duplicate adjacent letters. We denote by $\mathsf{Traces}(\mathcal{C})$ the set of all finite traces that can be obtained from runs of $\mathcal{C}$, and by $\mathsf{Traces}_\infty(\mathcal{C})$ the set of *infinite* traces. Define similarly $\mathsf{Traces}(\mathcal{C}_\alpha)$ and $\mathsf{Traces}_\infty(\mathcal{C}_\alpha)$ for the 01-CS. A *safety property* φ is a prefix-closed subset of C^*. We say that $\mathcal{C}$ satisfies the safety property φ, denoted $\mathcal{C} \models \varphi$, if $\mathsf{Traces}(\mathcal{C}) \subseteq \varphi$.

Existing Results. Many of the existing results for deciding trace properties are based on cutoffs [7,21,22,32]. That is, they view the system as a parallel composition $A \| B^n$ of controller and user processes, and derive a *cutoff* for n, i.e., a number c such that $A \| B^c \models \varphi \iff \forall n \geq c : A \| B^n \models \varphi$. This reduces the problem to a (decidable) model checking problem over a finite-state system. However, since the cutoff c is usually linear in $|B|$, the state space of this finite system is in the order of $O\left(|A| \times |B|^{|B|}\right)$.

As an improvement of these results, Aminof et al. [3] have shown that $\mathsf{Traces}_\infty(A)$ can be recognized by a Büchi-automaton of size $O(|A|^2 \cdot 2^{|B|})$, and the same for $\mathsf{Traces}_\infty(B)$, the infinite traces of a single user process in the parameterized system.

Deciding Trace Properties in the 01-CS. As a direct consequence of Lemma 2 we get:

Fig. 3. Example system with spurious loop

Lemma 6. *If CS $\mathcal{C}$ is fully $\preceq_0$-compatible and $\mathcal{C}_\alpha$ is its 01-CS, then* $\mathsf{Traces}(\mathcal{C}) = \mathsf{Traces}(\mathcal{C}_\alpha)$.

Note that the size of $\mathcal{C}_\alpha$ is $|C| \cdot 2^{|Q|}$, i.e., smaller than the Büchi automaton in the result of Aminof et al. [3]. On the other hand, our result in general only holds for finite traces. To see that 01-abstraction is not precise for infinite traces, consider the following example.

Example 5. Consider the CS $\mathcal{C}$ based on lossy broadcast depicted in Fig. 3. To its right we depict an infinite run of its 01-CS that executes a lossy broadcast on a infinitely often: on the first a, it moves from $\mathbf{v} = (1, 0)$ to $\mathbf{v}' = (1, 1)$, and then any further application loops in $\mathbf{v}' = (1, 1)$. However, such a behavior is not possible in $\mathcal{C}$: any concrete run of $\mathcal{C}$ will start with a fixed number n of processes, and therefore has to stop after n steps.

Despite this, we can extend Lemma 6 to infinite traces for disjunctive systems:

Lemma 7. *If $\mathcal{C}$ is a $\preceq_0$-compatible CS induced by disjunctive guard transitions, then* $\mathsf{Traces}_\infty(\mathcal{C}) = \mathsf{Traces}_\infty(\mathcal{C}_\alpha)$.

The proof for the lemma can be found in the extended version [20].

Note that it is easy to obtain a Büchi automaton B that recognizes the same language as $\mathcal{C}'_\alpha$: The states of B are the configurations of $\mathcal{C}_\alpha$, plus a special sink state $\bot$. Labels of transitions in B are from the set of minimal steps $\Delta_{\min}$. There is a transition between two automaton states with label $D \in \Delta_{\min}$ if both are configurations and there is a transition based on D between them in $\mathcal{C}'_\alpha$, and between a configuration and $\bot$ there is a transition labeled D if there is no transition based on D and starting in this configuration in $\mathcal{C}'_\alpha$. Finally, there is a self-loop with all labels from $\Delta_{\min}$ on $\bot$, and every state except $\bot$ is accepting.

Also note that we get corresponding results to Lemma 6 and Lemma 7 for traces of a user process, by encoding one copy of the user process into the controller (i.e., the controller simulates the product of the original controller and one user process), such that we can directly observe the traces of one fixed user process. The same construction works for any fixed number k of user processes. Table 1 summarizes our results on trace properties, and compares them to existing results from the literature.

Automata-Based Model Checking. Lemma 6 and Lemma 7 state language equivalences, but do not directly solve the PMCP. We assume that the specification φ is given in the form of an automaton $\mathcal{A}_\varphi$ that accepts the language φ. By Lemma 6, for safety properties it is then enough to check whether the product

Table 1. Decidability and Complexity of PMCP over finite and infinite traces, Comparison of Our Results to Existing Results

Our Results			Existing Results		
System Class	Traces	Result	System Class	Traces	Results
$\preceq_0$-compatible systems	finite	$\mathsf{Traces}(\mathcal{C}) = \mathsf{Traces}(\mathcal{C}_\alpha)$ (where $|\mathcal{C}_\alpha| = |\mathbf{C}| \cdot 2^{|Q|}$)	disjunctive systems	finite	$\mathsf{Traces}(\mathcal{C}) = \mathsf{Traces}(B)$ [3] (where $|B| = |C|^2 \cdot 2^{|Q|}$)
disjunctive systems	infinite	$\mathsf{Traces}_\infty(\mathcal{C}) = \mathsf{Traces}_\infty(\mathcal{C}_\alpha)$ (where $|\mathcal{C}_\alpha| = |\mathbf{C}| \cdot 2^{|Q|}$)	disjunctive systems	infinite	$\mathsf{Traces}(\mathcal{C}) = \mathsf{Traces}(B)$ [3] (where $|B| = |C|^2 \cdot 2^{|Q|}$)

$\mathcal{C}_\alpha \times \mathcal{A}_\varphi$ can reach a state which is non-accepting for $\mathcal{A}_\varphi$, and similarly for the PMCP over infinite traces based on Lemma 7.

6 Transition Counter Systems

In this section, we give a restriction on $\preceq_0$-compatible CSs, and show that CRP for $CC[\geq a]$ and $CC[\geq a, = 0]$ is PTIME- and NP-complete respectively. This restriction is inspired by Delzanno et al. [17], who study reconfigurable broadcast networks (RBN) without a controller process. Accordingly, they consider the CRP for cardinality constraints without the propositions $\mathsf{ctrl} = c$, $\mathsf{ctrl} \neq c$. They show that for RBN, CRP for $CC[\geq 1]$ is PTIME-complete and CRP for $CC[\geq 1, = 0]$ is NP-complete, where $CC[\geq 1]$ are the cardinality constraints in which atomic propositions are only of the form $\#q \geq 1$.

Transition Counter Systems. We consider CSs in which steps are based on local transitions between states, as is the case for the system models we have studied in this paper. Here, we do not consider a controller process, i.e., configurations are in $\mathbb{N}^Q$.

A CS *without controller* is $\mathcal{C} = (Q, \mathcal{T})$, where Q is a finite set of states, the step relation is $\mathcal{T} \subseteq \mathbb{N}^Q \times \mathbb{N}^Q$, and configurations are $\mathbf{v} \in \mathbb{N}^Q$. The results for CSs with controller in the previous sections still hold for CSs without controller: given $\mathcal{C} = (Q, \mathcal{T})$ without controller, add to it a set $C = \{c\}$ and consider configurations in which one process is in c. Since no steps of $\mathcal{T}$ involve C, this process cannot move and can be ignored.

Fix $\mathcal{C} = (Q, \mathcal{T})$ a CS without controller, and $\delta \subseteq Q^2$ a set of transitions between states (the transitions may have labels, but we ignore these for now). We denote transitions (p, p') by $p \to p'$. Given a multiset of transitions $D \in \mathbb{N}^\delta$, let $pre(D)$ be the multiset of states p such that $pre(D)(p) = m$ if there are m transitions in D of the form $p \to p'$ for some p'. Let $post(D)$ be the multiset of states p' such that $post(D)(p') = m$ if there are m transitions in D of the form $p \to p'$ for some p.

Let $\mathbf{v}, \mathbf{v}'$ be two configurations of $\mathbb{N}^Q$, and let $D \in \mathbb{N}^\delta$ be a multiset of transitions. We say $\mathbf{v}'$ is obtained by *applying* D to $\mathbf{v}$ if $\mathbf{v}' = \mathbf{v} - \sum_{i=1}^k \mathbf{p_i} + \sum_{i=1}^k \mathbf{p'_i}$, where $(p_1, p'_1), \ldots, (p_k, p'_k)$ are the transitions of D enumerated with multiplicity. Note that the result is only well-defined if $\mathbf{v}(p) \geq pre(D)(p)$ for all

$p \in Q$, and $[\![pre(D)]\!] \subseteq [\![\mathbf{v}]\!]$ (recall that $[\![\mathbf{m}]\!]$ is the support of a multiset $\mathbf{m}$), and that these conditions are ensured by our definitions of steps in Sect. 2.1.

A transition counter system is characterized by a finite set $\Delta_{\min}$ of "minimal steps", where a $D \in \Delta_{\min}$ is a multiset of transitions of δ such that each transition appears at most once in D, i.e., $D \in \{0,1\}^\delta$. Intuitively, D is a group of transitions that must be taken together in a step, and this group is of minimal size. All steps of a transition CS are based on a $D \in \Delta_{\min}$, by applying each transition of D one or more times.

Definition 3. *A CS without a controller* $\mathcal{C} = (Q, \mathcal{T})$ *is a transition counter system (TCS) if there exists a finite set of transitions $\delta \subseteq Q^2$ and a finite set of minimal steps $\Delta_{min} \subseteq \{0,1\}^\delta$ such that $\mathbf{v} \to \mathbf{v}'$ is a step of $\mathcal{C}$ if and only if $\mathbf{v}'$ is obtained by applying $D \in \mathbb{N}^\delta$ to $\mathbf{v}$, where D is a multiset of local transitions such that there exists a $D_0 \in \Delta_{min}$ with $[\![D]\!] = [\![D_0]\!]$, i.e. D and D_0 are non-zero on the same transitions.*

Notice that TCSs are entirely defined by the tuple $(Q, \delta, \Delta_{\min})$, and they are always polynomially abstractable: testing membership of a step in $\mathcal{T}$ is always in PTIME, and $B_\mathcal{C} \leq |Q|$.

Lemma 8.
1. *CSs without controller and with only lossy broadcast steps, or only disjunctive guard steps, are TCSs.*
2. *CSs without controller and with only synchronization steps are not TCSs.*

Proof. To prove (1), first consider a counter system $\mathcal{C} = (Q, \mathcal{T})$ without controller, and with only lossy broadcast steps based on broadcast and receive transitions from a set δ. These are still well defined, as the definition did not distinguish between controller and user processes. Define $\Delta_{\min}$ to be the set of $D \in \{0,1\}^\delta$ such that for each broadcast transition $t_0 = p_0 \xrightarrow{!a} p'_0 \in \delta$ and each subset of receive transitions $t_1 = p_1 \xrightarrow{?a} p'_1, \ldots, t_k = p_k \xrightarrow{?a} p'_k \in \delta$ for the same letter a, there is a $D = \mathbf{t}_0 + \mathbf{t}_1 + \ldots \mathbf{t}_k$. Then $\mathcal{C}$ is equivalent to the transition counter system $\mathcal{D} = (Q, \delta, \Delta_{\min})$ in the following sense: there is a step $\mathbf{v} \to \mathbf{w}$ in $\mathcal{C}$ if and only if there is a step $\mathbf{v} \to \mathbf{w}$ in $\mathcal{D}$.

Now, consider a counter system $\mathcal{C} = (Q, \mathcal{T})$ without controller, and with only disjunctive guard steps based on transitions from a set δ. These are still well defined, as the definition did not distinguish between controller and user processes. Define $\Delta_{\min}$ to be the set of $D \in \{0,1\}^\delta$ such that for each pair of transitions $t = p \xrightarrow{G_\exists} q \in \delta$ and $r \to r$ for $r \in G_\exists$, there is a $D = \mathbf{t} + \mathbf{r}$. Then $\mathcal{C}$ is equivalent to the transition counter system $\mathcal{D} = (Q, \delta, \Delta_{\min})$, in the same sense as above.

Regarding (2), consider a counter system $\mathcal{C} = (Q, \mathcal{T})$ without controller, and with only synchronization steps based on transitions from a set δ. These are still well defined, as the definition did not distinguish between controller and user processes. Assume there exists $\Delta_{\min}$ such that there is a step $(c, \mathbf{v}) \to (d, \mathbf{w})$ in $\mathcal{C}$ if and only if there is a step $(c, \mathbf{v}) \to (d, \mathbf{w})$ in the transition counter system $\mathcal{D} = (Q, \delta, \Delta_{\min})$. Assume there is a $D \in \Delta_{\min}$ containing a transition $p \xrightarrow{a} q$

and no transition $p \xrightarrow{a} p$. Consider a configuration $\mathbf{v}$ such that $\mathbf{v}(p) = pre(D)(p)$ for all $p \in Q$. Applying D to $\mathbf{v}$ defines a step $\mathbf{v} \to \mathbf{v}'$ in $\mathcal{D}$. Now consider configuration $\mathbf{v}'' = \mathbf{v} + \mathbf{p}$. By definition of a transition counter system, applying D to $\mathbf{v}''$ also defines a step in $\mathcal{D}$. However, this is not a step of $\mathcal{C}$ because there is a process of $\mathbf{v}''$ in p which takes no transition in the step. This is not possible in a synchronization step, where all processes in states with an a-labeled transition must take an a-labeled transition. Therefore counter systems without controller with only synchronization steps may not be equivalent to transition counter systems. □

It is known that RBN can simulate ASM systems [10], so they can indirectly be modeled as TCSs. We now show that TCSs are $\preceq_0$-compatible by design.

Lemma 9. *TCSs are fully $\preceq_0$-compatible.*

Proof. Let $\mathcal{C}$ be a transition counter system given by $(Q, \delta, \Delta_{\min})$. To prove forward $\preceq_0$-compatibility, assume there is a step $\mathbf{v} \to \mathbf{v}'$ and $\mathbf{v} \preceq_0 \mathbf{w}$. There exists a multiset of transitions D such that $\mathbf{v}'$ is obtained by applying D to $\mathbf{v}$.

For every state $q \in Q$, $\mathbf{v}(q) \leq \mathbf{w}(q)$. Therefore D can be applied to $\mathbf{w}$. Call $\mathbf{w}''$ the resulting configuration. We want to check $\mathbf{v}' \preceq_0 \mathbf{w}''$ or modify the step from $\mathbf{w}$ to make it true. This means we have to satisfy the following conditions:

1. $\mathbf{w}''(q) \geq \mathbf{v}'(q)$ for all $q \in Q$: the same transitions are taken from $\mathbf{w} \geq \mathbf{v}$, so this will hold.
2. $\mathbf{w}''(q) = 0$ if and only if $\mathbf{v}'(q) = 0$ for all $q \in Q$: if there are no such states then we are done; otherwise suppose $\mathbf{v}'(q) = 0$ and $\mathbf{w}''(q) > 0$. This entails $\mathbf{w}(q) > 0$ and thus also $\mathbf{v}(q) > 0$ by definition of $\preceq_0$. This means state q was emptied in the step $\mathbf{v} \to \mathbf{v}'$; one of the transitions in D is of the form $q \to p$. We call D' the multiset of transitions obtained by adding $\mathbf{w}(q) - \mathbf{v}(q)$ iterations of $q \to p$ to D, i.e., enough to empty q. The configuration $\mathbf{w}'$ obtained by applying D' to $\mathbf{w}$ is such that $\mathbf{v}' \preceq_0 \mathbf{w}'$.

Backward $\preceq_0$-compatibility can be proven in a similar way. □

TCSs are CSs, thus the definition of 01-CS carries over. In particular, a step in the 01-CS exists if there exists a corresponding step in the TCS. However, the 01-CS of a TCS can also be characterized in the following way.

Lemma 10. *Let $\mathcal{C}$ be a TCS given by $(Q, \delta, \Delta_{\min})$, and $\mathcal{C}_\alpha$ its 01-CS. There is a step $\mathbf{v}^\alpha \to \mathbf{w}^\alpha$ in $\mathcal{C}_\alpha$ if and only if there exists $D \in \Delta_{\min}$ such that $[\![pre(D)]\!] \subseteq [\![\mathbf{v}^\alpha]\!]$ and $\mathbf{w}^\alpha$ is such that (a) $\mathbf{w}^\alpha(q)$ equals 0 or 1 if $q \in [\![pre(D)]\!] \setminus [\![post(D)]\!]$, (b) $\mathbf{w}^\alpha(q)$ equals 1 if $q \in [\![post(D)]\!]$, and (c) $\mathbf{w}^\alpha(q)$ equals $\mathbf{v}^\alpha(q)$ otherwise.*

Proof. Let $\mathcal{C} = (Q, \delta, \Delta_{\min})$ be a transition counter system, $\mathcal{C}_\alpha$ its 01-CS and $\mathbf{v}^\alpha$ a configuration of $\mathcal{C}_\alpha$. Suppose there exists $D \in \Delta_{\min}$ such that $[\![pre(D)]\!] \subseteq [\![\mathbf{v}^\alpha]\!]$. Applying D to any $\mathbf{v}$ of $\mathcal{C}$ such that $\alpha(\mathbf{v}) = \mathbf{v}^\alpha$ and with enough processes so that D can be applied always yields a $\mathbf{w}$ whose image by α verifies points b) and c). The subtlety lies in point a).

Let $\mathbf{v}_1$ be the minimal configuration of $\mathcal{C}$ such that $\alpha(\mathbf{v}_1) = \mathbf{v}^\alpha$ and such that D can be applied to $\mathbf{v}_1$, i.e., $\mathbf{v}_1(p) = pre(D)(p)$ for all $p \in Q$. Let $\mathbf{w}_1$ be the configuration obtained by applying D to $\mathbf{v}_1$. Then $\alpha(\mathbf{w}_1) = \mathbf{w}_1^\alpha$ is such that $a), b), c)$ are verified, with $\mathbf{w}_1^\alpha(q) = 0$ for all $q \in [\![pre(D)]\!] \setminus [\![post(D)]\!]$. Step $\mathbf{v}_1 \to \mathbf{w}_1$ implies step $\mathbf{v}^\alpha \to \mathbf{w}_1^\alpha$ in $\mathcal{C}_\alpha$.

Let $\mathbf{v}_2$ be the configuration of $\mathcal{C}$ equal to $\mathbf{v}_1 + \mathbf{q_2}$ for a $q_2 \in [\![pre(D)]\!] \setminus [\![post(D)]\!]$. It is still that case that $\alpha(\mathbf{v}_2) = \mathbf{v}^\alpha$ and that D can be applied to $\mathbf{v}_2$. Let $\mathbf{w}_2$ be the configuration obtained by applying D to $\mathbf{v}_2$. Then $\alpha(\mathbf{w}_2) = \mathbf{w}_2^\alpha$ is such that $a), b), c)$ are verified, with $\mathbf{w}_2^\alpha(q) = 0$ for all $q \in [\![pre(D)]\!] \setminus [\![post(D)]\!]$ except for q_2, for which $\mathbf{w}_2^\alpha(q_2) = 1$. Step $\mathbf{v}_2 \to \mathbf{w}_2$ implies step $\mathbf{v}^\alpha \to \mathbf{w}_2^\alpha$ in $\mathcal{C}_\alpha$. We can repeat this proof idea for any configuration $\mathbf{v}_1 + \sum_{q \in S} \mathbf{q}$, for any subset S of $[\![pre(D)]\!] \setminus [\![post(D)]\!]$, to obtain all the $0, 1$ combinations for $\mathbf{w}$ to verify $a)$.

Now for the other direction, there exists a step $\mathbf{v}^\alpha \to \mathbf{w}^\alpha$ in $\mathcal{C}_\alpha$ if there exists a step $\mathbf{v} \to \mathbf{w}$ in $\mathcal{C}$ for $\mathbf{v}$ such that $\alpha(\mathbf{v}) = \mathbf{v}^\alpha$ and $\mathbf{w}$ such that $\alpha(\mathbf{w}) = \mathbf{w}^\alpha$. By definition of a transition counter system, there exists $D' \in \mathbb{N}^\delta$ and $D \in \Delta_{\min}$, such that $[\![D']\!] = [\![D]\!]$ and $\mathbf{w}$ is obtained by applying D' to $\mathbf{v}$. Multiset D' is such that $[\![pre(D')]\!] \subseteq [\![\mathbf{v}^\alpha]\!]$ since D' can be applied to $\mathbf{v}$ and $\alpha(\mathbf{v}) = \mathbf{v}^\alpha$ implies $[\![\mathbf{v}]\!] = [\![\mathbf{v}^\alpha]\!]$. Since $[\![pre(D')]\!] = [\![pre(D)]\!]$, we have $[\![pre(D)]\!] \subseteq [\![\mathbf{v}^\alpha]\!]$. It is clear that $\mathbf{w}$ obtained by applying D' is such that $\alpha(\mathbf{w})$ verifies the conditions $a), b), c)$. □

This lemma entails that one can check the existence of a step in the 01-CS of a TCS in polynomial time in the size of Q and $\Delta_{\min}$. This allows us to extend [17]'s results.

Theorem 4. *Given a TCS, deciding CRP for $CC[\geq a]$ is PTIME-complete.*

Proof. (Sketch). Let $\mathcal{C}$ be a TCS, $\mathcal{C}_\alpha$ its 01-CS and $\varphi \in CC[\geq a]$. By Lemma 5, the problem can be reduced to checking if there is a reachable configuration $\mathbf{v}^\alpha$ in $\mathcal{C}_\alpha$ that satisfies φ_α. Consider the following algorithm: start a run in the initial configuration $\mathbf{v}_0^\alpha$ containing the maximum number of ones, i.e. $\mathbf{v}_0^\alpha(q) = 1$ iff $q \in Q_0$. By Lemma 10 it is possible to only take steps that do not decrease the set of states with ones. This defines a maximal run $\mathbf{v}_0^\alpha \to \ldots \to \mathbf{v}_n^\alpha$ of length at most $|Q|$ such that $\mathbf{v}_n^\alpha(q) = 1$ for all q reachable in $\mathcal{C}_\alpha$. It then suffices to check whether $\mathbf{v}_n^\alpha \models \varphi_\alpha$. PTIME-hardness follows from PTIME-hardness of CRP for $CC[\geq 1]$ in RBN [17], which is a special case of this problem. The full proof can be found in the extended version [20]. □

In the following, we write $\mathbf{v}^\alpha \xrightarrow{D} \mathbf{w}^\alpha$ for a step as defined in the end of the last proof.

Theorem 5. *Given a TCS, deciding CRP for $CC[\geq a, = 0]$ is NP-complete.*

Proof. (Sketch). Let $\mathcal{C}$ be a TCS, $\mathcal{C}_\alpha$ its 01-CS and $\varphi \in CC[\geq a, = 0]$. By Lemma 5, it suffices to check if there is a $\mathbf{v}^\alpha \models \varphi_\alpha$ initially reachable in $\mathcal{C}_\alpha$. Consider the following (informal) non-deterministic algorithm: we guess a run $\mathbf{v}_0^\alpha \to \ldots \to \mathbf{v}_m^\alpha$ in two parts, first guessing a prefix that increases the set of

Table 2. Decidability and Complexity of the Constraint Reachability Problem (CRP)

Our Results			Existing Results		
System Class	Constraint Class	Result	System Class	Constraint Class	Results
$\preceq_0$-compatible systems	$CC[\text{ctrl}, \geq \mathbf{a}, = \mathbf{0}]$	PSPACE-complete (Theorem 3)	ASM disjunctive	$CC[\text{ctrl}]$ $CC[\text{ctrl}, \geq a]$ $CC[\text{ctrl}, \geq a, = 0]$	co-NP-complete [27] decidable [22] in EXPTIME [33]
TCS	$CC[\geq a]$	PTIME-complete (Theorem 4)	RBN disjunctive	$CC[\geq 1]$ $CC[\geq 1]$	PTIME-complete [17] in PTIME [10]
TCS	$CC[\geq a, = 0]$	NP-complete (Theorem 5)	RBN disjunctive	$CC[\geq 1, = 0]$ $CC[\geq 1, = 0]$	NP-complete [17] in NP [10]

states with ones, then guessing a suffix that decreases the set of states with ones. It then suffices to check whether $\mathbf{v}_m^\alpha \models \varphi_\alpha$, and the run is of length at most $2|Q|$. NP-hardness follows from NP-hardness of CRP for $CC[\geq 1, = 0]$ in RBN [17], which is a special case of this problem. The full proof can be found in the extended version [20]. □

Remark 5. Given a CS, the *deadlock* problem asks whether there is a reachable configuration from which no further step can be taken. In a TCS $\mathcal{C}$ where for all minimal steps $D \in \Delta_{\min}$ there is at most one transition starting in each state, i.e., $pre(D)(p) \leq 1, \forall p \in Q$, the deadlock problem is solvable in the abstract system $\mathcal{C}_\alpha$. Indeed, it can be expressed as a CRP problem with cardinality constraint $\bigwedge_{D \in \Delta_{\min}} \bigvee_{q \in pre(D)} \#q = 0$.

Table 2 summarizes our results on the CRP and compares them to existing results.

7 Conclusion

In this paper, we characterized parameterized systems for which $(0,1)$-counter abstraction is precise, i.e., a safety property holds in the parameterized system if and only if it holds in its 01-counter system. Several system models from the literature fall into this class, including reconfigurable broadcast networks, disjunctive systems, and asynchronous shared memory protocols. Our common framework for these systems provides a simpler explanation for existing decidability results, and also extends and improves them. We prove that the constraint reachability problem for the whole class of systems is PSPACE-complete (even without a controller process), and that lower complexity bounds can be obtained under additional assumptions.

Note that weaker versions of Lemmas 2 and 5 directly follow from the fact that $(\mathcal{C}, \preceq_0)$ is a well-structured transition system (cf. [2,30]): in these systems, infinite upward-closed sets (as defined by a constraint in $CC[\geq a, = 0]$) can be represented by a finite basis wrt. $\preceq_0$, resulting in a parameterized model checking algorithm with guaranteed termination. However, the complexity bound of the general algorithm is huge (e.g., for broadcast protocols [26] it has Ackermannian complexity [38]). Instead of relying only on this, we introduce a novel argument that directly connects $\preceq_0$-compatible systems to the 01-counter system.

In addition to the questions it answers, we think that this work also raises lots of interesting questions, and that it can serve as the basis of a more systematic study of the systems covered in our framework: while in this paper we have focused on reachability and safety properties, we conjecture that our framework can also be extended to liveness and termination properties, possibly under additional restrictions on the systems. Moreover, an extension to more powerful system models may be possible, for example to processes that are timed automata (like in [4]) or pushdown automata (like in [27]).

Acknowledgments. We thank Javier Esparza and Pierre Ganty for many helpful discussions at the start of this paper. P. Eichler carried out this work as a member of the Saarbrücken Graduate School of Computer Science. This research was funded in part by the German Research Foundation (DFG) grant GSP&Co (No. 513487900). C. Weil-Kennedy's work was supported by the grant PID2022-138072OB-I00, funded by MCIN, FEDER, UE and partially supported by PRODIGY Project (TED2021-132464B-I00) funded by MCIN and the European Union NextGeneration.

References

1. Abdulla, P.A., Cerans, K., Jonsson, B., Tsay, Y.: General decidability theorems for infinite-state systems. In: Proceedings, 11th Annual IEEE Symposium on Logic in Computer Science, New Brunswick, New Jersey, USA, 27–30 July 1996, pp. 313–321. IEEE Computer Society (1996). https://doi.org/10.1109/LICS.1996.561359
2. Abdulla, P.A., Cerans, K., Jonsson, B., Tsay, Y.: Algorithmic analysis of programs with well quasi-ordered domains. Inf. Comput. **160**(1–2), 109–127 (2000). https://doi.org/10.1006/INCO.1999.2843
3. Aminof, B., Kotek, T., Rubin, S., Spegni, F., Veith, H.: Parameterized model checking of rendezvous systems. Distrib. Comput. **31**(3), 187–222 (2018). https://doi.org/10.1007/S00446-017-0302-6
4. André, É., Eichler, P., Jacobs, S., Karra, S.L.: Parameterized verification of disjunctive timed networks. In: Dimitrova, R., Lahav, O., Wolff, S. (eds.) Verification, Model Checking, and Abstract Interpretation - 25th International Conference, VMCAI 2024, London, United Kingdom, 15–16 January 2024, Proceedings, Part I. Lecture Notes in Computer Science, vol. 14499, pp. 124–146. Springer, Cham (2024). https://doi.org/10.1007/978-3-031-50524-9_6
5. Angluin, D., Aspnes, J., Eisenstat, D., Ruppert, E.: The computational power of population protocols. Distrib. Comput. **20**(4), 279–304 (2007). https://doi.org/10.1007/S00446-007-0040-2
6. Apt, K.R., Kozen, D.: Limits for automatic verification of finite-state concurrent systems. Inf. Process. Lett. **22**(6), 307–309 (1986). https://doi.org/10.1016/0020-0190(86)90071-2
7. Außerlechner, S., Jacobs, S., Khalimov, A.: Tight cutoffs for guarded protocols with fairness. In: Jobstmann, B., Leino, K.R.M. (eds.) VMCAI 2016. LNCS, vol. 9583, pp. 476–494. Springer, Heidelberg (2016). https://doi.org/10.1007/978-3-662-49122-5_23

8. Balasubramanian, A.R., Bertrand, N., Markey, N.: Parameterized verification of synchronization in constrained reconfigurable broadcast networks. In: Beyer, D., Huisman, M. (eds.) TACAS 2018. LNCS, vol. 10806, pp. 38–54. Springer, Cham (2018). https://doi.org/10.1007/978-3-319-89963-3_3
9. Balasubramanian, A.R., Guillou, L., Weil-Kennedy, C.: Parameterized analysis of reconfigurable broadcast networks. In: FoSSaCS 2022. LNCS, vol. 13242, pp. 61–80. Springer, Cham (2022). https://doi.org/10.1007/978-3-030-99253-8_4
10. Balasubramanian, A.R., Weil-Kennedy, C.: Reconfigurable broadcast networks and asynchronous shared-memory systems are equivalent. In: Ganty, P., Bresolin, D. (eds.) Proceedings 12th International Symposium on Games, Automata, Logics, and Formal Verification, GandALF 2021, Padua, Italy, 20–22 September 2021. EPTCS, vol. 346, pp. 18–34 (2021). https://doi.org/10.4204/EPTCS.346.2
11. Baumeister, T., Eichler, P., Jacobs, S., Sakr, M., Völp, M.: Parameterized verification of round-based distributed algorithms via extended threshold automata. In: Platzer, A., Rozier, K.Y., Pradella, M., Rossi, M. (eds.) Formal Methods - 26th International Symposium, FM 2024, Milan, Italy, 9–13 September 2024, Proceedings, Part I. Lecture Notes in Computer Science, vol. 14933, pp. 638–657. Springer, Cham (2024). https://doi.org/10.1007/978-3-031-71162-6_33
12. Bertrand, N., Dewaskar, M., Genest, B., Gimbert, H., Godbole, A.A.: Controlling a population. Log. Methods Comput. Sci. **15**(3) (2019). https://doi.org/10.23638/LMCS-15(3:6)2019
13. Bertrand, N., Fournier, P., Sangnier, A.: Playing with probabilities in reconfigurable broadcast networks. In: Muscholl, A. (ed.) FoSSaCS 2014. LNCS, vol. 8412, pp. 134–148. Springer, Heidelberg (2014). https://doi.org/10.1007/978-3-642-54830-7_9
14. Bloem, R., et al.: Decidability of Parameterized Verification. Synthesis Lectures on Distributed Computing Theory, Morgan & Claypool Publishers (2015). https://doi.org/10.2200/S00658ED1V01Y201508DCT013
15. Bouyer, P., Markey, N., Randour, M., Sangnier, A., Stan, D.: Reachability in networks of register protocols under stochastic schedulers. In: Chatzigiannakis, I., Mitzenmacher, M., Rabani, Y., Sangiorgi, D. (eds.) 43rd International Colloquium on Automata, Languages, and Programming, ICALP 2016, 11–15 July 2016, Rome, Italy. LIPIcs, vol. 55, pp. 106:1–106:14. Schloss Dagstuhl - Leibniz-Zentrum für Informatik (2016). https://doi.org/10.4230/LIPIcs.ICALP.2016.106
16. Colcombet, T., Fijalkow, N., Ohlmann, P.: Controlling a random population. Log. Methods Comput. Sci. **17**(4) (2021). https://doi.org/10.46298/LMCS-17(4:12)2021
17. Delzanno, G., Sangnier, A., Traverso, R., Zavattaro, G.: On the complexity of parameterized reachability in reconfigurable broadcast networks. In: D'Souza, D., Kavitha, T., Radhakrishnan, J. (eds.) IARCS Annual Conference on Foundations of Software Technology and Theoretical Computer Science, FSTTCS 2012, 15–17 December 2012, Hyderabad, India. LIPIcs, vol. 18, pp. 289–300. Schloss Dagstuhl - Leibniz-Zentrum für Informatik (2012). https://doi.org/10.4230/LIPICS.FSTTCS.2012.289
18. Delzanno, G., Sangnier, A., Zavattaro, G.: Parameterized verification of ad hoc networks. In: Gastin, P., Laroussinie, F. (eds.) CONCUR 2010. LNCS, vol. 6269, pp. 313–327. Springer, Heidelberg (2010). https://doi.org/10.1007/978-3-642-15375-4_22
19. Delzanno, G., Sangnier, A., Zavattaro, G.: Verification of ad hoc networks with node and communication failures. In: Giese, H., Rosu, G. (eds.) FMOODS/FORTE

-2012. LNCS, vol. 7273, pp. 235–250. Springer, Heidelberg (2012). https://doi.org/10.1007/978-3-642-30793-5_15
20. Eichler, P., Jacobs, S., Weil-Kennedy, C.: Parameterized verification of systems with precise (0,1)-counter abstraction (2024). https://arxiv.org/abs/2408.05954
21. Emerson, E.A., Kahlon, V.: Reducing model checking of the many to the few. In: McAllester, D. (ed.) CADE 2000. LNCS (LNAI), vol. 1831, pp. 236–254. Springer, Heidelberg (2000). https://doi.org/10.1007/10721959_19
22. Emerson, E.A., Kahlon, V.: Model checking guarded protocols. In: 18th IEEE Symposium on Logic in Computer Science (LICS 2003), 22–25 June 2003, Ottawa, Canada, Proceedings, pp. 361–370. IEEE Computer Society (2003). https://doi.org/10.1109/LICS.2003.1210076
23. Emerson, E.A., Namjoshi, K.S.: Reasoning about rings. In: Cytron, R.K., Lee, P. (eds.) Conference Record of POPL'95: 22nd ACM SIGPLAN-SIGACT Symposium on Principles of Programming Languages, San Francisco, California, USA, 23–25 January 1995, pp. 85–94. ACM Press (1995). https://doi.org/10.1145/199448.199468
24. Emerson, E.A., Namjoshi, K.S.: On model checking for non-deterministic infinite-state systems. In: Thirteenth Annual IEEE Symposium on Logic in Computer Science, Indianapolis, Indiana, USA, 21–24 June 1998, pp. 70–80. IEEE Computer Society (1998). https://doi.org/10.1109/LICS.1998.705644
25. Esparza, J.: Keeping a crowd safe: on the complexity of parameterized verification (invited talk). In: Mayr, E.W., Portier, N. (eds.) 31st International Symposium on Theoretical Aspects of Computer Science (STACS 2014), STACS 2014, 5–8 March 2014, Lyon, France. LIPIcs, vol. 25, pp. 1–10. Schloss Dagstuhl - Leibniz-Zentrum für Informatik (2014). https://doi.org/10.4230/LIPICS.STACS.2014.1
26. Esparza, J., Finkel, A., Mayr, R.: On the verification of broadcast protocols. In: 14th Annual IEEE Symposium on Logic in Computer Science, Trento, Italy, 2–5 July 1999, pp. 352–359. IEEE Computer Society (1999). https://doi.org/10.1109/LICS.1999.782630
27. Esparza, J., Ganty, P., Majumdar, R.: Parameterized verification of asynchronous shared-memory systems. J. ACM **63**(1), 10:1–10:48 (2016). https://doi.org/10.1145/2842603
28. Esparza, J., Raskin, M., Weil-Kennedy, C.: Parameterized analysis of immediate observation petri nets. In: Donatelli, S., Haar, S. (eds.) PETRI NETS 2019. LNCS, vol. 11522, pp. 365–385. Springer, Cham (2019). https://doi.org/10.1007/978-3-030-21571-2_20
29. Finkel, A.: Reduction and covering of infinite reachability trees. Inf. Comput. **89**(2), 144–179 (1990). https://doi.org/10.1016/0890-5401(90)90009-7
30. Finkel, A., Schnoebelen, P.: Well-structured transition systems everywhere! Theor. Comput. Sci. **256**(1–2), 63–92 (2001). https://doi.org/10.1016/S0304-3975(00)00102-X
31. German, S.M., Sistla, A.P.: Reasoning about systems with many processes. J. ACM **39**(3), 675–735 (1992). https://doi.org/10.1145/146637.146681
32. Jacobs, S., Sakr, M.: Analyzing guarded protocols: better cutoffs, more systems, more expressivity. In: VMCAI 2018. LNCS, vol. 10747, pp. 247–268. Springer, Cham (2018). https://doi.org/10.1007/978-3-319-73721-8_12
33. Jacobs, S., Sakr, M., Völp, M.: Automatic repair and deadlock detection for parameterized systems. In: Griggio, A., Rungta, N. (eds.) 22nd Formal Methods in Computer-Aided Design, FMCAD 2022, Trento, Italy, 17–21 October 2022, pp. 225–234. IEEE (2022). https://doi.org/10.34727/2022/ISBN.978-3-85448-053-2_29

34. Kozen, D.: Lower bounds for natural proof systems. In: 18th Annual Symposium on Foundations of Computer Science, Providence, Rhode Island, USA, 31 October - 1 November 1977, pp. 254–266. IEEE Computer Society (1977). https://doi.org/10.1109/SFCS.1977.16
35. Kruskal, J.B.: The theory of well-quasi-ordering: a frequently discovered concept. J. Comb. Theory, Ser. A **13**(3), 297–305 (1972). https://doi.org/10.1016/0097-3165(72)90063-5
36. de Luca, A., Varricchio, S.: Well quasi-orders and regular languages. Acta Informatica **31**(6), 539–557 (1994). https://doi.org/10.1007/BF01213206
37. Pnueli, A., Xu, J., Zuck, L.: Liveness with (0,1, ∞)- counter abstraction. In: Brinksma, E., Larsen, K.G. (eds.) CAV 2002. LNCS, vol. 2404, pp. 107–122. Springer, Heidelberg (2002). https://doi.org/10.1007/3-540-45657-0_9
38. Schmitz, S., Schnoebelen, P.: The power of well-structured systems. In: D'Argenio, P.R., Melgratti, H. (eds.) CONCUR 2013. LNCS, vol. 8052, pp. 5–24. Springer, Heidelberg (2013). https://doi.org/10.1007/978-3-642-40184-8_2
39. Suzuki, I.: Proving properties of a ring of finite-state machines. Inf. Process. Lett. **28**(4), 213–214 (1988). https://doi.org/10.1016/0020-0190(88)90211-6

Formal Verification of Probabilistic Deep Reinforcement Learning Policies with Abstract Training

Junfeng Yang[1], Min Zhang[1], Xin Chen[2], and Qin Li[1(✉)]

[1] Shanghai Key Laboratory of Trustworthy Computing, East China Normal University, Shanghai, China
qli@sei.ecnu.edu.cn
[2] University of New Mexico, Albuquerque, NM, USA
chenxin@unm.edu

Abstract. Deep Reinforcement Learning (DRL), especially DRL with probabilistic policies, has shown great potential in learning control policies. In safety-critical domains, using probabilistic DRL policy requires strict safety assurances, making it critical to verify the probabilistic DRL policy formally. However, formal verification of probabilistic DRL policies still faces significant challenges. These challenges arise from the complexity of reasoning about the neural network's probabilistic outputs for infinite state sets and the state explosion problem during model construction. This paper proposes a novel approach based on abstract training for quantitatively verifying probabilistic DRL policies. Specifically, we abstract the infinite continuous state space into finite discrete decision units and train a deep neural network (DNN) policy on these decision units. This abstract training allows for the direct black-box computation of probabilistic decision outputs for a set of states, greatly simplifying the complexity of reasoning neural network outputs. We further abstract the execution of the trained DNN policy as a Markov decision model (MDP) and perform probabilistic model checking, obtaining two types of upper bounds on the probability of being unsafe. When constructing the MDP, we incorporate the reuse of abstract states based on decision units, significantly alleviating the state explosion problem. Experiments show that the proposed probabilistic quantitative verification can yield tighter upper bounds on unsafe probabilities over longer time horizons more easily and efficiently than the current state-of-the-art method.

Keywords: Probabilistic deep reinforcement learning · Formal quantitative verification · Safety-critical properties · Abstract training

1 Introduction

Deep Reinforcement Learning (DRL) is a subfield of machine learning focused on sequential decision-making problems in open environments. Through interaction with the environment, an agent learns a deep neural network (DNN) policy that determines its actions based on observed states to maximize cumulative reward [3,28]. Deep reinforcement learning has shown great potential in a variety of complex domains such as robot control [14], autonomous driving [21], gaming [24], healthcare [33] and dialogue generation [23]. DRL using probabilistic policies (choosing actions based on probability distributions outputted by learned DNNs) is particularly well-suited to managing complex scenarios. These include balancing the exploration-exploitation trade-off [12], coping with environmental uncertainty [13,25], diversity and stochasticity [16], and learning for continuous action spaces [24].

Before deploying probabilistic DRL policies in safety-critical domains, it is essential to strictly ensure their safety through formal methods. However, most current research has focused on the formal verification of DRL systems with deterministic policies, whether in deterministic environment [4,10,17,19,29,30,32], or uncertain environment [1,5,9], with minimal attention paid to the safety assurance of DRL systems utilizing probabilistic policies. The probabilistic policies are very different from the deterministic ones. For deterministic policies, the DNN policy forms a state path according to a unique decision sequence to achieve the control objective for fixed control tasks. The policy is safe if only the path is determined to be safe. However, when it comes to probabilistic policies, to maintain flexibility in a highly random environment, all actions have the possibility of being chosen. The policy can be executed following any state path within a tree-like structure formed by all possible action combinations. Therefore, ensuring the safety of a probabilistic policy faces significant challenges: (i) The infinite and uncountable state space makes it impractical to exhaustively analyze all relevant states and paths [20]. (ii) The complexity of deep neural networks makes it challenging to reason the output of a policy for a set of states, especially in the case of probabilistic policies, where the policy outcomes are not single values but probability distributions over available actions [7,18]. (iii) As the time horizon extends, the possible execution paths expand like a tree, causing the number of states to increase exponentially. (iv) Qualitative safety assessments become insufficient, necessitating a shift to quantitative safety assessments for probabilistic policies.

To our knowledge, no effective work has comprehensively solved these problems. The only two notable studies on probabilistic policy verification, Patak et al. in [26] addresses discrete state safety problems for probabilistic reinforcement learning without DNN embedded, and Bacci et al. in [6] made the first attempt to tackle probabilistic policy represented by DNN in continuous state spaces. Bacci et al. employed template polyhedron abstraction and mixed-integer linear programming (MILP) to reason the DNN outputs and model the execution of probabilistic policies as interval Markov decision processes, yielding upper bounds on the failure probabilities. However, this DNN inference method is constrained by the neural network's size and architecture, and their containment-check technique does not effectively address the challenge (iii). Consequently, their verification approaches are severely limited in scale, and the upper bound on the maximum probability of failure is too incomprehensive to ensure safety.

To address the challenges in verifying probabilistic policies for deep reinforcement learning (DRL) systems, we propose a new abstract training-based approach. This method produces more easily verifiable probabilistic policies through abstract training. It then models the trained policies from a continuous state space into a discrete Markov decision process (MDP) by grouping states into abstract states. Finally, it yields two upper bounds on the maximum and average unsafe probabilities of executing these policies within k steps using probabilistic model checking [22]. This approach builds on the concept of abstract training from [2] and our previous work [19,34], introducing the idea of *decision unit*, which is an abstract representation that aggregates a set of system states. During training, we replace concrete states with their corresponding decision units as inputs to the neural network. This ensures that the DNN policy computes identical action distributions for all states aggregated within the same decision unit. Then, in the MDP construction process, we can treat the trained neural network policy as an oracle, allowing us to directly obtain a probability distribution over actions for a set of states based on their decision units. Moreover, we can reuse many abstract states because the state space is discretized into a finite number of decision units, effectively mitigating the exponential state explosion problem. Finally, the MDP is solved and yields two probabilistic upper bounds: upper bounds on the maximum unsafe probability and upper bounds on the average unsafe probability.

Our ultimate goal is implementing a training-verification framework similar to the one in [31], integrating feedback from verification results to abstract train safer probabilistic policies. While their work focuses on deterministic policies, we target non-deterministic ones. In this work, however, we focus solely on the verification problem and the performance of the verification process, leaving the training component for future work.

We implemented a prototype tool for our approach and evaluated it using a series of benchmarks. The results show that the abstract-trained DRL systems have comparable performance to conventionally trained DRL systems while being more accessible to verify. Our approach delivers richer verification results more efficiently and on a larger scale than existing probabilistic policy verification methods.

In summary, this paper makes the following three major contributions:

- A probabilistic DRL policy abstract-training method based on decision units is proposed to alleviate the difficulty of reasoning DNN policy outputs and address the difficulties posed by infinite uncountable state spaces.
- A novel formal verification method is proposed for assessing and improving the abstract-trained probabilistic DRL policy. It provides upper bounds for both maximum and average probabilities of being unsafe within a certain horizon for specified initial states set.
- The results from extensive experiments demonstrate that our verification method outperforms the state-of-the-art by providing richer unsafe information, acquiring tighter error bounds, and realizing significant improvement in efficiency(up to over 90x speed up).

2 Preliminaries

2.1 Notation and Probabilistic Models

Dist(X) represents the set of discrete probability distributions over a set X, where the support set of the probability distribution $\mu \in \text{Dist}(X)$ is denoted as $supp(\mu)$. We use s^i to denote the i-th element of a vector s, $\mathcal{P}(X)$ to denote the powerset of X, and $|X|$ to denote the number of elements in X. $\mathbb{E}_X[Y(x)]$ is the average value of dependent variable $Y(x)$ over all elements x in the set X.

We consider the following probabilistic models in this paper: Discrete-Time Markov Processes (DTMPs), used to model DRL policy executions, and Markov Decision Processes (MDPs), used for abstractions and verification.

Definition 1 (Discrete-Time Markov Process). *A discrete-time Markov process (finite-branching) is a tuple $DTMP = (S, S_0, \mathbb{P}, AP, L)$, where S is the set of states; $S_0 \subseteq S$ is the set of initial states; $\mathbb{P} : S \times S \to [0, 1]$ is the transition probability function, where $\sum_{s' \in supp(\mathbb{P}(s,\cdot))} \mathbb{P}(s, s') = 1$ for all $s \in S$; AP is the set of atomic propositions; and $L : S \to \mathcal{P}(AP)$ is the labelling function.*

An infinite(finite) *path* through a DTMP is an infinite(finite) sequence of states $s_0 s_1 s_2...$ where $\mathbb{P}(s_i, s_{i+1}) > 0$ for all i which represents one possible execution of DTMP. We use $FPath(s)$ and $IPath(s)$ respectively, to denote the set of all finite and infinite paths starting from state s. We define a probability measure Pr_s over $IPath(s)$ in the usual way [11] and the probability of reaching a state labelled by l within k steps in the same way as in [6]:

$$Pr_s(\Diamond^{\leq k} l) = Pr_s(\{s_0... \in IPath(s) | \exists 0 \leq i \leq k, s_i \models l\}) \quad (1)$$

which can be computed recursively:

$$Pr_s(\Diamond^{\leq k} l) = \begin{cases} 1 & \text{if } s \models l \\ 0 & \text{if } s \not\models l \land k = 0 \\ \sum_{s' \in supp(\mathbb{P}(s,\cdot))} \mathbb{P}(s, s') Pr_{s'}(\Diamond^{\leq k-1} l) & \text{otherwise} \end{cases} \quad (2)$$

Definition 2 (Markov Decision Processes). *A Markov decision process is a tuple $MDP = (S, S_0, \mathbb{P}, AP, L)$, where S is the finite set of states; $S_0 \subseteq S$ is the set of initial states; $\mathbb{P} : S \times \mathbb{N} \times S \to [0, 1]$ is the transition probability function; AP is the set of atomic propositions; and $L : S \to \mathcal{P}(AP)$ is the labeling function.*

The MDP evolves from an initial state $s_0 \in S_0$ and forms paths like that in DTMP. But in each state $s \in S$, a choice $j \in \mathbb{N}$ must be non-deterministically chosen and transfer to successor $s' \in S$ with $\mathbb{P}(s, j, s')$. The choice is chosen following an advisor $\sigma \in Adv$. For an advisor σ we have the probability measure Pr_s^σ over $IPath(s)$ and we can compute the maximum or minimum probabilities for some measurable set of paths $\xi \subseteq IPath(s)$ (e.g., $\Diamond^{\leq k} l$):

$$Pr_s^{max}(\xi) = \sup_{\sigma \in Adv} Pr_s^\sigma(\xi) \quad Pr_s^{min}(\xi) = \inf_{\sigma \in Adv} Pr_s^\sigma(\xi) \quad (3)$$

2.2 Probabilistic DRL System and The Verification Problem

Probabilistic DRL. DRL is a technique for learning optimal control policies through sequential interaction with the environment and receiving feedback where the DNN is trained as a decision-maker. In particular, a probabilistic policy (or non—deterministic policy) processes the output of the DNN using functions like *softmax* to obtain a probability distribution $\mu \in Dist(A)$ over the actions A and samples an action to perform.

Indeed, a deep reinforcement learning system with a probabilistic policy is a 5-tuple PDRL = (S, S_0, A, π, f) where $S \subseteq \mathbb{R}^n$ is the set of n-dimensional system states over n continuous variables, $S_0 \subseteq S$ is the set of initial states, A is the set of available actions, $\pi : S \to Dist(A)$ is the DNN probabilistic policy, and $f : S \times A \to S$ describes the environment dynamics, indicating how states transition when an action is performed. Note: This paper assumes that the environmental dynamics f are prior known and deterministic-i.e., for any state-action pair, the system always deterministically transitions to the next unique state. The execution of this probabilistic DRL system can be modeled as a DTMP.

Definition 3 (PDRL Execution). *Given a PDRL = (S, S_0, A, π, f), the corresponding execution of this a PDRL is a DTMP=$(S, S_0, \mathbb{P}, AP, L)$ where the AP is the set of atomic propositions (Specifically in this paper, $AP = \{unsafe\}$). The labeling function L labels states such that, such that, $\forall s \in S$, $unsafe \in L(s)$ iff $s \in S_{unsafe} \subseteq S$. For $\forall s, s' \in S$: $\mathbb{P}(s, s') = \sum\{\pi(s)(a) | a \in A \text{ s.t. } f(s, a) = s'\}$.*

Verification Problem of PDRL. In a PDRL system, since all actions are likely to be chosen and some paths inevitably lead to unsafe states, the safety of PDRL cannot be verified by simply determining whether a path avoids entering an unsafe state. In this paper, we are concerned with quantitatively characterizing the safety of a PDRL: What is the probability that the PDRL enters an unsafe state within a certain time horizon from an initial state in a given non-singleton initial state set S_0?

However, it is impractical to compute such probabilities for each state in the infinite uncountable initial state set S_0 enumeratively. To practically describe the unsafe probabilities for the overall set, we compute the following two types of upper bounds on the probability of being unsafe for S_0.

Definition 4 (Upper Bound on Maximum Probability). *Given a PDRL Execution in Definition 3 and a horizon $k \in \mathbb{N}$, the upper bound p^+ on the maximum probability of being unsafe within k steps for all states in S_0 is defined as: $p^+ \geq \sup\{Pr_s(\lozenge^{\leq k} unsafe) | s \in S_0\}$*

Definition 5 (Upper Bound on Average Probability). *Given a PDRL Execution in Definition 3 and a horizon $k \in \mathbb{N}$, the upper bound p_E^+ on the average probability of of being unsafe within k steps for all states in S_0 is defined as: $p_E^+ \geq \mathbb{E}_{S_0}[Pr_s(\lozenge^{\leq k} unsafe)]$.*

The bound in Definition 4 aligns with the problem addressed by [6], describing the worst-case safety scenario for a PDRL executing from S_0. However, this metric might be too extreme: for a large state set S_0, even a small subset of states leading to unsafe conditions could yield excessively pessimistic bounds, resulting in a one-sided assessment. Therefore, we introduce a second, mean-based measure in Definition 5, where the average probability that the PRDL starts executing from random initial states in S_0, leading to unsafe conditions, does not exceed the verification bound.

3 Framework

We build on the success of the training-verification in loop framework we previously proposed for deterministic policies [19] and extend it to the verification of probabilistic policies. States in the continuous state space are pre-aggregated into *decision units (DUs)*, with both training and verification are performed directly on these DUs.

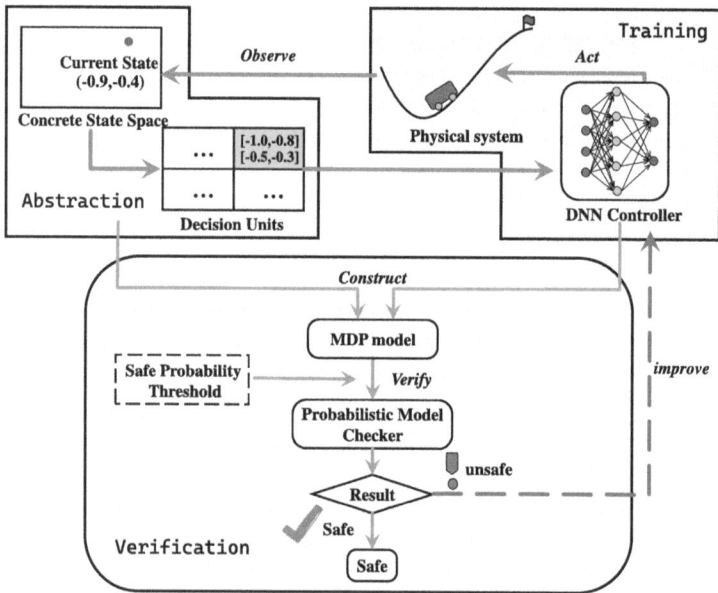

Fig. 1. The framework of abstraction-based training and verifying the probabilistic DRL policies. *Abstraction* module: states in the continuous state space are aggregated into decision units; *Training* module: observed states are mapped to DUs in the Abstraction module to train DNN policy. *Verification* module: the execution of the trained DNN policy is modeled as an MDP for model checking with the help of decision units.

As illustrated in Fig. 1, the overarching framework of our approach consists of three components: abstraction, training, and verification. In the **Abstraction** module, the infinite, uncountable states in the state space S are pre-grouped into a finite set of decision units in the form of high-dimensional interval boxes. A many-to-one mapping is established, linking a group of actual states to their corresponding DU. The **Training** module trains a probabilistic DNN policy using these pre-abstracted DUs as inputs. During interaction with the environment, observed states are mapped to their corresponding DUs, which are then used as inputs to the DNN for decision-making. Ultimately, a probabilistic DNN policy capable of controlling the actual system through DUs is learned. Finally, in the **Verification** module, an MDP abstraction of the PRDL execution controlled by the learned policy is constructed and verified using probabilistic model checking.

4 Decision Unit Abstraction and Abstract Training

We discretize the state space into a collection of decision units beforehand and conduct DNN policy training based on these DUs. In this way, decision-making for the infinite states in the state space is replaced by finite decision units, with decisions computed in a black-box manner by the DNN. This simplifies the verification and construction of the discrete probability model.

4.1 Decision Unit Abstraction

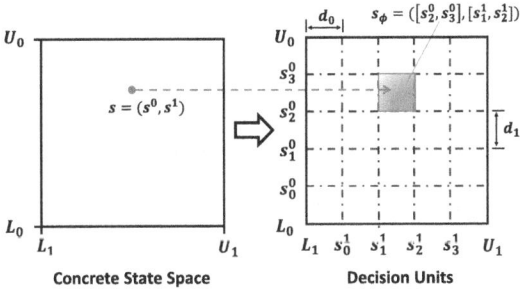

Fig. 2. An example of decision unit abstraction. The two-dimensional state space (left) is divided into a collection of decision units (right) following the granularity (d_0, d_1), where each small dashed rectangle represents a decision unit. State s is mapped to a corresponding s_ϕ with $s^0 \in [s_2^0, s_3^0]$ and $s^1 \in [s_1^1, s_1^1]$.

Recall that the state space $S \subseteq \mathbb{R}^n$ wherein each state s can be expressed as an n real-valued variables vector $(s^1, ..., s^n)$. Assuming that the variables of each dimension are bounded by an upper bound U_i and a lower bound L_i, the state space S is $\prod_{i=1}^{n}[L_i, U_i]$. We first define a n-dimensional vector $\gamma = (d^1, ..., d^n)$ as abstraction granularity. For each dimension i, we partition the state boundary $[L_i, U_i]$ into an interval set $\mathcal{I}_i$ following the i-th granularity $d^i \in \mathbb{R}$ $(0 < d^i <= U_i - L_i)$. The elements $I_i^1, ..., I_i^j$ $(j = 1, ..., U_i - L_i/d^i)$

in $\mathcal{I}_i$ are intervals of length d^i. The Cartesian product of all partitioned sets across all the dimensions, $\Omega = \mathcal{I}_1 \times ... \times \mathcal{I}_n$, forms a partition of the original concrete space S and constructs the collection of decision units. We call the $s_\phi = ((l_1, u_1), ..., (l_n, u_n)) \in \Omega$ *decision unit*, where $(l_i, u_i) \in \mathcal{I}_i$. All the concrete states $s \in \{(s^1, ..., s^n) | l_i \leq s^i \leq u_i, \forall 1 \leq i \leq n\}$ are aggregated into the decision unit $s_\phi = ((l_1, u_1), ..., (l_n, u_n))$ and written as $s \in s_\phi$. We use the DU mapping function $\phi : S \to \Omega$ to describe the aggregation relationship: $\forall s \in S$ and $\forall s_\phi \in \Omega$, $\phi(s) = s_\phi$ iff $s \in s_\phi$. Figure 2 shows an example that the state space $[L_0, U_0] \times [L_1, U_1]$ is uniformly partitioned into 25 DUs and the state $s = (s^0, s^1)$ is mapped to $s_\phi = ((s_2^0, s_3^0), (s_1^1, s_2^1))$. The mapping relation ϕ could be encoded as R-tree [15] for efficiently indexing multidimensional objects.

4.2 Abstract Training via Decision Units

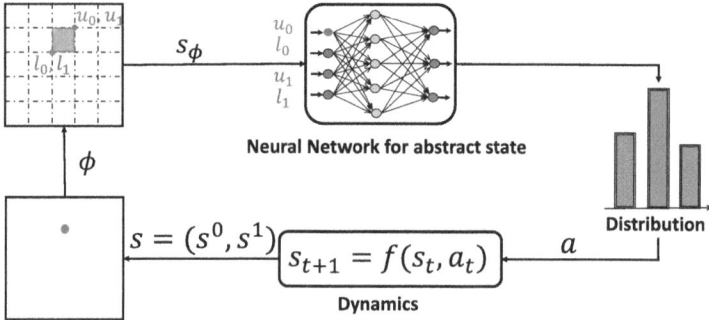

Fig. 3. An example of abstraction-based training. In each round of interaction, the observed state $s = (s^0, s^1)$ is mapped to its corresponding decision unit $s_\phi = ((l_0, u_0), (l_1, u_1))$. The upper and lower bounds of the decision unit, u_0, u_1, l_0, l_1, are then input into the neural network, replacing s for decision-making.

As shown in Fig. 3, we train a DNN policy that makes decisions based on decision units. The number of neurons in the input layer of the DNN is set to twice the number of state dimensions, with the upper and lower bounds of each dimension of the decision unit serving as the inputs. During training interactions, the DNN policy generates outputs for the DU s_ϕ corresponding to s, which are used as decisions for s.

Policies trained in this way achieve the control objective well in the original state space, as the training process is still essentially optimizing for state-generated control outputs. However, such a policy has a desirable property: the DNN generates identical output for all states within the same decision unit, simplifying the reasoning process for decision outputs across a set of states. These trained, DU-oriented DNN policies are formally defined as abstract policies:

Definition 6 (Abstract Policy). *Given a set of decision units $\Omega \subseteq \mathcal{P}(S)$, an abstract policy is a function $\hat{\pi} : \Omega \to Dist(A)$.*

The DRL system controlled by the abstract policy remains a PDRL, where $\forall s \in S, \pi(s) = \hat{\pi}(\phi(s))$.

5 Verification of Abstract-Trained PDRL

In this section, by aggregating states into abstract states, the execution of the abstract-trained policy over the uncountable state space is abstracted as a discrete Markov Decision Process. The MDP model is then solved to provide upper bounds on the original policy's maximum and average unsafe probabilities.

5.1 Constructing MDP Abstraction

An abstract state $\hat{s} \in \mathcal{P}(S)$ is a general subset of the state space representing the collection of possible system states at a given moment. The uncountable state set S_0 is abstracted into a finite set of initial abstract states $\hat{S}_0$ using interval boxes. The MDP is constructed through an iterative exploration that computes successor abstract states for each current abstract state, based on a breadth-first traversal, along with the corresponding transition probabilities.

We begin by implementing abstract state splitting based on decision units and defining abstract dynamics to compute the transitions for individual abstract states, as illustrated in Fig. 4. Finally, we outline the overall process of iterative MDP construction based on these individual transition calculations.

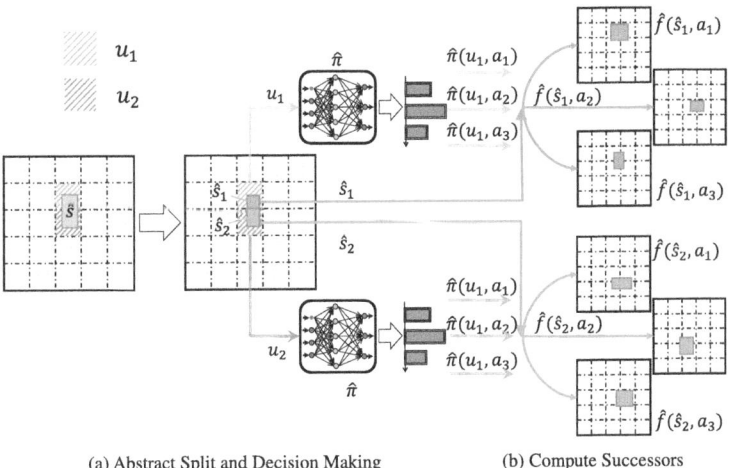

(a) Abstract Split and Decision Making   (b) Compute Successors

Fig. 4. An example of computing the transition for an individual abstract state. (a) Identify the decision units u_1 and u_2 that overlap with abstract $\hat{s}$. The $\hat{s}$ is then divided into two sub-abstract states $\hat{s}_1, \hat{s}_2$, using u_1 and u_2 compute the corresponding transition probability $\hat{\pi}(u_j, a_i)$ for each feasible action. (b) All actions are applied to the abstract dynamics to compute the successors of $\hat{s}_j$.

Abstract State Splitting and Decision Making. Each abstract state to be explored typically carries uncertainty due to potential overlaps with multiple decision units, resulting in various decision outputs for the states within it. This uncertainty hinders the analysis of the abstract state's behavior, but the non-deterministic choices in the MDP can effectively capture and address this property. Thus, the abstract state is divided into several sub-abstract states, each with a unique probability distribution output, modeled as different non-deterministic choices available for the abstract state within the MDP.

We identify the decision units $U_{\hat{s}} = \{u|u = \phi(s), s \in \hat{s}\}$ overlapping with $\hat{s}$. Next, we intersect the abstract state with each decision unit to obtain the sub-abstract states: $\{\hat{s}_j | \hat{s}_j = \hat{s} \cap u_j, u_j \in U_{\hat{s}}\}$. Each sub-abstract state belongs to a unique DU, ensuring a specific decision output: $\forall s \in \hat{s}_j, \pi(s) = \hat{\pi}(u_j)$. As illustrated in Fig. 4, we identify two overlapping decision units u_1 and u_2 for the abstract state $\hat{s}$. We then split $\hat{s}$ into the two sub-abstract states $\hat{s}_1, \hat{s}_2$ with u_1 governing decisions for all states on $\hat{s}_1$, and u_2 for $\hat{s}_2$.

Abstract Dynamics. As shown in Fig. 4, once the decisions for the abstract states are determined, we analyze how each sub-abstract state evolves when an action is performed. We define the abstract dynamics to formalize this process:

Definition 7 (Abstract Dynamics). *For a given environmental dynamics $f : S \times A \to S$ and a set of abstract states $\hat{S} \subseteq \mathcal{P}(S)$, the abstract dynamics is a function $\hat{f} : \hat{S} \to \hat{S}$ such that: for any $\hat{s} \in \hat{S}$, a concrete state $s \in \hat{s}$, and action $a \in A$, we have $f(s, a) \in \hat{f}(\hat{s}, a)$.*

Specifically, for each state dimension i, we compute the maximum and minimum boundaries that the state in the abstract state can reach:

$$u_i = \max_{s \in \hat{s}_j} v_i \cdot f(s, a) \quad l_i = \min_{s \in \hat{s}_j} v_i \cdot f(s, a) \tag{4}$$

Here, v_i is a one-hot coding vector, where the i-th component is 1 and all other components are 0. The interval box composed of these maximum and minimum boundaries provides the successive new abstract states, which overestimate the region to which all states on the abstract state can transition. Thus, the abstract dynamics is explicitly given by $\hat{f}(\hat{s}, a) = ((l_1, u_1), ..., (l_n, u_n))$ where each (l_i, u_i) is computed by Eq. (4).

Overall Process. Having combined the two operations to compute the transition of a single abstract state, we can construct the MDP in a breadth-first search (BFS) order, as outlined in Algorithm 1. For each abstract state to be explored, we first split it into several sub-abstract states (Line 5). Each sub-abstract state $\hat{s}_j$ represents one of the possible branches for decision-making, modeled as the j-th choice in MDP. For every action a_i in the action set A, we compute the decision probabilities $\hat{\pi}(u_j, a_i)$ for the j-th choice based on the corresponding DU u_j (Line 8) and compute the successor abstract state domains $\hat{s}'$ using the abstract dynamics $\hat{f}$ (Line 9). Subsequently, we obtain a transition entry in the transition matrix starting from the current abstract state (Line 10). The newly generated abstract state $\hat{s}'$ will be labeled by the labeling function $\hat{L}$ and added to the to-be-explored frontier to continue our breadth-first exploration (Line 12). The process is repeated until the explored depth reaches the desired horizon or no new abstract states can be generated.

Algorithm 1: Constructing MDP

Input: Initial abstract state $\hat{S}_0$, Action space A, Abstract Policy $\hat{\pi}$, mapping function ϕ, labeling function $\hat{L}$
Output: a MDP model $\mathcal{M}$

1 $\hat{\mathbb{P}} \leftarrow Empty\_Matrix(); \boldsymbol{Q} \leftarrow Initialise\_Queue(); labels \leftarrow \{\}$;
2 $\boldsymbol{Q} \leftarrow \{\hat{s}|\hat{s} \in \hat{S}_0\}$;
3 **while** $\boldsymbol{Q}$ *is not empty* **do**
4 $\quad$ $\hat{s} \leftarrow \boldsymbol{Q}.pop()$;
5 $\quad$ $\{\hat{s}_1, ..., \hat{s}_m\} = split(\hat{s}, \phi)$;
6 $\quad$ **for** $\hat{s}_j$ *in* $\{\hat{s}_1, ..., \hat{s}_m\}$ **do**
7 $\quad\quad$ **for** a_i *in* A **do**
8 $\quad\quad\quad$ $prob_j = \hat{\pi}(u_j, a_i)$;
9 $\quad\quad\quad$ $\hat{s}' = \hat{f}(\hat{s}_j, a_i)$;
10 $\quad\quad\quad$ $\hat{\mathbb{P}}.add(\hat{s}, j, \hat{s}', prob_j)$;
11 $\quad\quad\quad$ **if** $\hat{s}'$ *is not visited* **then**
12 $\quad\quad\quad\quad$ $labels \leftarrow labels \cup \{(\hat{s}', \hat{L}(\hat{s}'))\}$;
13 $\quad\quad\quad\quad$ $\boldsymbol{Q}.push(\hat{s}')$;

14 $\mathcal{M} \leftarrow (\hat{S}_0, \hat{\mathbb{P}}, labels)$;
15 return $\mathcal{M}$;

We end up with an MDP abstraction of the execution of PDRL formalized as follows:

Definition 8 (PDRL Execution Abstraction). *Given an abstract-trained PDRL, let $DTMP = (S, S_0, \mathbb{P}, AP, L)$ be the PDRL execution model, $\hat{\pi}$ the abstract policy, and ϕ the DU mapping function. Given also the abstract dynamics $\hat{f}$, the PDRL execution abstraction is a $MDP = (\hat{S}, \hat{S}_0, \hat{\mathbb{P}}, AP, \hat{L})$ such that:*

- $\hat{S} \subseteq \mathcal{P}(S)$ *is a set of abstract states, such that for all $s \in S$, there exists $\hat{s} \in \hat{S}$ with $s \in \hat{s}$;*
- *for all $s \in S_0$, there exists $\hat{s} \in \hat{S}_0$ with $s \in \hat{s}$;*
- *For every $\hat{s} \in \hat{S}$, $\hat{s}$ can be split into several sub-abstract states $\hat{s}_1, ..., \hat{s}_m$. Each sub-abstract state $\hat{s}_j$ is a subset of a DU u_j such that $\forall s \in \hat{s}_j, \phi(s) = u_j$ and represents a non-deterministic choice for making decisions. Thus a m-split $\hat{s}$ has m choices where $m \in \mathbb{N}$. For each choice $1 \leq j \leq m$, the $\hat{\mathbb{P}}(\hat{s}, j, \hat{s}')$ satisfying:*

$$\hat{\mathbb{P}}(\hat{s}, j, \hat{s}') = \sum \{\hat{\pi}(u_j, a) | a \in A \text{ s.t. } \hat{f}(\hat{s}_j, a) = \hat{s}'\}$$

- $\forall l \in AP, l \in \hat{L}(\hat{s})$ *iff $l \in L(s)$ for some $s \in \hat{s}$*

MDP Downsizing Through State Reuse. When constructing the MDP to verify the safety within k steps, the MDP unfolds like a tree of depth k. This structure leads to an exponential growth of the abstract state space following $(m|A|)^k$ (m is the maximum number of splitting that may occur in the abstract state). Such growth limits the time horizon for verification. Rapid growth arises because successor abstract states from each sub-abstract state may overlap significantly, but are still treated as distinct new states for further exploration. We eliminate this problem by directly using overlapped DUs $\{u_j|\hat{s} \cap u_j \neq \emptyset, u_j \in U_{\hat{s}}\}$ as the sub-abstract states to be applied in abstract dynamics. Then new abstract states are computed from a finite number of decision units, each with exactly $|A|$ successor states. Once a DU is repeatedly used as a sub-abstract state, its successors will already have been visited and will not be treated as new states for exploration again. Then, the state space of the MDP will never exceed $|A||\Omega \cup \hat{S}_0|$.

In this way, the size of the MDP can be controlled by the abstraction granularity of the decision units, significantly improving construction efficiency. Although this method introduces some over-approximation, it remains sound.

5.2 Verification and Upper Bound Computation

We use the existing probabilistic model checking tool to solve the MDP model and obtain $Pr_{\hat{s}_0}^{max}(\Diamond^{\leq k} unsafe)$ for all $\hat{s}_0 \in \hat{S}_0$. Based on these results, we compute the two types of bound of S_0 in Sect. 2 and state the soundness of the verified bounds.

For the maximum unsafe probability of an abstract state derived from solving the MDP, we present the following theorem, which will be proved in **The soundness** subsection.

Theorem 1. *Given a state $s \in S$ of an abstracted trained PDRL execution, and an abstract state $\hat{s} \in \hat{S}$ of the corresponding MDP abstraction for which $s \in \hat{s}$:*

$$Pr_s(\Diamond^{\leq k} unsafe) \leq Pr_{\hat{s}}^{max}(\Diamond^{\leq k} unsafe)$$

Upper Bound. From Theorem 1, it follows that $\sup\{Pr_{\hat{s}_0}^{max}(\Diamond^{\leq k} unsafe)|\hat{s}_0 \in \hat{S}_0\} \geq \sup\{Pr_s(\Diamond^{\leq k} unsafe)|s \in S_0\}$. Thus we take the maximum value of the verified probability of all abstract states on the region as an upper bound on the maximum probability of being unsafe of the initial state region S_0, i.e., $p^+ = \sup\{Pr_{\hat{s}_0}^{max}(\Diamond^{\leq k} unsafe)|\hat{s}_0 \in \hat{S}_0\}$.

For the upper bound on the average probability p_E^+, assuming that initial states are uniformly selected from the initial state set S_0 and that the S_0 is uniformly partitioned into an initial abstract states set $\hat{S}_0$, it is obtained that $\mathbb{E}_{S_0}[Pr_s(\Diamond^{\leq k} unsafe)] = \mathbb{E}_{\hat{S}_0}[\mathbb{E}_{\hat{s}_j}[Pr_s(\Diamond^{\leq k} unsafe)]]$. Additionally, for each abstract state $\hat{s}_j$, Theorem 1 and the relationship between the mean and maximum give $Pr_{\hat{s}_j}^{max}(\Diamond^{\leq k} unsafe) \geq \mathbb{E}_{\hat{s}_j}[Pr_s(\Diamond^{\leq k} unsafe)]$. Combining this with the previous equality, we obtain $\mathbb{E}_{\hat{S}_0}[Pr_{\hat{s}_j}^{max}(\Diamond^{\leq k} unsafe)] \geq \mathbb{E}_{S_0}[Pr_s(\Diamond^{\leq k} unsafe)]$. Here, $\mathbb{E}_{\hat{S}_0}[Pr_{\hat{s}_j}^{max}(\Diamond^{\leq k} unsafe)]$ is the average of the verification results for all abstract states, which is precisely p_E^+ we need.

Soundness. We now prove Theorem 1 to demonstrate the soundness of the probability upper bound derived from it.

Proof (Theorem 1). Based on Definition 2: $Pr_{\hat{s}}^{max}(\xi) = \sup_{\sigma \in Adv} Pr_{\hat{s}}^{\sigma}(\xi)$, to prove the Theorem 1, it is sufficient to show that for any $s \in \hat{s}$, there exists some σ such that $Pr_s(\Diamond^{\leq k} unsafe) \leq Pr_{\hat{s}}^{\sigma}(\Diamond^{\leq k} unsafe)$. We now demonstrate that for each $s \in \hat{s}$, such a σ satisfies the above inequality: σ selects the j_s-th choice for $\hat{s}$ if $s \in \hat{s}_{j_s}$, where $\hat{s}_{j_s}$ is the j_s-th sub-abstract state. The probability $Pr_{\hat{s}}^{\sigma}(\Diamond^{\leq k} unsafe)$ can be similarly computed like DTMP given such a advisor σ:

$$Pr_{\hat{s}}^{\sigma}(\Diamond^{\leq k} l) = \begin{cases} 1 & \text{if } \hat{s} \models l \\ 0 & \text{if } \hat{s} \not\models l \wedge k = 0 \\ \sum_{\hat{s}' \in supp(\hat{\mathbb{P}}(\hat{s}, j_{\hat{s}}, \cdot))} \hat{\mathbb{P}}(\hat{s}, j_{\hat{s}}, \hat{s}') Pr_{\hat{s}'}^{\sigma}(\Diamond^{\leq k-1} l) & \text{otherwise} \end{cases} \quad (5)$$

We employ strong induction over k for proving.

Base Case: When $k = 0$, we have $Pr_s(\Diamond^{\leq 0} unsafe) \leq Pr_{\hat{s}}^{\sigma}(\Diamond^{\leq 0} unsafe)$ due to Definition 8: (1) $\forall s \in S, s \in \hat{s}$ for some $\hat{s} \in \hat{S}$ and (2) $\forall l \in AP, l \in \hat{L}(\hat{s})$ iff $l \in L(s)$ for some $s \in \hat{s}$, such that $s \models unsafe$ implies $\hat{s} \models unsafe$.

Inductive Step: We assume Inductive Hypothesis(IH) that $Pr_{s'}(\Diamond^{\leq k-1} unsafe) \leq Pr_{\hat{s}'}^{\sigma}(\Diamond^{\leq k-1} unsafe)$ for $s' \in S$ and $\hat{s}' \in \hat{S}$ with $s' \in \hat{s}'$. Now prove for k: If $\hat{s} \models unsafe$, $Pr_s(\Diamond^{\leq k} unsafe) \leq Pr_{\hat{s}}^{\sigma}(\Diamond^{\leq k} unsafe) = 1$. Otherwise:

$$Pr_{\hat{s}}^{\sigma}(\Diamond^{\leq k} unsafe)$$
$$= \sum_{\hat{s}' \in supp(\hat{\mathbb{P}}(\hat{s}, j_{\hat{s}}, \cdot))} \hat{\mathbb{P}}(\hat{s}, j_{\hat{s}}, \hat{s}') Pr_{\hat{s}'}^{\sigma}(\Diamond^{\leq k-1} unsafe) \qquad - def \text{ of } Pr_{\hat{s}}^{\sigma}$$
$$= \sum_{a \in A} \hat{\pi}(u_{j_s}, a) Pr_{\hat{f}(\hat{s}_{j_s}, a)}^{\sigma}(\Diamond^{\leq k-1} unsafe) \qquad - def \text{ of } \hat{\mathbb{P}}(\hat{s}, j_{\hat{s}}, \hat{s}')$$
$$= \sum_{a \in A} \pi(s, a) Pr_{\hat{f}(\hat{s}_{j_s}, a)}^{\sigma}(\Diamond^{<k-1} unsafe) \qquad -s \in \hat{s}_{j_s} \subset u_{j}, \pi(s) = \hat{\pi}(u_j)$$
$$\geq \sum_{a \in A} \pi(s, a) Pr_{f(s,a)}^{\sigma}(\Diamond^{\leq k-1} unsafe) \qquad -by \text{ IH and } f(s, a) \in \hat{f}(\hat{s}, a)$$
$$= Pr_s(\Diamond^{\leq k} unsafe) \qquad - def \text{ of } Pr_s(\Diamond^{\leq k} unsafe)$$

The theorem is proved. □

6 Implementation and Experiments

We conducted experiments to evaluate our approach and compared it to similar state-of-the-art probabilistic DRL policy verification techniques. We aim to evaluate four aspects of our method through experimentation: **Q1**: The training performance of the abstract-trained PDRL. **Q2**: The correctness and generalizability of the upper bounds on maximum and average probabilities of unsafe (denoted as *Max-Bound* and *Avg-Bound*) generated by our verification method. **Q3**: The efficiency and tightness of our verification, along with its influencing factors. Finally, **Q4**: The advantages of our method over existing approaches. We discuss each issue in a separate subsection.

6.1 Implementation and Benchmarks

Implementation and Set Up. The framework is implemented using Python. All policies are trained using the Proximal Policy Optimization (PPO) algorithm [27]. Storm [8] is used as the backend tool for model checking. All experiments are conducted on a workstation running Ubuntu 22.04 with a 32-core 13th Gen Intel(R) Core(TM) i9-13900K @ 5800MHz and 128 GB RAM.[1]

Benchmarks. We carry out our experiments over the three control tasks used in [6] for evaluating their verification method: *Bouncing ball(BB)*, *Adaptive cruise control(ACC)* and *Inverted pendulum(PD)*. The respective time-step intervals for these tasks are 0.1 s, 0.1 s, and 0.05 s.

6.2 Training Performance

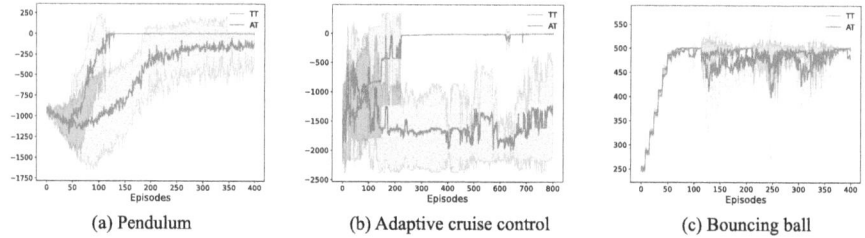

(a) Pendulum   (b) Adaptive cruise control   (c) Bouncing ball

Fig. 5. The trend of cumulative rewards (y-axis) of the systems controlled by abstract trained (red) and traditionally trained (green) using PPO. (Color figure online)

We compared the cumulative reward performance of abstract training versus traditional training methods. We trained five policies for each benchmark task using abstract and traditional training methods. We recorded the change in cumulative reward evaluation during the training sessions. Figure 5 compares the trend of cumulative rewards during training between these two training approaches in three tasks. The solid line indicates the mean reward, and the shading indicates the variance of the five policies, respectively. Red (legend AT) depicts the trend for abstract trained policies, while green (legend TT) depicts

---

[1] The experiments' code is available at https://github.com/Serendipity953/pdrlVerify.

the traditionally trained ones. The results indicate that the policies obtained through our abstract training achieve cumulative reward performance comparable to traditional methods, and they even reach good cumulative rewards earlier, suggesting that abstract training can effectively accomplish the control task.

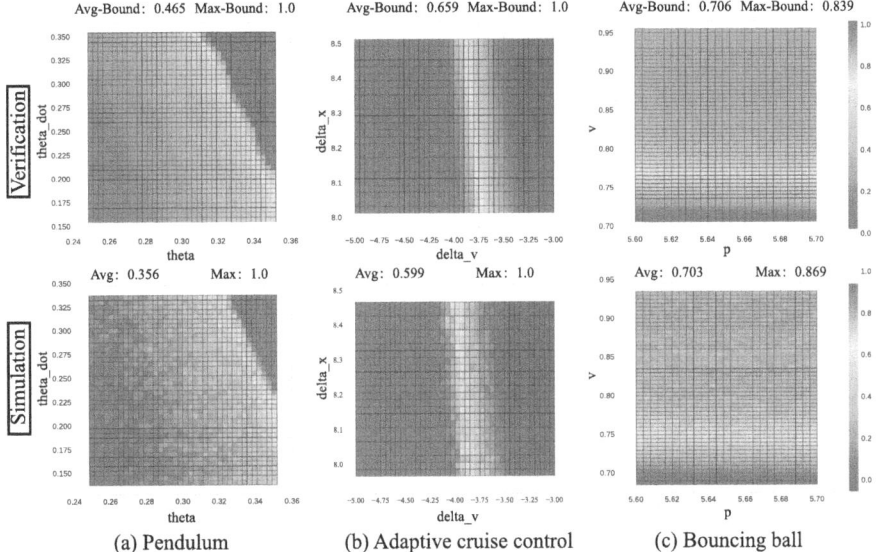

(a) Pendulum (b) Adaptive cruise control (c) Bouncing ball

Fig. 6. Verification and simulation results for the same initial state region and probabilistic policy. **Avg-Bound and Max-Bound**: the verified upper bound over the entire region; **Avg and Max**: the average and maximum unsafe probabilities from simulation over the entire region.

6.3 Correctness and Generalizability

Correctness. We conducted verification and obtained the Max-Bound and Avg-Bound for PD within 30 steps, ACC within 30 steps, and BB within 26 steps in their respective initial state regions. We ran simulations on the same initial regions for the verified abstract policy on each task. For each abstract state region, we sampled an initial state and repeated it 1000 times that run for k steps consistent with the verified horizon. This allows us to estimate the actual probability of being unsafe for the states within each abstract state. We calculated the maximum and average values of the simulated probabilities for all initial abstract states over the entire region. We plotted the verified and simulated results for all abstract states in the initial region using a heat map (Fig. 6). The entire colored area represents the initial state region, with each rectangle inside denoting an initial abstract state. The rectangle's color in *Verification* refers to the $Pr_{\hat{s}_0}^{max}(\lozenge^{\leq k} unsafe)$, and in *Simulation*, it represents the simulated probability. The verified Max-Bound, Avg-Bound, and the simulated maximum(Max) and average(Avg) probabilities for the entire initial state region are labeled at the top of each heatmap. From the result graphs, we observe that the verification results (first row in Fig. 6) consistently show a redder color in the corresponding abstract state regions compared to the simulation results (second row). This indicates that our verified $Pr_{\hat{s}_0}^{max}(\lozenge^{\leq k} unsafe)$ is indeed greater than the actual probability of being unsafe, aligning with our soundness proofs. Note that due to insufficient sampling, the simulated probabilities in certain abstract regions may exceed the verified maximum values. Thus, Avg-Bound and Max-Bound are the correct upper bounds on the average and maximum probabilities. Overall, the verification results closely match the simulation results across the entire initial region, demonstrating that our verification accurately reflects the safety trends.

Generalizability. We illustrate the generalizability of our verification by testing it under different initial state regions, policies, and verified horizons k. Figure 7 displays the verification results for the ACC example across nine different conditions.

Subfigures a1-a3 show the verification results of the same policy within 40 steps across three different initial state regions. From the verification results, we can see that in a1 to a3, as the initial velocity difference (delta_v) and the initial distance difference (delta_x) increase, the color of the heat map gradually shifts to purple-blue, indicating greater safety. The verification results for all three regions align with our intuition: at the beginning, the more and faster the leading car leads the ego car, the lower the likelihood of a collision becomes.

Subfigures b1-b3 present the verification results for three policies trained for 10^5, 3×10^5, and 5×10^5 steps on the same initial state region with a time horizon of $k = 30$. Although all three policies achieve convergence in cumulative reward, the verification results reveal variations in safety performance. As the training steps increase from 10^5 to 3×10^5 (b1 and b2), many abstract states transition to safer states (indicated by purple), and the Avg-Bound decreases from 0.596 to 0.538, suggesting that further training enhances the safety of the initial state region. However, some previously safe regions in b2 become unsafe again in b3, and the Avg-Bound rises from 0.538 to 0.588 after an additional 2×10^5 training steps. This indicates that prolonged training may introduce new safety issues. Unlike cumulative rewards, which only reflect overall task performance, our verification results provide insights into the safety of policies.

Subfigures c1-c3 show the verification of a policy under the same initial state region with time horizons of 20, 25, and 30 steps, respectively. As the time horizon increases, more abstract states shift towards red, indicating higher probabilities of being unsafe. Correspondingly, the overall probability of being unsafe for the PDRL increases, with the Avg-Bound rising from 0.226 in c1 to 0.437 in c2 and reaching 0.602 in c3. This trend confirms our expectation that the risk of entering unsafe states grows as the number of steps extends.

Our verification results effectively provide insights into the unsafeness of the PDRL system under different conditions and identify contributing factors. The Max-Bound helps identify the worst-case scenario, helping us pinpoint the most

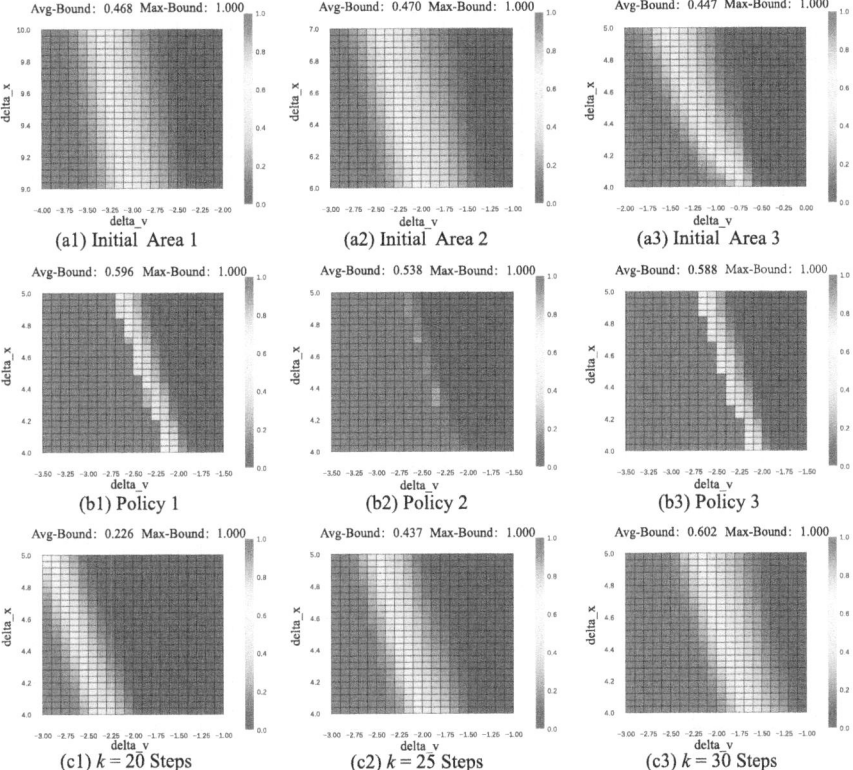

Fig. 7. Avg-Bound and Max-Bound generated for the ACC case under different initial state regions(subfigures a1-a3), various policies with different training levels(subfigures b1-b3), and different verified horizons k(subfigures c1-c3). The verified maximum unsafe probability of the abstract state within the region is visualized as heat maps.

critical state for improving the probabilistic policy. On the other hand, the Avg-Bound offers a comprehensive overview of the unsafe conditions across the region. We can further enhance this analysis by visualizing the safety situation through heat maps of the abstract states' verification results.

6.4 Tightness and Efficiency

We further confirm that this bound can be **sufficiently close** to the actual probability of being unsafe and that the verification can be completed within a **reasonable time**. We define *tightness* as the error between the verified upper bound and the actual probability, which is approximated through simulation experiments. More minor errors indicate more accurate verification results. We present the variation of tightness and elapsed verification time concerning the verified horizon for different levels of abstraction granularity.

For each control task-PD, ACC, and BB-we trained five policies at five distinct levels of abstraction granularity (L1-L5). Verification was conducted across multiple horizons under each granularity, employing the simulation approach outlined in the **Correctness** section to estimate the average and maximum probabilities in an initial region. We selected five distinct areas to verify and simulate for Avg-Bound for each task, recording the verification time. We then computed the mean values of the errors and recorded times. For the Max-Bound, we tested tightness only in a selected typical region that was less prone to extreme probability values. Figure 8 illustrates the time costs and bounds errors obtained over multiple horizons k from our experiments across the five granularity levels (L1-L5).

Tightness. The variations in tightness for Max-Bound and Avg-Bound are illustrated by the error curves in the first two rows of subplots in Fig. 8. As granularity becomes finer from L1 to L5, errors in both upper bounds decrease. Notably, once a sufficient level of refinement is reached, errors can remain really low for an extended period (e.g., Max-Bound errors for PD are below 0.04, Avg-Bound errors below 0.03; for ACC, they remain below 0.01 and 0.06; and for BB, they are below 0.10 and 0.12, when granularity is finer than L2), indicating reduced sensitivity to changes in granularity.

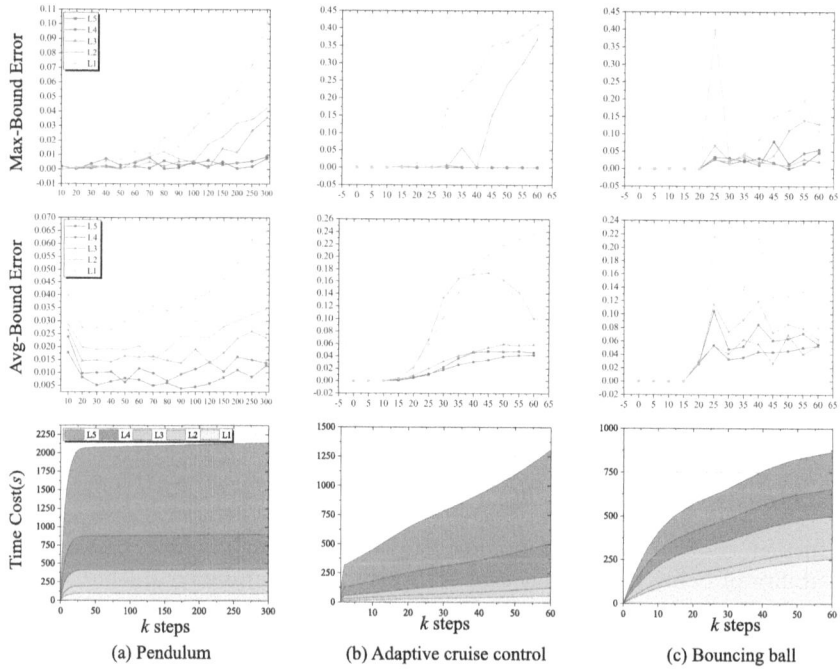

Fig. 8. The variation of tightness and time costs of verification concerning the horizon k across different granularity levels. **Row 1**: Max-Bound tightness; **Row 2**: Avg-Bound tightness; **Row 3**: Time costs. Granularity levels range from L1 (coarsest) to L5 (finest).

The verified time horizon also influences tightness, with errors rising as time steps increase, particularly at a coarser granularity. However, with sufficiently fine granularity, the error increase rate slows significantly. In the case of the BB task, the periodic nature causes the errors to fluctuate as time steps progress.

The deviation between upper bounds and simulated actual probabilities results from over-approximation from abstraction and MDP uncertainties. Coarser granularity produces coarser abstract states, leading to larger errors, which accumulate over time, reflected in the increasing error with k. For ACC, extreme situations (100% safe or unsafe) consistently exist, leading to alignment between the verified upper bound and simulation results. This highlights the extremity of Max-Bound.

Time Cost. Verification time is also influenced by granularity and verification horizons: it increases with finer granularity and larger horizons k. Finer granularity results in smaller abstract states, necessitating exploring a larger state space. The increase in time with k is intuitive, as more new abstract states must be explored. However, we observed that time increases very slowly after a particular horizon k due to our MDP downsizing technique mentioned in Sect. 5, significantly reducing the number of new abstract states. For tasks with small state spaces (e.g., PD and BB), the finite abstract state space can be nearly fully explored during progression, resulting in negligible increases in total elapsed time. Any subsequent minor increases in time are primarily due to the additional time required for solving MDPs. For instance, in the PD case, the time increase after 20 steps is almost imperceptible, demonstrating the effectiveness of our downsizing method.

Combining the conclusions regarding tightness, we can identify several granularity levels that achieve both verification accuracy and efficiency, such as L3.

6.5 Comparisons

The current state-of-the-art verification method for probabilistic reinforcement learning policies is presented in [6]. Our verification approach is more efficient and tighter and offers additional Avg-Bound compared to theirs. Since their verification time and results vary with parameters and sampling randomness, we compare them against their range of variation. We obtain our results at L3 granularity for each case, ensuring sufficient tightness and reasonable time consumption. The initial state area for verification aligns with their experimental setup.

First, we compare the efficiency and scalability of the verification methods. They verify the maximum horizons within a three-hour limit. Table 1 shows that they can only verify PD up to 6 steps, requiring 69–71 min (exceeding the three-hour limit at 7 steps). In comparison, our method efficiently verifies 300 steps in a maximum of 4 min, with low verification errors. For ACC, they verify up to 7 steps in 29–176 min, while we verify 60 steps in just 16 min. For BB, they achieve 20 steps in 30–111 min, whereas we complete 60 steps in approximately 20 min.

Table 1. comparison of our verification method with state-of-the-art techniques k: horizon of verification; **VT**: time cost of verification (in minutes); **impr.VT**: time speedup of our approach; MaxB/AvgB: the verified Max-Bound/Avg-Bound; **E-MAX**: tightness error of Max-Bound; **E-Avg**: tightness error of Avg-Bound. N/A: no data available due to excessive computation time. Our superior aspects are highlighted in bold.

Task	k	State-of-the-art			Our Method					
		VT(min)	Max	E-Max	VT(min)	**impr.VT**	Max	Avg	E-Max	E-Avg
PD	6	69-71	0.057	0.057	**0.76**	$\geq 90\times$	0.000	0.000	**0.000**	0.000
	150	N/A	N/A	N/A	4	\	0.042	0.041	0.008	0.015
	300	N/A	N/A	N/A	4	\	0.088	0.087	0.019	0.035
ACC	7	29-176	0.0-0.63	0.63	**0.76**	$\geq 38\times$	0.005	0.000	**0.005**	0.000
	30	N/A	N/A	N/A	2	\	1.000	0.161	0.000	0.024
	60	N/A	N/A	N/A	3	\	1.000	0.256	0.000	0.059
BB	20	30-111	0.031-0.62	0.62	**17**	$\geq 1.76\times$	0.000	0.000	**0.000**	0.000
	40	N/A	N/A	N/A	19	\	0.951	0.221	0.001	0.060
	60	N/A	N/A	N/A	20	\	1.0	0.722	0.000	0.112

The increase in time cost for our method as k grows is minimal (as illustrated in Fig. 6), allowing for the verification of an almost infinite horizon. In contrast, their method exhibits exponential growth in time cost with each additional step.

Regarding comprehensiveness and tightness, they provide only an upper bound on maximum probability, while we also include an Avg-Bound. Our simulations indicate that no unsafe situations arise within the horizons verified by their method, yielding an actual probability of 0. Their upper bounds for maximum failure probability are 0.057 for 6 steps (PD), 0-0.63 for 20 steps (BB), and 0.031-0.62 for 7 steps (ACC), showing significant not tightness and instability. In contrast, we generated an upper bound of 0.0 for all tasks within the same horizons. Our experiments maintain low errors over extended periods (Max-Bound errors below 0.02 and Avg-Bound errors below 0.04 for PD in 300 steps; 0.01 and 0.06 for ACC in 60 steps; and 0.01 and 0.12 for BB in 60 steps).

7 Conclusion and Future Work

We propose a formal verification method for verifying the probabilistic DRL policy using abstract-based training. Our method can efficiently provide two types of tight probability upper bounds on the probability of safety for probability policy over a long time horizon, providing rich information for analyzing the unsafe defects of probability policy.

Our future work will utilize the worst-case probability upper bounds identified through verification to trace unsafe paths and leverage them to improve the neural network policy. This optimization aims to enhance safety and ensure that safety requirements are met at minimal cost, thereby completing our framework for synthesizing safe probabilistic policies. Additionally, treating neural networks as black-boxes in probabilistic model checking opens up new possibilities for verifying probabilistic policies. This approach could pave the way for research into verifying more complex probabilistic temporal logic specifications, developing more efficient runtime verification methods, and tackling challenging high-dimensional reinforcement learning tasks using better abstraction.

Acknowledgments. This work is supported by the National Natural Science Foundation of China NSFC (No. 92370201, No. 62272165, No. 62372176), the "Digital Silk Road" Shanghai International Joint Lab of Trustworthy Intelligent Software (No. 22510750100) and Shanghai Trusted Industry Internet Software Collaborative Innovation Center.

References

1. Abate, A., Katoen, J.P., Lygeros, J., Prandini, M.: Approximate model checking of stochastic hybrid systems. Eur. J. Control. **16**(6), 624–641 (2010). https://doi.org/10.3166/ejc.16.624-641
2. Abel, D.: A theory of state abstraction for reinforcement learning. In: Proceedings of the AAAI Conference on Artificial Intelligence, AAAI 2019, vol. 33, pp. 9876–9877 (2019). https://doi.org/10.1609/aaai.v33i01.33019876
3. Arulkumaran, K., Deisenroth, M.P., Brundage, M., Bharath, A.A.: Deep reinforcement learning: a brief survey. IEEE Signal Process. Mag. **34**(6), 26–38 (2017). https://doi.org/10.1109/MSP.2017.2743240
4. Bacci, E., Giacobbe, M., Parker, D.: Verifying reinforcement learning up to infinity. In: Proceedings of the Thirtieth International Joint Conference on Artificial Intelligence, IJCAI-21 (2021). https://doi.org/10.24963/ijcai.2021/297
5. Bacci, E., Parker, D.: Probabilistic guarantees for safe deep reinforcement learning. In: Bertrand, N., Jansen, N. (eds.) FORMATS 2020. LNCS, vol. 12288, pp. 231–248. Springer, Cham (2020). https://doi.org/10.1007/978-3-030-57628-8_14
6. Bacci, E., Parker, D.: Verified probabilistic policies for deep reinforcement learning. In: NASA Formal Methods, NFM 2022, vol. 13260, pp. 193–212. Springer, Cham (2022). https://doi.org/10.1007/978-3-031-06773-0_10
7. Bunel, R.R., Turkaslan, I., Torr, P., Kohli, P., Mudigonda, P.K.: A unified view of piecewise linear neural network verification. In: Advances in Neural Information Processing Systems, NeurIPS 2018, vol. 31 (2018). https://proceedings.neurips.cc/paper_files/paper/2018/file/be53d253d6bc3258a8160556dda3e9b2-Paper.pdf
8. Dehnert, C., Junges, S., Katoen, J.-P., Volk, M.: A storm is coming: a modern probabilistic model checker. In: Majumdar, R., Kunčak, V. (eds.) CAV 2017. LNCS, vol. 10427, pp. 592–600. Springer, Cham (2017). https://doi.org/10.1007/978-3-319-63390-9_31
9. Dong, Y., Zhao, X., Huang, X.: Dependability analysis of deep reinforcement learning based robotics and autonomous systems through probabilistic model checking. In: 2022 IEEE/RSJ International Conference on Intelligent Robots and Systems (IROS), pp. 5171–5178. IEEE (2022). https://doi.org/10.1109/IROS47612.2022.9981794

10. Eliyahu, T., Kazak, Y., Katz, G., Schapira, M.: Verifying learning-augmented systems. In: Proceedings of the 2021 ACM SIGCOMM 2021 Conference, SIGCOMM'21, pp. 305–318 (2021). https://doi.org/10.1145/3452296.3472936
11. Forejt, V., Kwiatkowska, M., Norman, G., Parker, D.: Automated verification techniques for probabilistic systems. In: Bernardo, M., Issarny, V. (eds.) SFM 2011. LNCS, vol. 6659, pp. 53–113. Springer, Heidelberg (2011). https://doi.org/10.1007/978-3-642-21455-4_3
12. García, J., Fernández, F.: Probabilistic policy reuse for safe reinforcement learning. ACM Trans. Auton. Adapt. Syst. (TAAS) **13**(3), 1–24 (2019). https://doi.org/10.1145/3310090
13. Ghavamzadeh, M., Mannor, S., Pineau, J., Tamar, A., et al.: Bayesian reinforcement learning: a survey. Found. Trends® Mach. Learn. **8**(5-6), 359–483 (2015). https://doi.org/10.1561/2200000049
14. Gu, S., Holly, E., Lillicrap, T., Levine, S.: Deep reinforcement learning for robotic manipulation with asynchronous off-policy updates. In: 2017 IEEE International Conference on Robotics and Automation (ICRA), pp. 3389–3396. IEEE (2017). https://doi.org/10.1109/ICRA.2017.7989385
15. Guttman, A.: R-trees: a dynamic index structure for spatial searching. In: Proceedings of the 1984 ACM SIGMOD International Conference on Management of Data, SIGMOD 1984, pp. 47–57 (1984). https://doi.org/10.1145/602259.602266
16. Haarnoja, T., et al.: Soft actor-critic algorithms and applications. arXiv preprint arXiv:1812.05905 (2018). https://doi.org/10.48550/arXiv.1812.05905
17. Huang, C., Fan, J., Li, W., Chen, X., Zhu, Q.: Reachnn: reachability analysis of neural-network controlled systems. ACM Trans. Embed. Comput. Syst. (TECS) **18**(5s), 1–22 (2019). https://doi.org/10.1145/3358228
18. Ivanov, R., Carpenter, T., Weimer, J., Alur, R., Pappas, G., Lee, I.: Verisig 2.0: verification of neural network controllers using taylor model preconditioning. In: Silva, A., Leino, K.R.M. (eds.) CAV 2021. LNCS, vol. 12759, pp. 249–262. Springer, Cham (2021). https://doi.org/10.1007/978-3-030-81685-8_11
19. Jin, P., Tian, J., Zhi, D., Wen, X., Zhang, M.: Trainify: a CEGAR-driven training and verification framework for safe deep reinforcement learning. In: Computer Aided Verification, CAV 2022, vol. 13371, pp. 193–218. Springer, Cham (2022). https://doi.org/10.1007/978-3-031-13185-1_10
20. Johnson, T.T., et al.: Arch-comp20 category report: artificial intelligence and neural network control systems (AINNCS) for continuous and hybrid systems plants. EPiC Ser. Comput. **74** (2020). https://doi.org/10.29007/9xgv. https://par.nsf.gov/biblio/10195492
21. Kiran, B.R., et al.: Deep reinforcement learning for autonomous driving: a survey. IEEE Trans. Intell. Transp. Syst. **23**(6), 4909–4926 (2021). https://doi.org/10.1109/TITS.2021.3054625
22. Kwiatkowska, M., Norman, G., Parker, D.: Probabilistic model checking: advances and applications. In: Drechsler, R. (ed.) Formal System Verification, pp. 73–121. Springer, Cham (2018). https://doi.org/10.1007/978-3-319-57685-5_3
23. Li, J., Monroe, W., Ritter, A., Galley, M., Gao, J., Jurafsky, D.: Deep reinforcement learning for dialogue generation. arXiv preprint arXiv:1606.01541 (2016). https://doi.org/10.48550/arXiv.1606.01541
24. Mnih, V., et al.: Human-level control through deep reinforcement learning. Nature **518**(7540), 529–533 (2015). https://doi.org/10.1038/nature14236

25. Papoudakis, G., Christianos, F., Albrecht, S.: Agent modelling under partial observability for deep reinforcement learning. In: Advances in Neural Information Processing Systems, NeurIPS 2021, vol. 34, pp. 19210–19222 (2021). https://proceedings.neurips.cc/paper_files/paper/2021/file/a03caec56cd82478bf197475b48c05f9-Paper.pdf
26. Pathak, S., Pulina, L., Tacchella, A.: Verification and repair of control policies for safe reinforcement learning. Appl. Intell. **48**(4), 886–908 (2017). https://doi.org/10.1007/s10489-017-0999-8
27. Schulman, J., Wolski, F., Dhariwal, P., Radford, A., Klimov, O.: Proximal policy optimization algorithms. arXiv preprint arXiv:1707.06347 (2017). https://doi.org/10.48550/arXiv.1707.06347
28. Sutton, R., Barto, A.: Reinforcement learning: an introduction. IEEE Trans. Neural Networks **9**(5), 1054 (1998). https://doi.org/10.1109/tnn.1998.712192
29. Tian, J., Zhi, D., Liu, S., Wang, P., Chen, C., Zhang, M.: Boosting verification of deep reinforcement learning via piece-wise linear decision neural networks. In: Advances in Neural Information Processing Systems, NeurIPS 2023, vol. 36, pp. 10022–10037 (2023). https://proceedings.neurips.cc/paper_files/paper/2023/file/1f96b24df4b06f5d68389845a9a13ed9-Paper-Conference.pdf
30. Tian, J., Zhi, D., Liu, S., Wang, P., Katz, G., Zhang, M.: Taming reachability analysis of DNN-controlled systems via abstraction-based training. In: Verification, Model Checking, and Abstract Interpretation, VMCAI 2024, vol. 14500, pp. 73–97. Springer, Cham (2023). https://doi.org/10.1007/978-3-031-50521-8_4
31. Wang, Y., Huang, C., Wang, Z., Wang, Z., Zhu, Q.: Verification in the loop: correct-by-construction control learning with reach-avoid guarantees. arXiv preprint arXiv:2106.03245 (2021).https://doi.org/10.48550/arXiv.2106.03245
32. Xiong, Z., Jagannathan, S.: Scalable synthesis of verified controllers in deep reinforcement learning. arXiv preprint arXiv:2104.10219 (2021). https://doi.org/10.48550/arXiv.2104.10219
33. Yu, C., Liu, J., Nemati, S., Yin, G.: Reinforcement learning in healthcare: a survey. ACM Comput. Surv. (CSUR) **55**(1), 1–36 (2021). https://doi.org/10.1145/3477600
34. Zhang, Z., et al.: Boosting verified training for robust image classifications via abstraction. In: Proceedings of the IEEE/CVF Conference on Computer Vision and Pattern Recognition (CVPR), pp. 16251–16260 (2023). https://openaccess.thecvf.com/content/CVPR2023/html/Zhang_Boosting_Verified_Training_for_Robust_Image_Classifications_via_Abstraction_CVPR_2023_paper.html

A Real-Blasting Extension of cvc5 for Reasoning About Floating-Point Arithmetic

Daisuke Ishii(✉)

Japan Advanced Institute of Science and Technology, Nomi, Ishikawa, Japan
dsksh@jaist.ac.jp

Abstract. We present a new floating-point arithmetic (FPA) theory solver implemented in the cvc5 SMT solver. Differently from the major bit-blasting method, we use a *real-blasting* method that reasons about FPA formulas in a theory of real-integer arithmetic (RIA). It is based on an axiomatization of FPA operations with a set of conditional RIA formulas. The solver is implemented as an extended theory solver tightly coupled with cvc5 so that fragments of axioms are lazily instantiated regarding the solving context. Experimental results show that our solver performs better than existing solvers for several problem instances.

1 Introduction

Floating-point arithmetic (FPA) (Sect. 2.2) is one of the background theories (Sect. 2.1) of SMT solvers, essential for the rigorous analysis of numerical systems involving a machine-representation of real numbers (*FP numbers*) and the rounding operation from the infinite domain. The standard technique used for FPA in modern SMT solvers [2,6,23] is *bit-blasting* [8,9] that translates FPA formulas into those on bit vectors (BVs). However, the practical applications of FPA solvers are still limited in terms of scalability; for instance, if two arithmetic formulas involving the same number of arithmetic operators are handled in FPA and real arithmetic (RA), respectively, there can be a significant difference in the performance of the solving process.

In this paper, we present an FPA solver that uses an alternative method, *real-blasting*. This method represents each FP number as a rational and defines each operation with a set of axioms, i.e. lemmas in a real-integer arithmetic (RIA); then, the solving process benefits from an efficient RIA solver. Existing implementations [11,12,20] have reported performance issues in comparison with bit-blasting solvers. In this work, we investigate how a real-blasting FPA solver can be efficient by tightly coupling it with the DPLL(T) algorithm using an extension mechanism provided by cvc5. We propose a translation method for

---

This work was supported by JST, CREST Grant Number JPMJCR23M1, and JSPS KAKENHI Grant Number 23K11969.

FPA formulas in RIA and a set of axioms (Sect. 3); then, we implement a real-blasting theory solver by extending the RIA solver of cvc5; a main function of the extension is to instantiate the axioms either eagerly or lazily (Sect. 4). Section 5 describes an example of solving an FPA formula with the proposed method and the extended solver. In the experiments, we evaluate the performance in comparison with existing solvers including state-of-the-art bit-blasting solvers (Sect. 6). Our solver is available at https://github.com/dsksh/cvc5.

Related Work. The method we call "real-blasting" is proposed as an alternative approach to bit-blasting that uses theory solvers for real arithmetic (RA) [11,12,20]. There are tools REALIZER [20] and ALT-ERGO [10]; their functionality and performance are limited compared to recent solvers including ours; the former does not support subnormal numbers and non-real values; the latter can only check unsatisfiability, and in our preliminary experiment, it could only solve a few instances of the Griggio benchmark as reported in [11]. The axiomatization of FPA in RA required is closely related to the formalization for theorem proving [5,16] and program verification [1,13]. The axioms in this paper are modified from existing ones in e.g. [1,11]; they use additional symbols for special values but we represent them with rationals; we modify the axioms as templates to fit the learning process of cvc5 so that our extended theorem prover instantiates them efficiently. In other related work, the mixed-real FPA method [24,26,27] and constraint-programming-based methods [21,28] are proposed. The former proposes a process that transforms an FPA formula into an approximate RA formula and, after solving the formula, verifies the correctness of the approximation. In contrast, we do not approximate, as we convert using extended symbols and axioms. The latter represent FPA formulas with a set of constraints and solve with a constraint programming techniques. Since the problem formulation differs slightly between these methods and ours, a direct comparison of performance remains a future issue.

2 Preliminaries

2.1 Satisfiability Modulo Theories

We assume the concepts for SMT e.g. (sorted) signature, term, literal, formula and clause. See e.g. [4,19] for details. For a signature Σ, an Σ-*interpretation* over a set of variables V is a map from each symbol in Σ or each variable in V to an object of a corresponding sort. For a Σ-interpretation $\mathcal{M}$ and a Σ-formula φ, the *satisfiability* $\mathcal{M} \models \varphi$ is determined by an inductive interpretation of φ. A *theory* $(\Sigma, \boldsymbol{M})$ consists of a signature Σ and a class $\boldsymbol{M}$ of Σ-interpretations, also called *models* of the theory, that is closed under variable interpretation changes. A Σ-formula φ is T-*satisfiable* if $\mathcal{M} \models \varphi$ for some $\mathcal{M}$. A set C of Σ-formulas T-*entails* a Σ-formula φ, denoted $\mathsf{C} \models_T \varphi$, iff every Σ-interpretation in $\boldsymbol{M}$ that satisfies every $\varphi' \in \mathsf{C}$ satisfies φ. We say φ is T-*valid* iff $\emptyset \models_T \varphi$.

2.2 Floating-Point Arithmetic

We denote by $\mathbb{F}_{\epsilon,\sigma}$ a set (and sort) of *floating-point (FP) numbers* whose exponent and significand are ϵ and σ bits; we only consider the base 2 for the sake of simplicity. $\mathbb{F}_{\epsilon,\sigma}$ contains *normal* and *subnormal numbers*, which represent rational numbers; otherwise, the elements of $\mathbb{F}_{\epsilon,\sigma}$ are *special values* $-0_{\epsilon,\sigma}$, $+0_{\epsilon,\sigma}$, $-\infty_{\epsilon,\sigma}$, $+\infty_{\epsilon,\sigma}$ and $\mathrm{NaN}_{\epsilon,\sigma}$. We omit the precision $\cdot_{\epsilon,\sigma}$ when they do not matter in a context. *Rounding operations* from reals to FP numbers are parameterized by five *rounding modes*, e.g. RNE (*round to nearest, ties to even*), RTN and RTP (*round towards negative/positive infinity*); we denote by $\mathbb{M}$ the set (and sort) of rounding modes.

The theory of *floating-point arithmetic (FPA)* T_{FPA} is proposed for reasoning about formulas that describe numerical computation involving FP numbers and rounding modes. In this work, we consider the signature Σ_{FPA} defined by SMT-LIB [3] or [7], but do not include some symbols e.g. sqrt and toUbv, support for which is a future work. The syntax for atomic formulas A is as follows.

$$A ::= rop(P,P) \mid pop(P) \quad P ::= T \mid uop(P) \mid bop(rm,P,P) \quad T ::= id \mid c$$

The (meta-)symbol *rop* represents one of the six comparison operators (arity-2 predicate symbols) e.g. =, eq and gt; *pop* represents a property predicate e.g. isNormal; *uop* represents a unary function e.g. neg and abs; *bop* is one of the four arithmetic functions, i.e., add, sub, mul and div; *rm* represents the rounding modes. There are several notations for constants c; a basic notation $\mathrm{fp}(s,m,e)$ consists of three BVs, i.e., the sign bit s, the significand m and the exponent e. Two example atomic formulas are included in the Σ_{FPA}-formula φ in Sect. 5.

We omit the semantics, but the constants, functions and predicates are interpreted following the IEEE-754 standard [17] (see also [3,7,22]).

3 Representing FPA Formulas in RIA

We propose to extend the theory T_{RIA} with a signature that contains the symbols in Σ_{FPA} and a *translation* $[\![\cdot]\!]$, which maps FP numbers to $\mathbb{Q}$ and rounding modes to $[0,4] \subset \mathbb{Z}$; also, each Σ_{FPA}-term is mapped to a set of Σ_{RIA}-formulas. Thus, if φ is a Σ_{FPA}-formula, we can have a Σ_{RIA}-interpretation for $[\![\varphi]\!]$, where each term within is translated inductively. Given a solving context C (set of Σ_{FPA}-literals), we aim to make the inferences $\mathsf{C} \models_{T_{\mathrm{FPA}}} \varphi$ and $\mathsf{C} \cup \mathsf{C}' \models_{T} \varphi$ commute, where T is our extended theory and C' is an auxiliary set of Σ_{RIA}-literals.

For a constant term $\mathrm{fp}(s,m,e)$ of sort $\mathbb{F}_{\epsilon,\sigma}$ representing a finite (normal and subnormal) number, we translate it as an equivalent rational as $[\![\mathrm{fp}(s,m,e)]\!] := (-1)^s m 2^{e-\sigma+1}$. Translations may overlap for $\mathbb{F}_{\epsilon,\sigma}$ and $\mathbb{F}_{\epsilon',\sigma'}$ with different precisions, but they are mapped to two symbols sorted differently. We also map special values to rationals that do not interfere with any operation result (of the same precision). The constants representing the boundary values and translation for the special values are shown in the upper part of Fig. 1. Hereafter, we omit the parameter (ϵ,σ) for simplicity.

$$R_{\text{N,min}} := 2^{e_{\min}} \qquad R_{\text{N,max}} := (2 - 2^{1-\sigma})2^{e_{\max}} \qquad R_{\text{SN,min}} := 2^{e_{\min} - \sigma + 1}$$

$$[\![+0]\!] := 0 \qquad [\![-0]\!] := -\frac{1}{d+1} \qquad \text{(where } \tfrac{1}{d} = R_{\text{SN,min}})$$

$$[\![-\infty]\!] := -\tfrac{3R_{\text{N,max}}+1}{3} \qquad [\![+\infty]\!] := \tfrac{3R_{\text{N,max}}+1}{3} \qquad [\![\text{NaN}]\!] := -\tfrac{3R_{\text{N,max}}+2}{3}$$

$$[\![\circ]\!](m, x) := \begin{cases} x & \text{if } x \in \{[\![-0]\!], [\![+0]\!], [\![-\infty]\!], [\![+\infty]\!], [\![\text{NaN}]\!]\} \\ \odot_{[\![m]\!]}(x) & \text{otherwise} \end{cases} \quad (1)$$

$$[\![\text{mul}]\!](m, x, y) :=$$
$$\begin{cases} [\![\circ]\!](m, x \times y) & \text{if isFinite}(x) \wedge \text{isFinite}(y) \wedge (\neg\text{isZero}(x) \vee \neg\text{isZero}(y)) \wedge \\ & \text{isFinite}([\![\circ]\!](m, x \times y)) \\ [\![-0]\!] & \text{if isFinite}(x) \wedge \text{isFinite}(y) \wedge (\text{isZero}(x) \vee \text{isZero}(y)) \wedge m = [\![\text{RTN}]\!] \\ [\![-\infty]\!] & \text{if } (x = [\![-\infty]\!] \wedge \text{isFinite}(y) \wedge \neg\text{isZero}(y))) \vee \\ & (\text{isFinite}(x) \wedge \text{isFinite}(y) \wedge [\![\circ]\!](m, x \times y) = [\![-\infty]\!]) \vee \cdots \\ [\![\text{NaN}]\!] & \text{if } x = [\![\text{NaN}]\!] \vee y = [\![\text{NaN}]\!] \vee (\text{isZero}(x) \wedge y = [\![-\infty]\!]) \vee \cdots \\ \vdots & \end{cases} \quad (2)$$

$$[\![\text{gt}]\!](x, y) :\Leftrightarrow \begin{cases} x > y & \text{if isFinite}(x) \wedge \text{isFinite}(y) \wedge (\neg\text{isZero}(x) \vee \neg\text{isZero}(y)) \\ \bot & \text{if } (\text{isZero}(x) \wedge \text{isZero}(y)) \vee (x = [\![\text{NaN}]\!] \vee y = [\![\text{NaN}]\!]) \vee \cdots \\ \top & \text{if } (x = [\![+\infty]\!] \wedge \neg(y = [\![+\infty]\!]) \wedge \neg(y = [\![\text{NaN}]\!])) \vee \cdots \end{cases} \quad (3)$$

Fig. 1. Translation of Σ_{FPA}-terms. The constants and operators are parameterized by the precision (ϵ, σ) but we omit it for simplicity. Other symbols are defined as: $e_{\max} := 2^{\epsilon-1}$, $e_{\min} := 1 - e_{\max}$, $\text{isFinite}(r) :\Leftrightarrow (-R_{\text{N,max}} \leq r \leq R_{\text{N,max}})$, and $\text{isZero}(r) :\Leftrightarrow (r = [\![-0]\!] \vee r = [\![+0]\!])$.

Operators in Σ_{FPA} are represented explicitly as functions and predicates in the domain of $\mathbb{R}$ and $[0, 4] \subset \mathbb{Z}$. The rounding operator $\circ : \mathbb{M} \times \mathbb{R} \to \mathbb{F}$ is translated in two steps. We first define a function $\odot_m : \mathbb{R} \to \mathbb{R}$ that rounds with mode m into an FP number represented in $\mathbb{Q}$; it follows the IEEE-754 standard and is implemented using rational arithmetic as in [20]. Then, we define a wrapper that goes through special values (Eq. (1) in Fig. 1).

The translation of arithmetic operations requires consideration of a number of cases sorted by whether the arguments and results are special values or not. For example, the FPA multiplication operator mul is translated as Eq. (2). The first case is when the arguments x and y are both nonzero and finite and the result does not overflow; the multiplication of RA and then rounding are applied. The other branches provide conditions for each special value; note that the first case can also be $[\![-0]\!]$. Some of the conditions are omitted for reasons of space. Note that the arguments can be non-FP numbers; in that case, the result will be equivalent to the translation of $\text{mul}(m, \circ(m', x) \times \circ(m'', y))$ with certain rounding modes m' and m''.

$$\forall m, \forall y, \quad [\![y]\!] = [\![\circ]\!](m, [\![y]\!]) \tag{4}$$

$$\forall m, \forall x, \quad \text{isNormal}([\![\circ]\!](m, x)) \Rightarrow |x - [\![\circ]\!](m, x)| < 2^{1-\sigma}|x| \tag{5}$$

$$\forall m, \forall x_1, \forall x_2, \quad |x_1| \leq x_2 \Rightarrow |x_1 - [\![\circ]\!](m, x_1)| \leq 2^{\alpha} \tag{6}$$
$$\text{where } \alpha = \text{ilog2}(\max(x_2, R_{\text{N,min}})) - \sigma + 1$$

$$\forall m, \forall x_1, \forall x_2, \quad [\![\circ]\!](m, x_1) > [\![\circ]\!]([\![\text{RTN}]\!], x_2) \Rightarrow [\![\circ]\!](m, x_1) \geq [\![\circ]\!]([\![\text{RTP}]\!], x_2) \tag{7}$$

$$\forall m, \forall x, \quad (\neg(x = [\![\text{NaN}]\!]) \land x \leq -R_{\text{N,max}}) \Rightarrow$$
$$[\![\circ]\!](m, x) = -R_{\text{N,max}} \lor [\![\circ]\!](m, x) = [\![-\infty]\!] \tag{8}$$

$$\forall m, \forall x_1, \forall x_2, \quad x_1 \geq x_2 \Rightarrow [\![\circ]\!](m, x_1) \geq [\![\circ]\!](m, x_2) \tag{9}$$

$$\forall m, \forall x_1, \forall x_2, \quad [\![\circ]\!](m, x_1) > [\![\circ]\!](m, x_2) \Rightarrow [\![\circ]\!](m, x_1) > x_2 \land x_1 > [\![\circ]\!](m, x_2) \tag{10}$$

Fig. 2. Axioms for the rounding operator. We assume paramters m, x (or x_i) and y are bounded by $[0, 4]$, $\mathbb{R}$ and the set of Σ_{FPA}-terms, respectively. isNormal(r) represents $R_{\text{N,min}} \leq |r| \leq R_{\text{N,max}}$ and ilog2 is an integer logarithmic function with base 2.

Likewise, the FPA comparison operator gt is translated as Eq. (3). If the arguments are finite, it is delegated to the operator of RA. The two branches below specify the conditions for some cases involving special values.

3.1 Axiomatization

We define each operator with a set of axioms, so that it can be defined in stages rather than explicitly at once during the solving process. Our axioms are similar to those in [1,11], but they are modified for our translation method.

Some axioms for the rounding operator are shown in Fig. 2. Equation (4) claims the idempotency of ∘ on an FP number $[\![y]\!]$. Equation (5) bounds relative errors for normal numbers, and Eq. (6) bounds absolute errors, given a bound x_2 on the argument x_1. Equation (7) gives a relation between FP numbers $[\![\circ]\!]([\![\text{RTN}]\!], x_2)$ and $[\![\circ]\!]([\![\text{RTP}]\!], x_2)$, which are different but adjacent in most cases. Equation (8) states how large negative values are rounded. Equation (9) and (10) are axioms related to the comparison operators. There are 17 axioms in total.

Axioms for other operators are basically deduced from the explicit representation shown in Fig. 1. Given a translated formula $\psi(m, x, y)$ for a Σ_{FPA}-operator op guarded by a condition $\chi(m, x, y)$, we can have an axiom

$$\forall m, \forall x, \forall y, \ \chi(m, x, y) \Rightarrow [\![\text{op}]\!](m, x, y) = \psi(m, x, y).$$

For example, there are 10 and 9 axioms for mul and gt, respectively.

3.2 Correctness

The proposed method is expected to be correct in the following sense, and we will partially confirm this in the experiment. Let T be a theory extended from

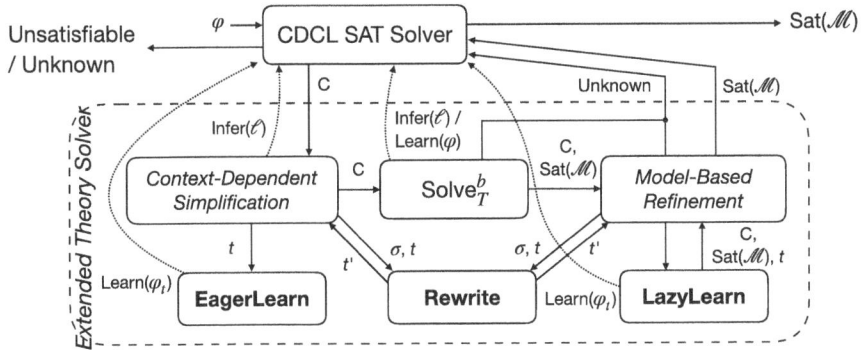

Fig. 3. Overview of the solving procedure of cvc5.

T_{RIA} with the signature Σ_{FPA} and the Σ_{RIA}-interpretations composed with $[\![\cdot]\!]$. Also, let $\varphi(\boldsymbol{m}, \boldsymbol{x})$ be a quantifier-free Σ_{FPA}-formula over variables $\boldsymbol{m}$ of sort $\mathbb{M}$ and $\boldsymbol{x}$ of sort $\mathbb{F}$. We prepare variables $\boldsymbol{z}_m$ of sort $\mathbb{Z}$ and $\boldsymbol{z}_x$ of sort $\mathbb{R}$ according to $\boldsymbol{m}$ and $\boldsymbol{x}$. Then, the formula $\varphi(\boldsymbol{m}, \boldsymbol{x})$ is T_{FPA}-satisfiable iff

$$\Big(\bigwedge_{i=1}^{|z_m|} 0 \leq z_{m,i} \leq 4\Big) \wedge \Big(\bigwedge_{i=1}^{|z_x|} z_{x,i} = \circ(c_m, z_{x,i})\Big) \wedge [\![\varphi]\!](\boldsymbol{z}_m, \boldsymbol{z}_x) \text{ is } T\text{-satisfiable,}$$

where c_m is an arbitrary rounding mode constant.

4 Implementation

We have implemented a theory solver for T_{FPA} in cvc5 that can be used in place of the standard one. It checks T-satisfiability where T is an extended theory of the non-linear RIA (NLRIA). We review the extension mechanism of cvc5 (Sect. 4.1) and then describe our implementation (Sects. 4.2 and 4.3). Our implementation consists of around 5,000 LOC.

4.1 cvc5 and Extended Theory Solvers

cvc5 (version 1.2.0) [2][1] is an SMT solver implemented in C++. It is based on the DPLL(T) algorithm that tightly integrates a SAT solver and theory solvers. Briefly, an input formula φ is fed to the SAT solver (after Boolean encoding) and it enumerates propositional assignments (*context*) C, i.e., a set of theory literals. Then, a theory solver for T validates or refines a T-context C by interpreting the Σ_T-literals contained.

For an extended theory T whose signature is extended as $\Sigma = \Sigma^b \cup \Sigma^e$, an *extended theory solver (ETS)* [25] provides a means to reason about Σ-formulas

---

[1] https://cvc5.github.io/.

using a built-in Σ^b-theory solver. In the process, we represent every Σ^e-term t with a fresh *purification variable* z_t. Accordingly, we extend a context C^b with a set C^e of equalities $z_t = t$ for all purification variables. The procedure involving ETS is shown in Fig. 3. The output of ETS is either of: (1) $\mathsf{Sat}(\mathcal{M})$ where $\mathcal{M}$ is a T-model such that $\mathcal{M} \models \mathsf{C}$ (where $\mathsf{C} = \mathsf{C}^b \cup \mathsf{C}^e$); (2) $\mathsf{Learn}(\varphi)$ where φ is a *lemma*, a T-valid clause not propositionally entailed by the context; (3) $\mathsf{Infer}(\ell)$ where ℓ is a T-literal entailed by the context; (4) $\mathsf{Unknown}$ indicating that it is unable to determine the satisfiability.

The following two components are important in the reasoning about extended theories using ETS.

Context-Dependent Simplification. As a lightweight reasoning process, ETS computes a simplified form of each Σ^e-term t under an extended context C. If $\mathsf{C}^b \models_T y = s$ for a variable y, we can have a substitution σ that assigns as $y \mapsto s$. A subroutine REWRITE first receives t and σ constructed from C^b, and computes a simplified Σ^b-term t'. Then, it introduces $z_t = t'$ in C^b.

Model-Based Refinement. ETS first tries to obtain a model $\mathcal{M}$ such that $\mathcal{M} \models \mathsf{C}^b$ based on the bare Σ^b-theory. If there is $\mathcal{M}$, then ETS checks whether it also satisfies C^e by interpreting Σ^e-terms t. To do so, we can obtain a substitution σ entailed by $\mathcal{M}$, and check $z_t \sigma = \text{REWRITE}(\sigma, t)$ for every t. If an inconsistency is found, a subroutine LAZYLEARN is invoked for a Σ^e-term t to obtain a *refinement lemma*, a Σ^b-clause φ_t such that $\mathsf{C}^e \models_T \varphi_t$ (i.e., it is entailed by the axioms, etc.) and $\mathcal{M} \not\models \varphi_t$ (i.e., it could refine the model).

Apart from the lazy learning of refinement lemmas, ETS can have an EAGERLEARN routine to learn the basic T-valid facts about Σ^e-terms in the initial phase.

4.2 Preprocessing for Intermediate Representations

We implemented an additional preprocessing pass that intercepts the intermediate representation of the input Σ_{FPA}-formula and converts each term into a representation for the extended theory. In this process, FP number constants are translated as rationals and rounding mode constants are converted to integers in $[0, 4]$. Variables v of sort $\mathbb{F}$ are re-sorted as $\mathbb{R}$ and associated with a constraint Eq (4) instantiated for a rounding mode constant and v. Similarly, variables v of sort $\mathbb{M}$ are re-sorted as $\mathbb{Z}$ with a constraint $0 \le v \le 4$. These conversions are exemplified in Sect. 5.

Also, the intermediate representation for each Σ_{FPA}-operation is converted to a representation of extended symbols (in Σ^e) that is sorted according to the translation described in Sect. 3. Some examples of the extended symbols (parameterized with precisions) and their sorts are shown below.

- Rounding operator: (_ rfp.round ϵ σ) : $\mathbb{Z} \times \mathbb{R} \to \mathbb{R}$.
- Arithmetic operators e.g.: (_ rfp.neg ϵ σ) : $\mathbb{R} \to \mathbb{R}$ and
  (_ rfp.mul ϵ σ) : $\mathbb{Z} \times \mathbb{R} \times \mathbb{R} \to \mathbb{R}$.
- Comparison operators e.g.: (_ rfp.gt ϵ σ) : $\mathbb{R} \times \mathbb{R} \to \mathbb{Z}$.

Note that we represent the comparison operators as arithmetic function with the codomain $\{0, 1\} \subset \mathbb{Z}$. The extension mechanism of CVC5 assigns purification variables (sorted as the codomain sort) to these Σ^e-terms.

4.3 Real-Blasting Extension for FPA

This section describes the implementation of the three subroutines REWRITE, EAGERLEARN and LAZYLEARN required for the theory solver extension for FPA.

Rewriter. Given a substitution σ and a Σ^e-term t, REWRITE checks whether σ gives constants to the arguments of t, and if so, simplifies the term by performing the corresponding FPA operation according to the definition shown in part in Fig. 1. If the arguments are not concretized, nothing is done, except in a few cases e.g. $\mathrm{mul}(m, 1, x) = x$. In the case of comparison operators, we first transform to standard forms (e.g. $\mathrm{lt}(x_1, x_2)$ to $\mathrm{gt}(x_2, x_1)$) and then simplify, e.g. with Eq. (3).

Implementation of Axioms. For each Σ^e-term t, EAGERLEARN can instantiate every corresponding axiom for t and append it to the context (as a set of Σ-clauses). However, the context can grow quickly in this way, since each operation needs to consider a number of cases as we saw in Sect. 3.

The more limited the solving context, the fewer cases to consider. In many cases, it is more efficient to issue a fragment of axioms as a refinement lemma with LAZYLEARN, while analyzing the model $\mathcal{M}$. Here, we illustrate some such lemma implementations.

For a rounding term $t := \circ(m, x)$, if a model $\mathcal{M}$ interprets its purification variable z_t or the argument x as a finite FP number, and the error $|z_t - x|$ is large, we can issue axioms on the absolute (Eq. (6)) and relative errors (Eq. (5)) as refinement lemmas (preliminary experiments show that it is more efficient to issue both). Similarly, if z_t is interpreted as a rational number that does not correspond to FP numbers, we issue an axiom e.g. Equation (7) to exclude non-FP numbers in the neighborhood. Otherwise, we can issue a value-based refinement lemma. If $\mathcal{M}$ interprets m, x and z_t as constants c_m, c_x and c_1, and c_1 differs from $c_2 := [\![\circ]\!](c_m, c_x)$, then we can issue a lemma $(m = c_m \wedge x = c_x) \Rightarrow z_t = c_2$. However, since there can be infinitely many value-based lemmas, we restrict issuance of lemmas for most operators; we basically issue them if the argument values are special values. While this restriction reduces the search space, it also leads to unknown results, as will be shown in the experimental results.

For a multiplication term $t := \mathrm{mul}(m, x, y)$, if x and y are assigned finite constants and the operation does not overflow (the value is in the finite range), we instantiate an axiom that corresponds to the first case of Eq. (2) to interpret t as $[\![\circ]\!](m, x \times y)$. Lazy generation of this axiom is important because the multiplication term can be immediately simplified as a linear term or processed as a nonlinear term, depending on the context.

Our implementation issues axioms with both EagerLearn and LazyLearn. To make the solving process efficient, we need to adjusted how each axiom is issued. After trying out multiple settings for which routines issue which axioms in preliminary experiments, we implemented two of them. The number of lemma templates for some operators in the default setting is as follows. ◦: $1+16$, add: $6+1$, mul: $2+8$, and gt: $2+7$, where $i+j$ means i and j templates are used by Eager- and LazyLearn, respectively.

Interaction with Nonlinear Term Handling. The NLRIA theory solver of CVC5 on which the real-blasting method relies is an extended solver, which assigns purification variables to nonlinear int/real terms (monomials) and instantiates axioms during solving. To improve the efficiency, some modifications were made to its implementation. First, we suppressed the generation of axioms that were considered to be redundant for our method. For example, the NLRIA solver propagates each boundary constraint (i.e. inequalities) on variables x to those on the monomials e.g. $x \times y$; to suppress the number of such propagations, we modified to ignore the boundary constraints of the form $x = \circ(m, x)$ derived from Eq. (4). Second, we facilitated reasoning about inequality terms over monomials. For example, if the term $x \times z \geq y \times z$ appears in the NLRIA solver, we instantiate Eq. (9) accordingly so as to relate the learning clauses on monomials to those on the rounding terms.

5 Example

Let us consider a formula of an 8-bit FPA

$$\varphi \Leftrightarrow \mathrm{isFinite}(x) \wedge \mathrm{mul}(\mathrm{RNE}, x, \mathrm{fp}(0_2, 0001_2, 111_2)) = \mathrm{fp}(0_2, 1001_2, 001_2),$$

where the constants of the form $\mathrm{fp}(s, m, e)$ represent rational numbers $\frac{15}{512}$ and $\frac{9}{2}$. We then solve the problem by translating it in RIA as follows.

$$[\![\varphi]\!] \Leftrightarrow -240 \leq z_x \leq 240 \wedge z_{\mathrm{mul}} = \tfrac{9}{2} \wedge \varphi_x \wedge \varphi_{\mathrm{mul}},$$

where (1) the first predicate describes the definition of isFinite, (2) z_x and z_{mul} are purification variables in RIA representing the variable x and the multiplication term, and (3) φ_x and φ_{mul} are formulas encoding the axiomatization of FPA. Σ_{RIA}-formula φ_x describes the domain of the variable ($x \in \mathbb{F}_{4,4}$); due to the idempotency of the rounding operator ◦, we can put as $\varphi_x :\Leftrightarrow [\![x = \circ(\mathrm{RNE}, x)]\!]$, which is translated in RIA as $z_x = z_{\circ(x)} \wedge \varphi_{\circ(x)}$. Likewise, φ_{mul} encodes the axiom for the FPA operator mul, which can be an equation $z_{\mathrm{mul}} = [\![\mathrm{mul}]\!](0, z_x, \frac{15}{512})$ where the function $[\![\mathrm{mul}]\!]$ is defined in Eq. (2) in Fig. 1 (the cases can be combined into an ite term).

Instead of defining mul as above with EagerLearn, φ_{mul} can be gradually refined with LazyLearn. Initially, we can start solving with $\varphi_{\mathrm{mul}}^1 \Leftrightarrow \top$. The SMT solver enumerates models e.g. $\mathcal{M}^1 = \{z_x \mapsto 0, x_{\circ(x)} \mapsto 0, z_{\mathrm{mul}} \mapsto \frac{9}{2}\}$.

Then, we can verify whether the literals in φ are consistent using REWRITE of our extended solver that simplifies Σ_{FPA}-terms assuming a model. For $\mathcal{M}^1$, the term $[\![\text{mul}]\!](0, z_x, \frac{15}{512})$ simplifies to 0, so it is inconsistent with the value of z_{mul}.

To search for another model, assuming the inconsistency in $\mathcal{M}^1$ and the fact isFinite(x), φ_{mul} can be refined (by instantiating the first case of Eq. (2)) as

$$\varphi_{\text{mul}}^2 :\Leftrightarrow \bigl(\text{isFinite}(z_x) \wedge \text{isFinite}(\tfrac{15}{512} z_x) \wedge \cdots \bigr) \Rightarrow z_{\text{mul}} = z_{\circ(\times)},$$

where isFinite abbreviates the range constraint and $z_{\circ(\times)}$ is a variable for $\circ(\text{RNE}, x \times \text{fp}(\frac{15}{512}))$. Accordingly, the solver obtains a model

$$\mathcal{M}^2 = \{z_x \mapsto \tfrac{1}{64}, z_{\circ(x)} \mapsto \tfrac{1}{64}, z_{\text{mul}} \mapsto \tfrac{9}{2}, z_{\circ(\times)} \mapsto \tfrac{9}{2}\}.$$

However, there are again inconsistencies between the assignment for $z_{\text{mul}}/z_{\circ(\times)}$ and the simplified value of the original Σ_{FPA}-term. To make the values for z_x and $z_{\circ(\times)}$ consistent, we need lemmas on the rounding operator. Thus, $\varphi_{\circ(\times)}$ can be refined (by instantiating Eq. (5) and Eq. (7) in Fig. 2) as

$$\varphi_{\circ(\times)}^3 :\Leftrightarrow \bigl(\text{isNormal}(\tfrac{15}{512} z_x) \Rightarrow |z_{\circ(\times)} - \tfrac{15}{512} z_x| \leq \tfrac{1}{16}|\tfrac{15}{512} z_x|\bigr) \wedge$$
$$\bigl(z_{\circ(x)} > 144 \Rightarrow z_{\circ(x)} \geq 160\bigr),$$

where the first part describes the relative error bound of the rounding operation and the second part prohibits the assignment of non-FP numbers in (144, 160). Finally, the refined formulas yield the model

$$\mathcal{M}^3 = \{z_x \mapsto 160, z_{\circ(x)} \mapsto 160, z_{\text{mul}} \mapsto \tfrac{9}{2}, z_{\circ(\times)} \mapsto \tfrac{9}{2}\},$$

and the satisfiability is verified by the REWRITE routine that evaluates the Σ_{FPA}-terms in φ for the assigned values.

6 Experiment

We have conducted experiments to answer the following research questions. **RQ1**: Whether the method and the implementation work correctly? **RQ2**: Is our solver more efficient and scalable than existing solvers? **RQ3**: How effective is the model-based refinement with LAZYLEARN compared to the lemma instantiation with EAGERLEARN?

Experiments were run on a 2.2 GHz Intel Xeon E5-2650v4 with a memory limit of 4 GB. The timeout for each run was set to 1 min.

Benchmark Instances. We prepared the following three FPA benchmarks. **(1)** To answer RQ1, we experimented with the Wintersteiger benchmark[2] that consists

---
[2] https://smt-lib.org/benchmarks.shtml.

Table 1. Results for the Wintersteiger-UC and Griggio benchmarks.

	Wintersteiger-UC		Griggio	
Solver	# solved	avg time	# solved	avg time
CVC5-RB	$7,909+9,700(87.7\%)$	0.02 s	$16+40(26.2\%)$	1.9 s
C5-RB-EL	$7,555+9,206(83.5\%)$	**0.01** s	$11+33(20.6\%)$	**1.34** s
CVC5	$\mathbf{10,036+10,038(100\%)}$	0.29 s	$56+80(63.6\%)$	12.2 s
BITWUZLA	$\mathbf{10,036+10,038(100\%)}$	0.11 s	$\mathbf{56+83(65.0\%)}$	7.7 s
Z3	$9,836+\mathbf{10,038}(99.0\%)$	1.8 s	$19+35(25.2\%)$	13.7 s
Total	10,036+10,038		71+106 (37 unsolved)	

of simple instances to exercise the basic operations. We only used instances on four arithmetic operators and operators gt, geq and eq. Since original instances can be solved by simply rewriting them, we modified them so that the solver would use the axioms. For example, an UNSAT instance of the form $\mathrm{add}(c_m, c_x, c_y) = c_r$ was modified into $\mathrm{geq}(\mathrm{add}(c_m, x, c_y), c_r) \wedge \mathrm{lt}(x, c_x)$. **(2)** We used the Griggio [15][3] benchmark designed to stress bit-blasting solvers. **(3)** We prepared encoded formulas for bounded model checking (BMC) of simple numerical systems, intended to evaluate the scalability; three systems are taken from [18] and slightly modified. We encoded a safety property of each system with either of 64bit, 128bit or 256bit FP numbers and every operation was with mode RNE; the formulas involve some uncertain variables representing input signals or initial values; the properties are valid, so the expected results are UNSAT.

Solvers. We experimented on our real-blasting solver (named CVC5-RB) together with a version that instantiates most of the axioms eagerly (named C5-RB-EL). We made a comparison with the bit-blasting solvers built-in to CVC5 (v. 1.2.0; with the `--fp-exp` option to handle non-standard precisions), BITWUZLA (v. 0.5.0) [23] and Z3 (v. 4.13.0) [14].

6.1 Experimental Results

Table 1 shows the results for the two benchmarks, Wintersteiger-UC (under constrained) and Griggio. Each section consists of the numbers $m+n$ of solved UNSAT and SAT instances and the average time taken for the conclusive runs (those resulted in UNSAT and SAT). None of the results were unsound. Uncounted results were either out of time or inconclusive.

Table 2 shows the results of BMC. Each column corresponds to a system and a precision of FP numbers (32, 64, 128 or 256 bits) used during encoding. We verified that the executions of lengths 1–k do not satisfy the specified property, with a time limit of 1min for each. Each number k in the table represents the maximum length of executions that could be verified.

---

[3] https://smt-lib.org/benchmarks.shtml.

Table 2. Results for the BMC instances.

Solver	Integrator				Rotation				Filter			
	$k(32)$	$k(64)$	$k(128)$	$k(256)$	$k(32)$	$k(64)$	$k(128)$	$k(256)$	$k(32)$	$k(64)$	$k(128)$	$k(256)$
CVC5-RB	12	12	**12**	6	1	1	1	1	4	3	3	2
C5-RB-EL	2	2	2	2	**6**	**5**	**5**	**2**	1	1	1	1
CVC5	22	10	7	4	2	1	1	1	8	4	3	**3**
BITWUZLA	**32**	**13**	10	**7**	2	2	1	1	**10**	**5**	**4**	**3**
Z3	15	7	4	3	2	2	1	1	6	3	2	2

6.2 Discussion

Answer to RQ1: Since there were no unsound results, we consider that our solver works correctly, at least in reasoning about the basic operations.

Answer to RQ2 (1): Although the number of instances solved was less than CVC5 and BITWUZLA for Wintersteiger-UC and Griggio, the results of BMC using our solver was comparable to those using other solvers. In the BMC of the rotation system, our solver C5-RB-EL performed well. Because the encoding formula for this system contains many nonlinear terms, we believe that the lemmas about monomials were effective. The eager setting of C5-RB-EL sometimes results in faster inferences, and this may have been the case here. The result of many BMC runs with our solver was less efficient than other solvers probably due to a large number of axiom instantiations for the nonlinear terms. With other solvers, the number of steps decreased according to the precision, but it was often decreased only slightly with our solver.

Answer to RQ2 (2): The proposed solver seemed to be good at problems with different characteristics than bit-blasting solvers as observed by the BMC experiment; namely, problems that address error bounds after a number of arithmetic operations and that used large-precision FP numbers. In Griggio, our solver solved 3/4/24 instances involving many arithmetic operations that were not solved by CVC5/BITWUZLA/Z3.

Answer to RQ2 (3): Our solver was typically faster on the same instances of Wintersteiger-UC and Griggio that were also solved by other solvers (and not just in aggregate). However, it tended not to terminate when it took time. We suppose that the reason is the explosion in the number of learning clauses as is common with other axiomatization methods; future implementation of a restarting strategy or an abstraction process (e.g. with intervals [11,18]) might improve this.

Answer to RQ2 (4): Our solver resulted in unknown or timeout for a number of instances of Wintersteiger-UC. The process to solve them resulted in enumerating many assignments of non-FP numbers satisfying the rounding error constraint. This phenomena can be avoided by using more axioms on arithmetic operators; however, we had not implemented some useful axioms e.g. for backward operations because they would complicate the process with additional case analyses (currently, we only have such lemmas for the operations with the "to nearest" rounding modes).

Answer to RQ3: For Wintersteiger-UC, Griggio, and for two of the BMC experiments, model-based learning of axioms with LAZYLEARN was more efficient, but there were cases where EAGERLEARN was better. As the number of learning clauses increases (e.g. in the BMC involving more nonlinear terms), LAZYLEARN is expected to become less efficient. We consider there is a tradeoff between the speed of adding axioms and the size of the solving context.

7 Conclusion

A real-blasting FPA solver implemented as an extended theory solver in CVC5 was presented. In the experiments, our solver was able to perform better on some practical BMC examples. One issue for the future is to fully support the FPA vocabulary (e.g. accurate handling of sqrt [11]). Because of the complexity of its implementation, for further development, a mechanism to support the specification of axiom templates would be helpful; also, formal verification and certificate generation will be a future work.

References

1. Ayad, A., Marché, C.: Multi-prover verification of floating-point programs. In: Giesl, J., Hähnle, R. (eds.) IJCAR 2010. LNCS (LNAI), vol. 6173, pp. 127–141. Springer, Heidelberg (2010). https://doi.org/10.1007/978-3-642-14203-1_11
2. Barbosa, H., et al.: cvc5: a versatile and industrial-strength SMT solver. In: TACAS 2022. LNCS, vol. 13243, pp. 415–442. Springer, Cham (2022). https://doi.org/10.1007/978-3-030-99524-9_24
3. Barrett, C., Fontaine, P., Tinelli, C.: The SMT-LIB Standard (Version 2.6) (2021)
4. Barrett, C., Tinelli, C.: Satisfiability modulo theories. In: Handbook of Model Checking, pp. 305–343. Springer, Cham (2018). https://doi.org/10.1007/978-3-319-10575-8_11
5. Boldo, S., Melquiond, G.: Flocq: a unified library for proving floating-point algorithms in Coq. In: ARITH, pp. 243–252 (2011). https://doi.org/10.1109/ARITH.2011.40
6. Brain, M., Schanda, F., Sun, Y.: Building better bit-blasting for floating-point problems. In: Vojnar, T., Zhang, L. (eds.) TACAS 2019. LNCS, vol. 11427, pp. 79–98. Springer, Cham (2019). https://doi.org/10.1007/978-3-030-17462-0_5
7. Brain, M., Tinelli, C., Rüemmer, P., Wahl, T.: An automatable formal semantics for IEEE-754 floating-point arithmetic. In: ARITH, pp. 160–167 (2015). https://doi.org/10.1109/ARITH.2015.26

8. Brillout, A., Kroening, D., Wahl, T.: Mixed abstractions for floating-point arithmetic. In: FMCAD, pp. 69–76. IEEE (2009)
9. Clarke, E., Kroening, D., Lerda, F.: A tool for checking ANSI-C programs. In: Jensen, K., Podelski, A. (eds.) TACAS 2004. LNCS, vol. 2988, pp. 168–176. Springer, Heidelberg (2004). https://doi.org/10.1007/978-3-540-24730-2_15
10. Conchon, S., Coquereau, A., Iguernlala, M., Mebsout, A.: Alt-Ergo 2.2. In: SMT Workshop, pp. 1–11 (2018)
11. Conchon, S., Iguernlala, M., Ji, K., Melquiond, G., Fumex, C.: A three-tier strategy for reasoning about floating-point numbers in SMT. In: Majumdar, R., Kunčak, V. (eds.) CAV 2017. LNCS, vol. 10427, pp. 419–435. Springer, Cham (2017). https://doi.org/10.1007/978-3-319-63390-9_22
12. Conchon, S., Melquiond, G., Roux, C.: Built-in treatment of an axiomatic floating-point theory for SMT solvers. In: SMT Workshop, pp. 12–21 (2012)
13. Daumas, M., Melquiond, G.: Certification of bounds on expressions involving rounded operators. ACM Trans. Math. Softw. **37**(1), 1–20 (2010). https://doi.org/10.1145/1644001.1644003
14. de Moura, L., Bjørner, N.: Z3: an efficient SMT solver. In: Ramakrishnan, C.R., Rehof, J. (eds.) TACAS 2008. LNCS, vol. 4963, pp. 337–340. Springer, Heidelberg (2008). https://doi.org/10.1007/978-3-540-78800-3_24
15. Haller, L., Griggio, A., Brain, M., Kroening, D.: Deciding floating-point logic with systematic abstraction. In: FMCAD, pp. 131–140 (2012)
16. Harrison, J.: Floating-point verification using theorem proving. In: Bernardo, M., Cimatti, A. (eds.) SFM 2006. LNCS, vol. 3965, pp. 211–242. Springer, Heidelberg (2006). https://doi.org/10.1007/11757283_8
17. IEEE: 754-2008 – IEEE Standard for Floating-Point Arithmetic (2008)
18. Ishii, D., Tomita, T., Aoki, T.: Approximate translation from floating-point to real-interval arithmetic. In: NFM, LNCS, vol. 13260, pp. 733–751. Springer, Heidelberg (2022). https://doi.org/10.1007/978-3-031-06773-0_39
19. Kroening, D., Strichman, O.: Decision Procedures, 2nd edn. Springer, Heidelberg (2016). https://doi.org/10.1007/978-3-662-50497-0
20. Leeser, M., Mukherjee, S., Ramachandran, J., Wahl, T.: Make it real: effective floating-point reasoning via exact arithmetic. In: DATE, pp. 1–4 (2014). https://doi.org/10.7873/DATE2014.130
21. Marre, B., Bobot, F., Chihani, Z.: Real behavior of floating point numbers. In: SMT Workshop, pp. 1–12 (2017)
22. Muller, J.M., et al.: Handbook of Floating-Point Arithmetic, 2 edn. Birkhäuser (2018)
23. Niemetz, A., Preiner, M.: Bitwuzla. In: CAV. LNCS, vol. 13965, pp. 3–17. Springer, Heidelberg (2023). https://doi.org/10.1007/978-3-031-37703-7_1
24. Ramachandran, J., Wahl, T.: Integrating proxy theories and numeric model lifting for floating-point arithmetic. In: FMCAD, pp. 153–160 (2016). https://doi.org/10.1109/FMCAD.2016.7886674
25. Reynolds, A., Tinelli, C., Jovanović, D., Barrett, C.: Designing theory solvers with extensions. In: Dixon, C., Finger, M. (eds.) FroCoS 2017. LNCS (LNAI), vol. 10483, pp. 22–40. Springer, Cham (2017). https://doi.org/10.1007/978-3-319-66167-4_2
26. Salvia, R., Titolo, L., Feliú, M.A., Moscato, M.M., Muñoz, C.A., Rakamarić, Z.: A mixed real and floating-point solver. In: Badger, J.M., Rozier, K.Y. (eds.) NFM 2019. LNCS, vol. 11460, pp. 363–370. Springer, Cham (2019). https://doi.org/10.1007/978-3-030-20652-9_25

27. Zeljić, A., Backeman, P., Wintersteiger, C.M., Rümmer, P.: Exploring approximations for floating-point arithmetic using UppSAT. In: Galmiche, D., Schulz, S., Sebastiani, R. (eds.) IJCAR 2018. LNCS (LNAI), vol. 10900, pp. 246–262. Springer, Cham (2018). https://doi.org/10.1007/978-3-319-94205-6_17
28. Zitoun, H., Michel, C., Rueher, M., Michel, L.: Search strategies for floating point constraint systems. In: Beck, J.C. (ed.) CP 2017. LNCS, vol. 10416, pp. 707–722. Springer, Cham (2017). https://doi.org/10.1007/978-3-319-66158-2_45

Two-Way Collaboration Between Flow and Proof in SPARK

Claire Dross, Joffrey Huguet, and Johannes Kanig[✉]

AdaCore, Paris, France
kanig@adacore.com

Abstract. There are various kinds of static (and dynamic) verification frameworks and tools, all with weaknesses and strengths. Combining them manually or automatically is common by making different tools interoperate or by integrating various analyses in the same tool. In the SPARK verification tool for the Ada language, two static analyses work together: deductive verification and data+information flow analysis. Deductive verification offers strong guarantees at the cost of manual effort: users need to write contracts, and the guarantees are subject to a number of assumptions. Data+information flow analysis is used to alleviate some of these difficulties, thus making the tool easier to use. This article presents how the verification is shared between these analyses so that each one is used at its full potential and how they collaborate to improve the efficiency and precision of the SPARK tool while retaining soundness. If most of these collaborations are automated, a feature of SPARK lets the user choose which analysis should be used to verify initialization in a fine-grained manner, depending on their preferred trade-off.

Keywords: Combination · Static analysis · Deductive verification

1 Introduction

Software verification is pervasive nowadays both in critical systems where safety is paramount (avionics, trains...) and in everyday-life programs, with the growth in the number of connected devices and the risks of cyber attacks. Depending on the domain, this verification can involve various techniques and tools, more or less automated, giving different levels of guarantees. It can range from mostly manual testing to full-blown formal verification, through automated fuzzing and various classes of quality checkers and bug finders. These verification frameworks and tools all come with their weaknesses and strengths. To get a better trade-off, it is common to combine them manually or automatically by making different tools interoperate or by integrating various analyses in the same tool.

The SPARK tool [9,17] is an open-source[1] industrial-strength deductive verification framework, used in various industries to implement safety- or security-

---
[1] The source code of SPARK is freely available under the GPL licence at https://github.com/AdaCore/spark2014/.

critical software[2]. It allows users to formally verify their programs at the source level by generating a set of logical formulas whose validity implies the validity of the program. Deductive verification offers strong guarantees, ranging from complete freedom of some kinds of bugs or exploits such as buffer overflows or division by zero to full functional correctness with respect to requirements written as contracts (typically preconditions and postconditions). These guarantees however come at a cost. They are subject to a number of assumptions, more or less difficult to discharge, like the correctness of the compiler (as the verification is done at the level of the source code). Also, running the tool requires manual intervention, in particular to supply contracts for all functions in the program.

In order to alleviate some of these difficulties and to make the tool easier to use, SPARK combines a data+information flow analysis with deductive verification. This article presents how these analyses collaborate to improve the efficiency and precision of the SPARK tool while retaining soundness. Section 2 presents how the work is shared between the two analyses so that each one works at its full potential. In Sects. 3 and 4, we explain more complex parts of the design where the two analyses collaborate more closely: how flow analysis computes additional information for deductive verification is presented in Sect. 3 while Sect. 4 focuses on checks which are automatically handed over from flow analysis to deductive verification to increase precision. Finally, Sects. 5 and 6 dwell on a particular feature of SPARK where the user can choose on a case-by-case basis which analysis they want to use to enforce proper initialization of data before it is read.

2 Flow and Proof - Divide and Conquer

SPARK is a subset of the Ada programming language with the goal of making formal verification accessible to non-experts. As mentioned in the introduction, SPARK provides strong guarantees while requiring some effort from the developer, mainly providing annotations in the form of contracts. However, several techniques are applied to reduce the cost while keeping the same guarantees.

The SPARK subset is relatively large and contains many advanced features, but some features are excluded (such as dynamic creation of parallel threads or "tasks") or their use is restricted by additional limitations and checks. Some of these *simplifying assumptions* are discussed in this section in more detail.

SPARK subprograms[3] can be annotated with additional contracts, for the most part preconditions and postconditions, as well as a summary of effects on global variables. In Ada, effects on parameters are part of the signature of the subprogram, and do not require separate annotations. This necessity to annotate subprograms is shared with other deductive verification tools.

---

[2] The SPARK Pro product page can be found on the AdaCore website: https://www.adacore.com/sparkpro.
[3] Ada differentiates functions which return a result from procedures which work by side effect. Functions and procedures are called subprograms.

```
type Array_Type is array (1 .. 10) of Integer;

procedure Init_Arr (X : out Array_Type) is
begin
    for I in X'Range loop
        X (I) := Compute_Something;
    end loop;
end Init_Arr;

function Consume_Arr (X : Array_Type) is
begin
    Do_Something (X (5));
end Consume_Arr;
```

Fig. 1. The SPARK initialization policy requires accessed data to be fully initialized at subprogram boundaries. The procedure Init_Arr needs to fully initialize the array output parameter before returning. The function Consume_Arr can assume that all array cells are initialized, including the fifth cell.

One major difference with respect to most deductive verification tools is that many annotations follow the semantics of Ada and are executable. Preconditions and postconditions act as additional assertions which can be enabled or disabled at compile time. If assertions are disabled, contracts are stripped by the compiler, otherwise they are executed alongside the regular code. Executable contracts can be easier to write for programmers who are not experts in formal verification, as the semantics for the program and annotations are the same, and there is no need to learn a different annotation language. Also, most proof obligations[4] directly correspond to runtime checks, such as the check that a value used as an array index is indeed in the range of the bounds of the array. This further reduces the number of concepts for users to learn. As a consequence, SPARK has no concept of *logical* functions - any function can be called from anywhere, including from preconditions and postconditions. This choice also comes at a cost: the translation of SPARK programs to a logic language for proof requires additional considerations to ensure soundness. This issue will be discussed in Sect. 3.

The simplifying assumptions introduced by SPARK have the simultaneous goal of reducing the annotation burden and making the tool more efficient. One such assumption is that variables must be initialized entirely at subprogram boundaries, including arrays and records. For example, when analyzing a subprogram that has an array as an input parameter, we can assume that all cells of the array are initialized, and the values of the cells respect the properties of their types. Conversely, when analyzing a subprogram that has an array as a return type or output parameter, we must check that the subprogram initializes the array entirely before returning (See Fig. 1). This assumption eliminates the need

---

[4] Proof obligations are logical formulas generated by deductive verification tools. Their satisfiability implies the correctness of the program.

```ada
procedure Increment_Two (X, Y : in out Integer)
   with Pre => X < Integer'Last and Y < Integer'Last
is
begin
   X := X + 1;
   Y := Y + 1;
end Increment_Two:
```

Fig. 2. If X and Y could alias, the second increment operation could overflow. SPARK checks that Increment_Two is never called with aliased parameters, so the overflow check is proved. No separation condition is needed in the precondition, it is implicit.

for "validity" predicates and reduces the proof search space (as invalid values do not need to be considered).

Another assumption is the absence of aliasing. If a procedure has two parameters, at least one of which is writable, it can assume that they are distinct, see Fig. 2. The same is true for a procedure which has a parameter and also accesses a global variable. This assumption is enforced for pointers through a borrow checker inspired by Rust [11]. It eliminates the need for "separation" predicates and simplifies the memory model for the translation to logical formulas, which ultimately results in proof obligations that are simpler and easier to prove.

Ada defines three modes for parameters of subprograms: in parameters (the default) can only be read, whereas out and in out parameters can be written too. The initial value of an out parameter like the parameter X of Set_X_To_Zero in Fig. 3 should never be used. As deductive verification requires knowing the global effects of all subprograms, SPARK introduces a special Global annotation to allow users to specify global effects of their subprograms. This information is used by deductive verification to bound the effect of calls to these subprograms. Global contracts must be complete, so the contracts also implicitly state that there are no other global effects.

```ada
procedure Set_X_To_Zero (X : out Integer);

Total : Integer := 0;

procedure Add_To_Total (Incr : in Integer) with
   Global => (In_Out => Total);
```

Fig. 3. Example of Ada parameter modes and SPARK Global contracts. The annotation on Add_To_Total specify that it reads and writes the Total variable.

In general, SPARK is only concerned with partial correctness: procedures might not terminate, their postcondition is only enforced when they do. However, it is possible for users to specify the termination behavior of a procedure via an annotation if they want, see Fig. 4. SPARK checks the termination annotations for correctness.

```
procedure P with Always_Terminates ⇒ True;
-- procedure P must terminate

procedure Q with Always_Terminates ⇒ False;
-- procedure Q is allowed to not terminate

procedure R with Always_Terminates ⇒ Condition;
-- procedure R is required to terminate when Condition is True
```

Fig. 4. Examples of various `Always_Terminates` annotations.

These points, the simplifying assumptions as well as the need for annotations, gave rise to the initial (simple) structure of the SPARK tool. It consists of two different analyses:

Flow - A flow analysis algorithm checks `Always_Terminates` and `Global` annotations, as well as initialization and aliasing rules. The analysis is very fast and deterministic, but it is imprecise as it does not track values of variables.

Proof - Using the simplifying assumptions mentioned above, proof carries out deductive verification of the SPARK program in a modular way. This analysis is precise, but it necessarily requires manual annotations, is relatively slow, and incomplete as automatic provers may not be able to discharge all proof obligations.

Users can run the fast Flow analysis only, or both Flow and Proof, but not Proof alone, as Proof generally relies on the results of Flow. The remainder of the article presents improvements upon this initial structure. Section 3 explains how Flow, which is interprocedural, generates additional data for proof. Sections 4-6 describe the other direction: some value dependent parts of verification activities done by Flow are left as assumptions that are discharged later by Proof. Figure 5 gives an overview of this improved structure.

3 Use Flow to Compute Information for Proof

As Flow is fast and can be run on the whole codebase, it can be used to compute information needed by Proof for the precision and the soundness of its analysis. This mechanism can be used both to accommodate missing contracts and to generate complex soundness-related information.

3.1 Compute Global Contracts and Termination Annotations

`Global` contracts are necessary for Proof to produce a sound analysis. As presented in Sect. 2, they provide the tool and the user information about the global effects of the subprogram. Proof can then use these contracts to precisely model the effect of statements in the code. For example, in Fig. 6, Proof is able to prove that the value of V has not been modified by the call to `Add_V_To_G`, using its

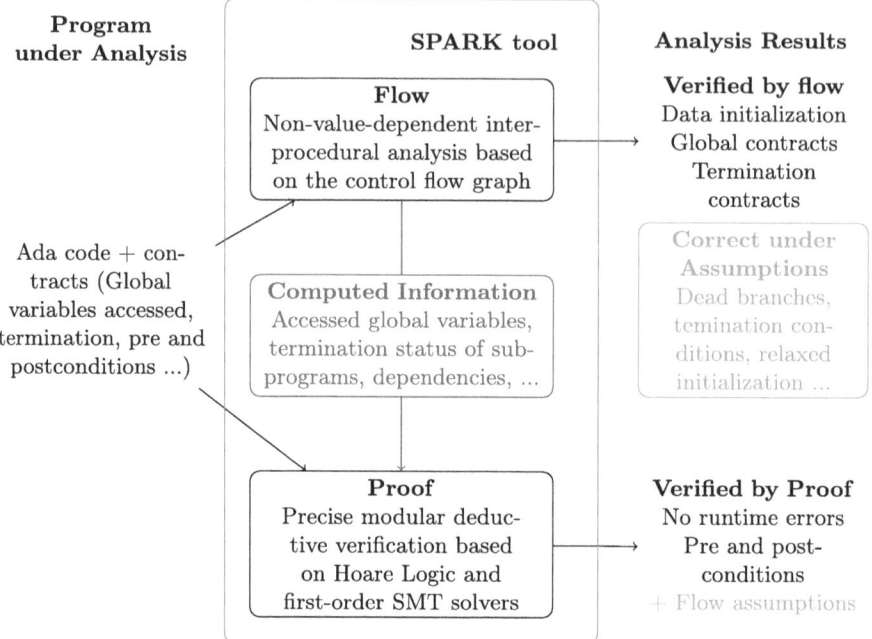

Fig. 5. The analysis of a SPARK program is done in two phases, Flow runs first, and then Proof, each verification activity being done by the analysis which is the best suited for it. Flow, which is interprocedural, generates additional data for proof (in red). On the other direction, some value dependent parts of verification activities done by Flow are left as assumptions that are discharged later by Proof (in green). (Color figure online)

Global contract. Providing correct global contracts is also necessary for soundness. As an example, knowing that Read_G accesses the value of G is necessary for Proof to determine that Read_G = Old_Read_G might not hold after the call to Add_V_To_G because the value of G can be modified during the call.

Manually writing the global contracts and the termination annotations is considered as a pain point for users. First, they need to completely describe the effects of the subprogram. As a consequence, they grow with the call graph. Second, maintenance of global contracts is also costly. Changes in the global contracts of a subprogram propagate to the callers of said subprogram. Users should manually edit all global contracts to reflect the changes in the callees. As a result, the time required to maintain global contracts across the entire program grows with the call graph. In addition, users might not even be interested in global contracts. They might only want to prove the absence of runtime errors for their program for example. Therefore, making these contracts optional is an important step toward making the tool easier to use and decreasing the amount of required manual effort.

```
procedure Add_V_To_G with Global => (Input => V, In_Out => G);
function Read_G return Integer with Global => (Input => G);

V := 1;
G := 0;
Old_Read_G := Read_G;
Add_V_To_G;
pragma Assert (V = 1);
pragma Assert (Read_G = Old_Read_G);
```

Fig. 6. After the call to `Add_V_To_G`, Proof is able to prove that the value of `V` is still equal to 1, but it is unable to prove that `Read_G = Old_Read_G`, as `G` has been modified.

Since Proof needs the contracts for soundness and precision, it is necessary for Flow to generate them when they are not provided. This is done in two phases. In the first phase, all information about the direct references to global variables and calls to other subprograms is stored for each subprogram. In the second phase, all this information is combined to generate global contracts for all subprograms containing all the variables accessed by the subprogram along with the mode in which they are accessed. In general, the generated global contracts, and in particular the modes, can be an over-approximation of the actual contracts (for example, the mode of a global item can be computed to be `In_Out` instead of `Output`).

The situation is similar for termination annotations. As explained in Sect. 2, termination of procedures is not required for Proof as it is only concerned with the partial correctness of the program: the postcondition is proven in the case where the subprogram terminates. However, it is not the same for functions, which can be called in contracts. In other deductive verification tools, there is often a distinction between *program* and *logical* functions, also called *pure* functions. The former have contracts and a body made of statements and are used only in the program. When they are called, the tool proves the precondition, abstracts the body of said function, and assumes the postcondition to prove the rest of the program. Logical functions, called in contracts, are mathematical functions in the underlying logic. In particular, they are complete[5] and do not have contracts. Due to the executability of contracts in Ada, SPARK does not make such a distinction between functions called in programs and in contracts. In particular, functions that are used in contracts can have contracts themselves. To take their contract into account when these functions are called inside other contracts, it is turned into an axiom: it states that for all valid inputs of the function, the precondition implies the postcondition. However, such an axiom might not be correct if there are some valid inputs of the function for which there cannot be any valid result. In this case, the axiom is said to not be realizable and it can compromise soundness. To make sure that the axiom is realizable,

---

[5] A function is said to be *complete* if it returns a result for every input.

it is in general enough to verify the function itself as it should return a valid result for all valid inputs. However, if a function does not terminate, its axiom might not be realizable even if the function is proved. Figure 7 gives an example of this. The function Unrealizable does not terminate because it calls another non-terminating subprogram. Therefore, the postcondition of Unrealizable is proved, as the end of the function is never reached. Adding the axiom of Unrealizable in the context given to provers could introduce unsoundness. Hence, SPARK needs to prove termination of functions so that it produces a sound analysis.

```
function Unrealizable (X : Natural) return Boolean with
   Post ⇒ False is
begin
   Never_Terminates;   --  call does not terminate
   return True;
end Unrealizable;
```

Fig. 7. As the end of the function is never reached, a postcondition of False is proved in partial correctness. The axiom of the corresponding logic function states that for all valid parameters, True implies False, which makes the axiom unsound.

As with global contracts, using the same two-phase mechanism, Flow analyzes loops in subprograms and an over-approximation of the call graph to prove termination of functions. In particular, it also computes termination of all procedures called by functions, so that the user is not required to annotate such procedures. In Fig. 7, it identifies an issue on the call to Never_Terminates in Unrealizable, stating that it might not terminate.

3.2 Verify Assumptions

As explained in Sect. 3.1, Proof generates axioms for postconditions of functions called in contracts. These axioms are verified when the body of the corresponding function is analyzed. This method might be unsound when there are circular dependencies between such functions, as an axiom can indirectly be used to prove itself. A simple example of circular dependency is recursion. In Ada, subprograms are not explicitly marked as recursive. Recursion can occur across different compilation units, which makes it difficult to track and maintain for the users. Consider the recursive variant of Unrealizable defined in Fig. 8. When verifying it, Proof should make sure to not include its own axiom or its unsound postcondition could be proved easily, and so, even if Unrealizable terminates. In general, SPARK removes the axiom of all mutually recursive functions when verifying a subprogram. The detection of recursion is done with the help of the call graph generated by Flow to compute termination annotations discussed in Sect. 3.1.

Note that Proof keeps the postcondition on recursive calls occurring in the program (not in contracts). In Fig. 8, SPARK can prove the postcondition of

Add. Indeed, the postcondition is assumed on the parameters of the recursive call, after the call occurs in the program. This handling, which is standard practice in deductive verification, is compliant with partial correctness. Indeed, if the subprogram terminates, then the exact same recursive call cannot appear twice on the same execution path and the postcondition can be verified inductively assuming that it holds for all earlier calls.

```
function Unrealizable (X : Natural) return Boolean with
  Post =>
    False       -- postcondition might fail, cannot prove False
    and then (if X /= 0 then Unrealizable (X - 1));

function Add (N, M : Natural) return Natural with
  Post => Add'Result = N + M
is
begin
   if M = 0 then
      return N;
   else
      -- Proof introduces the postcondition of Add in the
      -- context, but only with the actual parameters.
      return Add (N, M - 1) + 1;
   end if;
end Add;
```

Fig. 8. When proving its postcondition, Proof knows that `Unrealizable` is recursive and does not use its axiom to prove itself. When proving the postcondition of `Add`, Proof does not have its postcondition available as a global axiom. However, it does assume the postcondition on the parameters of the recursive call after the call.

It can happen for the proof of a non-recursive subprogram to depend on itself. As an example, in Fig. 9, the type `My_Natural` is annotated with a strong type invariant, called a *subtype predicate* in Ada, which calls the function `Is_Natural`. Declaring a constant of this type in `Is_Natural` has the effect of creating a dependency: Proof sees a call to `Is_Natural` when considering the properties of `Zero`. Such dependencies should be detected to avoid introducing soundness issues as with recursive subprograms.

Finding, tracking and maintaining such dependencies is cumbersome and difficult for the users, because the sources of proof dependencies are numerous. Besides predicates, several annotations added for SPARK can create circular dependencies in the program. To ensure soundness, Flow detects these dependencies automatically and Proof removes the axioms of interdependent functions called in contracts.

```ada
function Is_Natural (X : Integer) return Boolean with
   Post ⇒ False;

subtype My_Natural is Integer with
   Predicate ⇒ Is_Natural (My_Natural);

function Is_Natural (X : Integer) return Boolean is
   Zero : constant My_Natural := 0;
   --  Implicit use of Is_Natural
begin
   return X ≥ Zero;
end Is_Natural;
```

Fig. 9. When Proof processes the declaration of Zero, it would normally pull the contract of Is_Natural into the context as an axiom, which would make it possible to prove its False postcondition. Flow detects this circular dependency and the axiom is not introduced.

4 Use Proof to Gain Precision in Flow

The fact that Flow does not track values of variables and its resulting imprecision can be frustrating for users. In this section, we discuss two mechanisms where some checks originally performed by Flow are automatically handed over to Proof: Flow checks the code assuming some properties, then Proof verifies these assumptions during its run.

4.1 Termination Proofs

Section 2 introduced termination annotations, while Sect. 3 explained how Flow computes a termination status for subprograms. In this section, we show how Proof can help when Flow cannot establish termination.

When a subprogram only uses constructs that are syntactically known to terminate, like a simple loop over a range, Flow can check its termination status on its own. However, the termination of several language constructs cannot be determined without the use of value-dependent reasoning. In particular, the verification of the termination of loops and recursive calls is usually handled in deductive verification tools by requiring the user to provide a *variant*, usually a non-negative integer value which decreases at each iteration or recursive call. Checking that a provided expression is indeed a variant is out of reach of Flow, so it needs to be done by Proof. This is implemented using an *automated handover* to Proof: Flow assumes that the construct is correct and Proof emits checks for it afterward.

As an example, consider a *while* loop statement. If it has no annotations, Flow considers it as potentially non-terminating, possibly resulting in messages on user-provided termination annotations and functions called in contracts. If it has a loop variant, Flow assumes that it terminates. As a result, the subprogram containing this loop can be determined to be terminating by Flow. The check

```ada
procedure Simple_Loop (X : Array_Type) with
   Always_Terminates => True is
begin
   for I in X'Range loop  -- termination proved (flow)
      X (I) := X (I) / 2;
   end loop;
end Simple_Loop;

procedure Variant_Loop (C : Positive) with
   Always_Terminates => True is
   Z : Integer := 0;
begin
   while Z < C loop  -- termination assumed (flow)
      pragma Loop_Variant (Increases => Z);
      -- loop variant proved (proof)
      Z := Z + 1;
   end loop;
   for I in Integer range 0 .. C loop
   -- termination proved (flow)
      Z := Z - 1;
   end loop;
end Variant_Loop;

procedure Variant_Loop_Fail (C : Positive)
   with Always_Terminates => True is
   Z : Integer := 0;
begin
   while Z < C loop  -- termination assumed (flow)
      pragma Loop_Variant (Increases => C);
      -- loop variant might fail (proof)
      Z := Z + 1;
   end loop;
   while Z > 0 loop  -- loop might not terminate (flow)
      Z := Z - 1;
   end loop;
end Variant_Loop_Fail;
```

Fig. 10. Various examples of termination proofs, discharged by Flow or Proof.

for the variant is handed over to Proof, which emits a message in case of failure. Variants for (mutually) recursive subprograms and calls to subprograms with a non-static termination condition (see Sect. 2) are handled similarly: Flow trusts that the construct terminates and Proof verifies it when it is run and emits a message in case of failure.

Consider Fig. 10. The procedure Simple_Loop can easily be checked as terminating by Flow, as the loop iterates over a fixed range. For the procedure Variant_Loop, a collaboration of Flow and Proof is required. Flow can check that the second loop terminates (a fixed range again) and trusts that the first loop terminates, because it has a variant. It computes an overall status of "terminating" for this procedure, pending the proof of the loop variant. The loop variant is easily proved by Proof. Note that in SPARK, where machine Integer types are bounded, a variant can strictly increase at each loop iteration like Z here. In `Loop_Variant_Fail`, the loop variant is wrong. While Flow still assumes termination of the loop, Proof emits a check on the loop variant.

4.2 Dead Branches

Consider the subprogram in Fig. 11. It has a precondition stating that no errors have been detected. However, it still contains defensive code which for example can contain cleanup operations. Despite being dead, such code can still be meaningful if not all callers of the subprogram are verified, or the programmer wants to protect against hardware failures or other abnormal situations.

```
procedure Proceed (Result : out Boolean) with
  Pre => not Has_Errors is
begin
  if Has_Errors then
      S := Get_Status;
      Cleanup (S);
      pragma Assert (False);
  else
      ...
  end if;
end Proceed;
```

Fig. 11. The subprogram `Proceed` contains defensive code that is intended for abnormal operation. The `pragma Assert (False)` indicates this intent. However, Flow should still check for flow errors in the dead branch leading up to this assertion. Outputs of the subprogram do not need to be initialized in this branch.

Ideally, Flow should not consider such dead execution branches as potentially reaching the end of a subprogram, so that the data flow happening in such branches does not contribute to the overall data flow of the subprogram. As an example, in Fig. 11, the fact that `Result` is not set if `Has_Errors` is True is fine as it should never happen on calls where the precondition of `Proceed` is respected. For this, Flow needs to recognize such branches and *cut* them to compute the overall data flow. However, it is not possible in general as deciding whether a branch is dead usually depends on the values of variables.

A solution to this issue is the notion of *error-signaling statements*. These statements are statically recognized by Flow as meaning that the branch is dead. In particular, it is the case of `pragma Assert (False)` in Fig. 11. It is recognized specifically by Flow which, after checking the code on the branch using the regular rules, does not consider this branch as flowing to the exit of the subprogram. Assignments on this branch do not contribute to the overall data flow of the subprogram, and the branch does not need to initialize outputs of the subprogram.

Aside from `pragma Assert (False)`, other error-signaling statements are recognized by the tool. In particular, raising an exception is a typical pattern for defensive code in Ada. As SPARK requires users to specify the exceptions that a subprogram is allowed to propagate, raising an exception which is neither handled locally nor declared in the contract of the subprogram can be considered as an error-signaling statement. To make sure that this handling is correct, Flow relies on Proof to make sure that the branch really cannot be reached. When it is run, Proof checks that *error-signaling statements* cannot be executed by replacing them with an obligation to prove `False`. In practice, this means that the branch is dead.

The soundness argument for the collaboration described in Sect. 4 is based on the fact that Flow and Proof only use information on statements that occur earlier in the execution of the program. It means that a proof by induction can be carried out, where the correctness of each statement is based on the correctness of previous statements.

5 Handover from Flow to Proof as a Feature - Relaxed Initialization

As explained in Sect. 2, subprograms have to abide by a default initialization policy. This policy has been designed to simplify user-written annotations by precisely describing what should be initialized on entry and exit of each subprogram - all inputs are initialized on entry and all outputs are initialized on exit. This simplified model makes it possible for Flow to check initialization efficiently in general. However, it can sometimes be insufficient.

First, Flow might not be precise enough to verify initialization, if it requires value-dependent reasoning. In particular, this happens for arrays. The initialization policy itself can also be too constraining. For example, mandating all outputs of a subprogram to be entirely initialized disallows valid use cases in which some outputs of a subprogram are only initialized when they can be used. As an example, procedure `Set_V` in Fig. 12 might fail to properly initialize its parameter `V`. In this case, it sets the boolean flag `Error` to True and the value of `V` should never be read. This is not supported by the initialization policy of SPARK which mandates the initialization of all outputs on all paths. The initialization policy also prevents initializing a structure by part, using different procedures. As the partial initialization procedures `Init_F1` and `Init_F2` in Fig. 12 preserve the parts of the structure they do not initialize, they need their parameter `X` to be an input as well as an output. As per the initialization policy of SPARK, it means that they cannot be called on an uninitialized object.

Even if, in general, it is possible to work around these restrictions by refactoring the program or introducing useless initialization, it might not be desirable in terms of efficiency and readability. It might even lead to missing verification in the tool. For example, if `Set_V` in Fig. 12 is modified to always initialize its

```
procedure Set_V (S : String; V : out R; Error : out Boolean);
-- Read V from S. V is only initialized if Error is False.

type R is record
   F1 : T1;
   F2 : T2;
end record;

procedure Init_F1 (X : in out R; V : T1);
-- Initialize X.F1, initialization status of X.F2 is preserved
procedure Init_F2 (X : in out R; V : T2);
-- Initialize X.F2, initialization status of X.F1 is preserved
```

Fig. 12. Set_V sets the boolean flag Error to False if it was not able to properly initialize its parameter V. This pattern is disallowed by the initialization policy of SPARK as it mandates all outputs to be entirely initialized. The procedures Init_F1 and Init_F2 initialize a part of the record X, leaving the rest as it was. They cannot be called on an uninitialized variable as their parameter X needs to be entirely initialized before the call, as per the initialization policy of SPARK.

parameter V using a default value, the verification tool will not be able to detect an incorrect usage of the API where V is read even if Error is True.

As an alternative, it would be possible to always use Proof to ensure proper initialization of objects on read. This would solve the problem of value dependency, as Proof tracks precisely the value of all objects. In addition, it would be possible to use preconditions and postconditions to specify precisely the initialization status of inputs and outputs and relax the initialization policy. As an example, Fig. 13 gives variants of the procedures from Fig. 12 with contracts stating precisely the expected initialization status of all parameters. To state that an object or a subcomponent is entirely initialized, it uses a specific attribute named Initialized. The postcondition of Set_V enforces that Error is always initialized after the call and that V is entirely initialized if Error evaluates to False. It is enough to verify user code if it does not read V when Error is set. The precondition of Init_F1 does not constrain the initialization status of X so it can be called on an entirely uninitialized object. Its postcondition states that the procedure has initialized X.F1 and left X.F2 as it was. Together with a similar contract for Init_F2 which expresses the opposite postcondition, it is enough to verify that calling Init_F1 and then Init_F2 initializes the object completely.

However, using Proof to verify initialization and to relax the initialization policy has drawbacks. First, it requires users to manually write additional contracts for initialization. These contracts need to mention all inputs and outputs of the subprogram as initialization is no longer specified implicitly by the initialization policy. This can increase the complexity of the contracts as can be seen in Fig. 13. In addition to contracts, other annotations might be needed to help

```
procedure Set_V (S : String; V : out R; Error : out Boolean) with
   Pre  => S'Initialized,
   Post => Error'Initialized and (Error or else V'Initialized);
-- Read V from S. V is only initialized if Error is False.

procedure Init_F1 (X : in out R; V : T1) with
   Pre  => V'Initialized,
   Post => X.F1'Initialized
      and (X.F2'Initialized = X.F2'Initialized'Old);
-- Initialize X.F1, initialization status of X.F2 is preserved
```

Fig. 13. Procedures Set_V and Init_F1 with contracts making the initialization status of their parameters before and after the call explicit. The Initialized attribute is used to state that an object is entirely initialized.

Proof. In particular, if an object is initialized by a loop, a loop invariant[6] will need to be supplied to achieve the proof. What is more, using Proof to verify initialization might make verifying other properties harder. Indeed, it increases the size of the search space by making it possible for objects to not be initialized when read.

So users can get the best trade-off depending on their needs, SPARK allows them to choose on a case-by-case basis whether they want to use Flow or Proof to check initialization. By default, the initialization policy applies and Flow is used to verify initialization. This avoids requiring additional annotations from the user and reduces the size of the contracts while retaining efficient provability. When this is not enough, it is possible for the user to explicitly ask for a relaxation of the initialization policy and a handover to Proof by using the Relaxed_Initialization annotation. In general, it is specified on the object for which we want to relax initialization. Other objects are not impacted.

For subprograms, it is possible to decide separately to relax the initialization policy for some parameters and not others. A single Relaxed_Initialization annotation is supplied on the subprogram which lists all the relaxed parameters, and possibly the function result if it can be partially initialized. It simplifies the necessary contracts by retaining the initialization policy whenever possible. In Fig. 14, the Relaxed_Initialization aspect is used to relax the initialization policy on the parameter V of Set_V and the parameter X of Init_F1. We can see that the postcondition of Set_V has been simplified compared to Fig. 13 because the Error parameter is still subject to the initialization policy which makes its initialization at the end of Set_V implicit. The preconditions of Set_V and Init_F1 are no longer necessary for a similar reason.

For some use cases, relaxing the initialization policy at the level of the object might be too coarse. In particular, when designing data structures, there might be components on which the initialization policy should be relaxed and components on which it should not. As an example, consider a stack implemented

---

[6] Loop invariants are assertions which have to be supplied manually by users to verify loops using deductive verification.

```
procedure Set_V (S : String; V : out R; Error : out Boolean) with
   Relaxed_Initialization ⇒ V,
   Post ⇒ (if not Error then V'Initialized);
-- Read V from S. V is only initialized if Error is False.

procedure Init_F1 (X : in out R; V : T1) with
   Relaxed_Initialization ⇒ X,
   Post ⇒ X.F1'Initialized
      and (X.F2'Initialized = X.F2'Initialized'Old);
-- Initialize X.F1, initialization status of X.F2 is preserved
```

Fig. 14. The `Relaxed_Initialization` aspect can be used on subprograms to specify which parameters should be exempted from the initialization policy. The initialization status of these parameters needs to be described explicitly using contracts. It is not necessary for other parameters as the initialization policy applies.

using a fixed-sized array of elements and the index of the last inserted element. The definition of such a data structure is given in Fig. 15. The `Top` index should always be initialized, as it will be used for example to push an element on the stack. However, because the API of the stack only allows accessing elements at the top of the stack, it is never possible to read the value of an element stored after this index in the `Content` array. Therefore, it should not be necessary to give a value to `Content` when a stack is created. It will be initialized gradually when elements are pushed. In addition to being inefficient, requiring such an initialization would weaken the checks performed by the verification. Indeed, it would not ensure that the values stored at these invalid indexes are never read, relying instead on the dummy values supplied at initialization.

```
subtype Top_Range is Natural range 0 .. Max;
type Int_Array is array (1 .. Max) of Integer;

type Stack is record
   Content : Int_Array;
   Top     : Top_Range := 0;
end record;
```

Fig. 15. A stack implemented using a fixed-sized array of elements and the index of the last inserted element if there is one.

Even if components cannot be annotated with `Relaxed_Initialization`, mixed structures can be created by annotating types. If a type is annotated with `Relaxed_Initialization`, then the annotation will implicitly apply to all objects and subcomponents of the type. Figure 16 presents a variant of the type `Stack` defined in Fig. 15 where the type of the array `Content` is annotated with `Relaxed_Initialization`. It makes it possible to leave the `Content` field uninitialized when creating an object of type `Stack`. Proof becomes responsible for checking that elements of Content are initialized correctly when accessed. For it to be possible, it is necessary to supply a strong type invariant, called subtype predicate in Ada, stating that all values stored in Content up to Top are initialized at all times[7].

```
type Relaxed_Int_Array is array (1 .. Max) of Integer with
   Relaxed_Initialization;

type Stack is record
     Content : Relaxed_Int_Array;
     Top     : Top_Range := 0;
end record with
   Ghost_Predicate =>
     (for all I in 1 .. Top => Content (I)'Initialized);
   -- The content of a stack needs to be initialized up to Top
```

Fig. 16. Definition of a stack using a type annotated with Relaxed_Initialization for the content of the stack. A predicate is used to enforce that values in Content are initialized up to the Top index.

6 Relaxed Initialization - How Does It Work?

As explained in Sect. 5, the initialization of objects or subcomponents can be checked either by Flow or by Proof on a case-by-case basis, depending on user-supplied annotations. To make sure that no initialization check is missed, the routine used to determine which analysis is responsible for the initialization of an object or subcomponent is shared between Flow and Proof. If an object is checked by Proof, Flow always considers it as entirely initialized when read, and vice versa.

For the sake of simplicity, the two analyses do not attempt to communicate mixed initialization status with each other in a precise manner. To be copied from an object annotated with Relaxed_Initialization to an object with no annotation (or conversely), a value has to be entirely initialized. In the example in Fig. 17, only the components F1 of the variables X and Y are initialized. This mixed status is handled precisely by both analyses which can verify that reading X.F1 and Y.F1 in the assignment to I is allowed. However, copying X into Y leads to a failed initialization check. Indeed, Proof makes sure that the object is entirely initialized so Flow can safely assume it.

For Proof to be able to check initialization of composite structures precisely, it is necessary to encode the initialization status of each scalar subcomponent

---

[7] The Initialized attribute is considered as ghost code - code which is only used for verification - as it cannot be executed safely in all cases. As a result, it cannot be used in a standard predicate which is not ghost by default in SPARK.

```
type T2 is record
   G1, G2 : Integer;
end record;

type R is record
   F1 : Integer;
   F2 : T2;
end record;

X : R with Relaxed_Initialization;
Y : R;
X.F1 := 12;
Y.F1 := 12;
I := X.F1 + Y.F1;   -- Both X.F1 and Y.F1 are initialized
Y := X;   -- Error: X should be completely initialized
```

Fig. 17. Flow and Proof can deal with partially initialized structures and determine that the reads of X.F1 and Y.F1 in the assignment to I are allowed. Mixed initialization status is not communicated between Flow and Proof, so X needs to be entirely initialized for its value to be copied into Y.

separately. More precisely, a boolean flag needs to be introduced per scalar subcomponent of objects annotated with the Relaxed_Initialization annotation. For example, an object of type R defined in Fig. 17 requires three boolean flags, one for F1, one for F2.G1, and one for F2.G2, while an array needs as many boolean flags as there are components in the array. To make it simpler, the flags are embedded in the data structure, so the encoding of an object of type R has two fields: the first one containing a boolean flag and an integer value for F1 and the second one containing itself two fields with a boolean flag and an integer value for F2.G1 and F2.G2. When a scalar object or scalar subcomponent annotated with Relaxed_Initialization is evaluated, Proof makes sure that it is initialized by checking the associated boolean flag. For example, when X.F1 is evaluated in the assignment to I in Fig. 17, Proof checks that its associated boolean flag is true. This is enough to make sure that all read scalar values are necessarily in the bounds of their types and that no random value can be read in memory.

Using these additional flags, the Initialized attribute can be defined as the conjunction of all the initialization flags of the subcomponents in its prefix. As an example, a reference to X.F1'Initialized evaluates to the boolean flag associated with X.F1 while X'Initialized is the conjunction of the three boolean flags associated with X.F1, X.F2.G1, and X.F2.G2. Using this definition, it is possible to verify that X'Initialized evaluates to true after an initialization by part as in Fig. 12.

Let us consider the case of mixed structures with components subject to Relaxed_Initialization. We can remark that, when an object with such components is taken as an input by a subprogram call, it is not necessary for its annotated components to be initialized. Indeed, the called subprogram will expect the annotated components to be partially initialized. Inside the subprogram body, if reads of uninitialized parts of the parameter occur, Proof will emit checks accordingly. As an example, consider an object of type `Stack` defined in Fig. 16. Calling `Push` on a partially initialized stack cannot lead to reads of uninitialized values as `Push` considers its parameter as partially initialized as per the annotation on the `Relaxed_Int_Array` type. More generally, a composite object with annotated components being initialized should not entail the initialization of its annotated components. Therefore, neither Flow nor Proof attempts to verify that components subject to `Relaxed_Initialization` are initialized when a composite object is taken as input by a subprogram call.

7 Related Work

Starting with the influential work around code contracts in Eiffel [18] and JML [8], contracts can be interpreted as both formal specifications for proof and executable specifications for testing and runtime monitoring [15]. Various methods and tools taking advantage of them have emerged in the verification ecosystem, some static, some dynamic, all with their strengths and weaknesses [3,7].

Running several verification tools on the same program to take advantage of the strengths of different methods and reach a better trade-off is common practice. However, the collaboration between the tools is in general reduced to applying them separately to verify different (implicit or explicit) assertions [16] or the verification of other properties like relational ones [4]. Fine-grained combination of verification tools is made more difficult by the fact that they need to agree on the semantics of annotations. Small variations in its interpretation can lead to differences in the evaluation of the validity of the program [5].

Combining several analyses in the same tool has the advantage of allowing a better handle on the interpretation of the annotation language. Some works have targeted the combination of proof and runtime checking to gradually introduce program verification [1] or to verify information flow [2]. Frama-C is a well-known platform for verification of C programs that supports the combination of different static analyses to prove properties of programs [3]. As opposed to what we present in this article, the combination framework used in Frama-C is more general [10]. It allows each analysis to state which properties it has verified under which assumptions in a common format. The kind of interactions we presented in Sects. 1 and 3 could be encoded in this framework. However, it cannot accommodate more involved interactions where an analysis uses general information computed by the other (as opposed to a check being discharged), as is done in Sect. 2. This kind of combination of verification techniques is also present in Frama-C but in a more ad hoc manner, similar to what we do in this article [6].

As writing contracts is a costly activity in proof, many works have proposed static [13] or dynamic techniques [12] to generate such contracts, in particular invariants for loops ([14], see Sect. 5). However, no general method for generating contracts and loop invariants exists for now and deductive verification tools require users to supply them by hand.

8 Conclusion and Perspectives

Deductive verification is a powerful verification technique which provides strong guarantees. However, its usage can be complicated as it requires manual interaction to annotate programs with contracts. For verification and annotation to be tractable, it generally introduces simplifying assumptions which should be justified separately to ensure soundness. To tackle these limitations, the SPARK tool includes a data-flow analysis as a separate pass. It is generally used to verify simplifying assumptions and to generate some annotations for the user. The interaction between the analyses goes both ways as deductive verification can be used to gain precision in checks usually done by flow analysis. This handover can be automated or requested by the user on a case-by-case basis, as done for the `Relaxed_Initialization` feature.

As future work, it could be possible to extend this collaboration further, in particular to handover more checks from flow analysis to proof. Non-aliasing checks would be good candidates as they can be value-dependent, in the presence of array accesses in particular. The same kind of handover could also be applied to the verification of complex transfers of ownership when using pointers to relax some of the restrictions related to the ownership policy of SPARK.

Though this article contains some discussion about the soundness of our combination, a mechanized proof would definitely provide some confidence here. This remains for now a long-term goal as it would require a formalization of the Ada language which does not exist yet.

The combination mechanism experienced for SPARK could also be used by other teams developing similar tools for other languages. Properties common to most programs can be specified as a simplifying assumption which can be separately analyzed and enforced. An escape hatch for when the property is intentionally violated can be provided either automatically, or using a specific annotation. This mechanism could be used for analysis of C/C++ code or Rust unsafe code to support aliasing or initialization more efficiently.

References

1. Bader, J., Aldrich, J., Tanter, É.: Gradual program verification. In: VMCAI 2018. LNCS, vol. 10747, pp. 25–46. Springer, Cham (2018). https://doi.org/10.1007/978-3-319-73721-8_2
2. Barany, G., Signoles, J.: Hybrid information flow analysis for real-world C code. In: Gabmeyer, S., Johnsen, E.B. (eds.) TAP 2017. LNCS, vol. 10375, pp. 23–40. Springer, Cham (2017). https://doi.org/10.1007/978-3-319-61467-0_2
3. Baudin, P., et al.: The dogged pursuit of bug-free C programs: the Frama-C software analysis platform. Commun. ACM **64**(8), 56–68 (2021)

4. Blatter, L., Kosmatov, N., Le Gall, P., Prevosto, V., Petiot, G.: Static and dynamic verification of relational properties on self-composed C code. In: Dubois, C., Wolff, B. (eds.) TAP 2018. LNCS, vol. 10889, pp. 44–62. Springer, Cham (2018). https://doi.org/10.1007/978-3-319-92994-1_3
5. Boerman, J., Huisman, M., Joosten, S.: Reasoning about JML: differences between KeY and OpenJML. In: Furia, C.A., Winter, K. (eds.) IFM 2018. LNCS, vol. 11023, pp. 30–46. Springer, Cham (2018). https://doi.org/10.1007/978-3-319-98938-9_3
6. Bouillaguet, Q., Bobot, F., Sighireanu, M., Yakobowski, B.: Exploiting pointer analysis in memory models for deductive verification. In: Enea, C., Piskac, R. (eds.) VMCAI 2019. LNCS, vol. 11388, pp. 160–182. Springer, Cham (2019). https://doi.org/10.1007/978-3-030-11245-5_8
7. Burdy, L., et al.: An overview of JML tools and applications. Int. J. Softw. Tools Technol. Transf. **7**, 212–232 (2005)
8. Chalin, P., Kiniry, J.R., Leavens, G.T., Poll, E.: Beyond assertions: advanced specification and verification with JML and ESC/Java2. In: de Boer, F.S., Bonsangue, M.M., Graf, S., de Roever, W.-P. (eds.) FMCO 2005. LNCS, vol. 4111, pp. 342–363. Springer, Heidelberg (2006). https://doi.org/10.1007/11804192_16
9. Chapman, R., Dross, C., Matthews, S., Moy, Y.: Co-developing programs and their proof of correctness. Commun. ACM **67**(3), 84–94 (2024)
10. Correnson, L., Signoles, J.: Combining analyses for C program verification. In: Stoelinga, M., Pinger, R. (eds.) FMICS 2012. LNCS, vol. 7437, pp. 108–130. Springer, Heidelberg (2012). https://doi.org/10.1007/978-3-642-32469-7_8
11. Dross, C., Kanig, J.: Recursive data structures in SPARK. In: Lahiri, S.K., Wang, C. (eds.) CAV 2020. LNCS, vol. 12225, pp. 178–189. Springer, Cham (2020). https://doi.org/10.1007/978-3-030-53291-8_11
12. Ernst, M.D., et al.: The Daikon system for dynamic detection of likely invariants. Sci. Comput. Program. **69**(1–3), 35–45 (2007)
13. Fähndrich, M., Logozzo, F.: Static contract checking with abstract interpretation. In: Beckert, B., Marché, C. (eds.) FoVeOOS 2010. LNCS, vol. 6528, pp. 10–30. Springer, Heidelberg (2011). https://doi.org/10.1007/978-3-642-18070-5_2
14. Furia, C.A., Meyer, B., Velder, S.: Loop invariants: analysis, classification, and examples. ACM Comput. Surv. **46**(3), 1–51 (2014)
15. Hatcliff, J., Leavens, G.T., Leino, K.R.M., Müller, P., Parkinson, M.: Behavioral interface specification languages. ACM Comput. Surv. **44**(3) (2012)
16. Kosmatov, N., Marché, C., Moy, Y., Signoles, J.: Static versus dynamic verification in Why3, Frama-C and SPARK 2014. In: 7th International Symposium on Leveraging Applications, p. 16. Springer, Corfu (2016). https://hal.inria.fr/hal-01344110
17. McCormick, J.W., Chapin, P.C.: Building High Integrity Applications with SPARK. Cambridge University Press, Cambridge (2015)
18. Meyer, B.: Object-Oriented Software Construction, 2nd edn. Pearson College Div, Boston (2000)

Abstract Interpretation

Affine Disjunctive Invariant Generation with Farkas' Lemma

Jingyu Ke[1], Hongfei Fu[1(✉)], Hongming Liu[1], Zhouyue Sun[1], Liqian Chen[2], and Guoqiang Li[1]

[1] Shanghai Jiao Tong University, Shanghai, China
{Windocotber,jt002845,hm-liu,sunzhouyue,li.g}@sjtu.edu.cn
[2] National University of Defense Technology, Changsha, China
lqchen@nudt.edu.cn

Abstract. In the verification of loop programs, disjunctive invariants are essential to capture complex loop dynamics such as phase and mode changes. In this work, we develop a novel approach for the automated generation of affine disjunctive invariants for affine while loops via Farkas' Lemma, a fundamental theorem on linear inequalities. Our main contributions are two-fold. First, we combine Farkas' Lemma with a succinct control flow transformation to derive disjunctive invariants from the conditional branches in the loop. Second, we propose an invariant propagation technique that minimizes the invariant computation effort by propagating previously solved invariants to yet unsolved locations as much as possible. Furthermore, we resolve the infeasibility checking in the application of Farkas' Lemma which has not been addressed previously, and extend our approach to nested loops via loop summary. Experimental evaluation over more than 100 affine while loops (mostly from SV-COMP 2023) demonstrates that our approach is promising to generate tight linear invariants over affine programs.

1 Introduction

An invariant at a program location is an assertion that over-approximates the set of program states reachable to that location, i.e., every reachable program state to the location is guaranteed to satisfy the assertion. Since invariants provide an over-approximation for reachable program states, they play a fundamental role in program verification and can be used for safety [2,56,62], reachability [3,4,10,16,20,29,63] and time-complexity [14] analysis. Invariant generation targets the automated generation of invariants which can be used to aid the verification of critical program properties.

Automated approaches for invariant generation have been studied for decades and there have been an abundance of literature along this line of research. From different program objects, invariant generation targets numerical values (e.g., integers or real numbers) [5,9,15,19,65,76], arrays [51,78], pointers [12,52], algebraic data types [46], etc. By different methodologies, invariant generation can be solved by abstract interpretation [9,22,24,37], constraint solving [15,19,21,39],

© The Author(s), under exclusive license to Springer Nature Switzerland AG 2025
K. Shankaranarayanan et al. (Eds.): VMCAI 2025, LNCS 15529, pp. 187–213, 2025.
https://doi.org/10.1007/978-3-031-82700-6_9

inference [12,30,31,34,35,57,71,77,86], recurrence analysis [33,49,50], machine learning [36,41,67,88], data-driven approaches [17,26,52,58,64,73], etc. Most results in the literature consider a strengthened version of invariants, called *inductive invariants*, that requires the inductive condition that the invariant at a program location is preserved upon every execution back and forth to the location.

In this work, we consider the automated generation of disjunctive invariants, i.e., invariants that are in the form of a disjunction of assertions. Compared with conjunctive invariants, disjunctive invariants capture disjunctive features such as multiple phases and mode transitions in loops. Although extensive research has been conducted on conjunctive invariant generation, verification of programs with complex disjunctive loops still demands a more precise and scalable approach, rather than merely generating conjunctive invariants at the loop entry point to summarize the entire loop.

Moreover, most of the existing disjunctive invariant analyses rely on specific program patterns, such as alternating loop paths [72] or periodic regular loops [54,85], which cannot be effectively generalized to real-world programs with arbitrary execution traces. Methods based on abstract interpretation or symbolic execution often fail to converge, or converge to low-precision invariants, when dealing with loops that are deeply nested or have large constant iteration counts. To address this, we propose a modular approach capable of handling complex loops and efficiently mitigating the computational explosion caused by the interleaving of paths in disjunctive loops.

We use constraint solving to generate real-valued affine disjunctive invariants. A typical constraint solving method is via Farkas' Lemma [19,45,55,69] that provides a complete characterization for affine invariants. However, the application of Farkas' Lemma is mostly limited to the conjunction of affine inequalities. The question on how to leverage Farkas' Lemma to affine disjunctive invariants remains to be a challenge. In this paper, we focus on the generation of affine disjunctive invariants in affine loops. An affine loop is a while loop where all conditional and assignment statements are in the form of linear expressions.

Our Contributions. First, we introduce a novel control flow transformation that extracts loop paths (from entry to exit) as standalone locations in a transition system and establishes transitions between them. Second, to alleviate the exponential computational overhead introduced by the control flow transformation, we propose an invariant propagation technique that propagates already-computed invariants to locations whose invariants yet need to be computed as much as possible. Third, we fully resolve the infeasible situation in the application of Farkas' Lemma [55,69] and extend our approach to nested loops through loop summary. Fourth, we implement our approach as a prototype DInvG[1]. Experimental evaluation with various state-of-the-art verification tools using over 100 benchmarks from SV-COMP 2023 [80] and [9], shows that our app-

---

[1] The tool implementation is available on GitHub: https://github.com/WindOctober/DInvG.

roach is both tight (in the accuracy of the generated invariant) and is time efficient for real-valued disjunctive affine invariant generation.

The remainder of this paper is structured as follows. Section 2 revisits the fundamental definitions of affine transition systems and invariants, providing a variant of *Farkas' Lemma* as well as the basic definitions of polyhedra. Section 3 offers an overview of how our tool DInvG transforms programs and solves for disjunctive invariants. In Sect. 4, we formalize the definition of control flow transformation and extract the corresponding affine transition systems, present the pseudocode for invariant propagation as well as optimizations for nested loop and infeasible traces, and prove that the disjunctive invariants generated are inductive. Section 5 demonstrates the efficiency and precision advantages of DInvG compared to several state-of-the-art tools and conducts an ablation study on invariant propagation. Section 6 compares our method with related verification approaches, elaborating on the conceptual and implementation differences between invariant propagation and control flow transformation. A full version of this paper can be found at [48].

2 Preliminaries

Below we revisit affine transition systems [69] and their associated invariants, elucidate Farkas' Lemma, and outline fundamental principles from polyhedra theory. It is important to note that, within the scope of this paper, we treat linear and affine concepts equivalently.

2.1 Affine Transition Systems and Invariants

An *affine inequality* over a set $V = \{x_1, \ldots, x_n\}$ of real-valued variables is of the form $a_1 x_1 + \cdots + a_n x_n + b \geq 0$, where a_i's and b are real coefficients. An *affine assertion* over V is a conjunction of affine inequalities over V.

An affine transition system possesses a finite number of locations as well as real-valued variables, and specifies transitions between locations with affine guards and affine updates on the values of the variables.

Definition 1 (Affine Transition Systems [69]). *An affine transition system (ATS) is a tuple $\Gamma = \langle X, X', L, \mathcal{T}, \ell^*, \theta \rangle$:*

- *X is a finite set of real-valued variables and $X' = \{x' \mid x \in X\}$ is the set of primed variables.*
- *L is a finite set of locations and $\ell^* \in L$ is the initial location.*
- *$\mathcal{T}$ is a finite set of transitions where each transition τ is a triple $\langle \ell, \ell', \rho \rangle$ from location ℓ to location ℓ' with the guard affine assertion ρ over $X \cup X'$.*
- *θ is a disjunction of affine assertions over X that specifies the initial condition at ℓ^*.*

The directed graph $DG(\Gamma)$ of the ATS Γ is defined as the graph where the vertices are the locations of Γ and there is an edge (ℓ, ℓ') if and only if there is a transition $\langle \ell, \ell', \rho \rangle$ with source location ℓ and target location ℓ'.

The intuition of an ATS $\Gamma = \langle X, X', L, \mathcal{T}, \ell^*, \theta \rangle$ is as follows. Each variable $x \in X$ represents the current value of the variable and each primed variable $x' \in X'$ represents the next value of its unprimed variable $x \in X$ after one step of transition. The transition $\langle \ell, \ell', \rho \rangle$ specifies the jump from the current location ℓ to the next location ℓ' with the guard condition ρ specifying the condition to enable the transition. The guard condition involves both the current values (represented by X) and the next values (by X'), so that it can specify the relationship between the current and next values.

Below we describe the semantics of an ATS. A *valuation* over a finite set V of variables is a function $\sigma : V \to \mathbb{R}$ that assigns to each variable $x \in V$ a real value $\sigma(x) \in \mathbb{R}$. We mostly consider valuations over the variables X of an ATS and simply abbreviate "valuation over X" as "valuation" (i.e., omitting X). Given an ATS, a *configuration* is a pair (ℓ, σ) with the intuition that ℓ is the current location and σ is a valuation that specifies the current values for the variables.

Given an affine assertion φ and a valuation σ over a variable set V, we write $\sigma \models \varphi$ to mean that σ satisfies φ, i.e., φ is true when one substitutes the corresponding values $\sigma(x)$ into all the variables x in φ. Given an ATS Γ, two valuations σ, σ' and an affine assertion φ over $X \cup X'$, we write $\sigma, \sigma' \models \varphi$ to mean that φ is true when one substitutes every variable $x \in X$ by $\sigma(x)$ and every variable $x' \in X'$ with $\sigma'(x)$ in φ. Moreover, given two affine assertions φ, ψ over a variable set V, we write $\varphi \models \psi$ to mean that φ implies ψ, i.e., for every valuation σ over V we have that $\sigma \models \varphi$ implies $\sigma \models \psi$. The case of disjunction of affine assertions is similar.

The semantics of an ATS Γ is given by its paths. A *path* π of the ATS Γ is a finite sequence of configurations $(\ell_0, \sigma_0) \ldots (\ell_k, \sigma_k)$ such that

- **(Initialization)** $\ell_0 = \ell^*$ and $\sigma_0 \models \theta$, and
- **(Consecution)** for every $0 \leq j \leq k-1$, there exists a transition $\tau = \langle \ell, \ell', \rho \rangle$ such that $\ell = \ell_j$, $\ell' = \ell_{j+1}$ and $\sigma_j, \sigma_{j+1} \models \rho$.

We say that a configuration (ℓ, σ) is *reachable* if there exists a path $(\ell_0, \sigma_0) \ldots (\ell_k, \sigma_k)$ such that $(\ell_k, \sigma_k) = (\ell, \sigma)$. An *invariant* at a location ℓ of an ATS is an assertion φ such that for every path $\pi = (\ell_0, \sigma_0) \ldots (\ell_k, \sigma_k)$ of the ATS and each $0 \leq i \leq k$, it holds that $\ell_i = \ell$ implies $\sigma_i \models \varphi$. An invariant φ is *affine* if φ is an affine assertion over the variable set X, and is *disjunctively affine* if φ is a disjunction of affine assertions.

In invariant generation, one often investigates a strengthened version of invariants called *inductive invariants*. In this work, we present affine inductive invariants in the form of inductive affine assertion maps [19,55,69] as follows.

An *affine assertion map* (AAM) over an ATS is a function η that maps every location ℓ of the ATS to an affine assertion $\eta(\ell)$ over the variables X. An AAM η is called *inductive* if the following holds:

- **(Initialization)** $\theta \models \eta(\ell^*)$;
- **(Consecution)** For every transition $\tau = \langle \ell, \ell', \rho \rangle$, we have that $\eta(\ell) \wedge \rho \models \eta(\ell')'$, where $\eta(\ell')'$ is the affine assertion obtained by replacing every variable $x \in X$ in $\eta(\ell')$ with its next-value counterpart $x' \in X'$.

By a straightforward induction on the length of a path under an ATS, one could verify that every affine assertion in an inductive AAM is indeed an invariant.

2.2 Farkas' Lemma and Polyhedra

Farkas' Lemma [32] is a classical theorem in the theory of affine inequalities and previous results [19,55,69] have applied the theorem to affine invariant generation. In these results, the form of Farkas' Lemma follows [70, Corollary 7.1h].

Theorem 1 (Farkas' Lemma). *Consider an affine assertion φ over a set $V = \{x_1, \ldots, x_n\}$ of real-valued variables as in Fig. 1a. When φ is satisfiable (i.e., there is a valuation over V that satisfies φ), it implies an affine inequality ψ as in Fig. 1b (i.e., $\varphi \models \psi$) if and only if there exist non-negative real numbers $\lambda_0, \lambda_1, \ldots, \lambda_m$ such that (i) $c_j = \sum_{i=1}^{m} \lambda_i \cdot a_{ij}$ for all $1 \leq j \leq n$, and (ii) $d = \lambda_0 + \sum_{i=1}^{m} \lambda_i \cdot b_i$ as in Fig. 1c. Moreover, φ is unsatisfiable if and only if the inequality $-1 \geq 0$ (as ψ) can be derived from above.*

$$\varphi : \begin{array}{c} a_{11} \cdot x_1 + \cdots + a_{1n} \cdot x_n + b_1 \geq 0 \\ \vdots \qquad \vdots \qquad \vdots \\ a_{m1} \cdot x_1 + \cdots + a_{mn} \cdot x_n + b_m \geq 0 \end{array}$$

(a) φ in Farkas' Lemma

$$\psi : c_1 \cdot x_1 + \cdots + c_n \cdot x_n + d \geq 0$$

(b) ψ in Farkas' Lemma

$$\begin{array}{c|c}
\lambda_0 & 1 \geq 0 \\
\lambda_1 & a_{11} \cdot x_1 + \cdots + a_{1n} \cdot x_n + b_1 \geq 0 \\
\vdots & \vdots \qquad \vdots \qquad \vdots \\
\lambda_m & a_{m1} \cdot x_1 + \cdots + a_{mn} \cdot x_n + b_m \geq 0 \\
\hline
 & c_1 \cdot x_1 + \cdots + c_n \cdot x_n + d \geq 0 \leftarrow \psi \\
 & -1 \geq 0 \leftarrow \text{false}
\end{array} \right\} \varphi$$

(c) The Tabular Form for Farkas' Lemma

Fig. 1. The φ, ψ and Tabular Form for Farkas' Lemma [19,69]

Farkas' Lemma simplifies the inclusion of a polyhedron inside a halfspace into the satisfiability of a system of affine inequalities. We refer to the case of unsatisfiable φ with $\psi := -1 \geq 0$ in the statement of Theorem 1 as *infeasible implication*. The application of Farkas' Lemma can be visualized by the tabular form in Fig. 1c (taken from [19]), and we multiply $\lambda_0, \lambda_1, \ldots, \lambda_m$ with their inequalities in φ and sum up them together to get ψ. For $1 \leq j \leq m$, we require $\lambda_j \geq 0$.

A subset P of $\mathbb{R}^n$ is a *polyhedron* if $P = \{\mathbf{x} \in \mathbb{R}^n \mid \mathbf{A} \cdot \mathbf{x} \leq \mathbf{b}\}$ for some real matrix $A \in \mathbb{R}^{m \times n}$ and real vector $\mathbf{b} \in \mathbb{R}^m$, where $\mathbf{x}$ is treated as a column vector and the comparison $\mathbf{A} \cdot \mathbf{x} \leq \mathbf{b}$ is defined in the coordinate-wise fashion. A polyhedron P is a *polyhedral cone* if $P = \{\mathbf{x} \in \mathbb{R}^n \mid \mathbf{A} \cdot \mathbf{x} \leq \mathbf{0}\}$ for some real matrix $A \in \mathbb{R}^{m \times n}$, where $\mathbf{0}$ is the m-dimensional zero column vector. It is well-known from the Farkas-Minkowski-Weyl Theorem [70, Corollary 7.1a] that any polyhedral cone P can be represented as $P = \{\sum_{i=1}^{k} \lambda_i \cdot \mathbf{g}_i \mid \lambda_i \geq 0 \text{ for all } 1 \leq i \leq k\}$ for some real vectors $\mathbf{g}_1, \ldots, \mathbf{g}_k$, where such vectors $\mathbf{g}_i$'s are called a collection of *generators* for the polyhedral cone P.

```
int x = 0, y = 50;
                         while ( x < 100 ){
while ( x < 100 ){          case x > 49 :
    x = x + 1;                  x = x + 1;
    if ( x > 50 )               y = y + 1;
        y = y + 1;          case x ≤ 49 :
}                               x = x + 1;
                         }
       P₁                          P₂
```

(a) Source P_1 and Its Transformation P_2.

$X = \{x, y\}, L = \{\ell_1, \ell_2^*\}, \mathcal{T} = \{\tau_1, \tau_2, \tau_3, \tau_4\},$
$\theta : x = 0 \wedge y = 50, \tau_1 : \langle \ell_1, \ell_1, \rho_1 \rangle,$
$\tau_2 : \langle \ell_1, \ell_2, \rho_2 \rangle, \tau_3 : \langle \ell_2, \ell_2, \rho_3 \rangle, \tau_4 : \langle \ell_2, \ell_1, \rho_4 \rangle,$

$\rho_1 : \begin{bmatrix} 50 \leq x \leq 99 \\ 50 \leq x' \leq 99 \\ x' = x + 1 \\ y' = y + 1 \end{bmatrix}, \rho_2 : \begin{bmatrix} 50 \leq x \leq 99 \\ x' \leq 49 \\ x' = x + 1 \\ y' = y + 1 \end{bmatrix}$

$\rho_3 : \begin{bmatrix} x \leq 49 \\ x' \leq 49 \\ x' = x + 1 \\ y' = y \end{bmatrix}, \rho_4 : \begin{bmatrix} x \leq 49 \\ 50 \leq x' \leq 99 \\ x' = x + 1 \\ y' = y \end{bmatrix}$

(b) The ATS Corresponding to P_2

Fig. 2. An example from [72] and its transformed form and corresponding ATS

3 An Overview of Our Approach

Consider the affine program P_1 in Fig. 2a. Our approach has three parts, namely control flow transformation, invariant computation and invariant propagation.

Control Flow Transformation. For the non-nested loop P_1, we extract each execution path from the loop entry to the exit and transform it into the form of loop P_2. Each **case** statement corresponds to a possible path in the original loop with a path condition ϕ of taking the path specified by the conjunction of a conditional formula ϕ_c and an assignment formula ϕ_a. For example, in the first case of P_2 that corresponds to the case of entering the if-branch in P_1, we have ϕ_a is $(x' = x + 1 \wedge y' = y + 1)$, ϕ_c is $x > 49$, and ϕ is $\phi_a \wedge \phi_c$.

Then, an ATS in Fig. 2b is directly derived from the transformed program. In this example, each **case** statement is considered as an independent location in the ATS, e.g., the location ℓ_1 stands for the first **case** in P_2. The transitions between the locations are derived from the jumps between different paths, where each transition guard ρ for a transition τ, can be obtained through the formula:

$$\rho := \phi_c \wedge \phi_c'[x'/x] \wedge \phi_a \wedge G \wedge G[x'/x]$$

where, ϕ_a and ϕ_c represent the assignment and conditional formulas at the transition's start location, ϕ_c' represents the conditional formula at the transition's end location, and G is the loop guard. After the ATS is constructed, we apply the approaches [55,69] in Farkas' Lemma combined with the technique of invariant propagation to obtain invariants at all locations, and group them disjunctively together to obtain result invariants for original loop. Note that the construction of the ATS is the key connection to apply Farkas' Lemma.

Invariant Computation. We first establish affine invariant templates at each location in the ATS by setting:

$$\eta(\ell_i) := c_{\ell_i,1} x + c_{\ell_i,2} y + d_{\ell_i} \geq 0 \,, \forall i \in \{1, 2\}$$

$$\begin{array}{c|cc}
\lambda_0 & 1 \geq 0 & \\
\lambda_1 & a_{11}x_1 + \cdots + a_{1n}x_n + b_1 \bowtie_1 0 & \\
\vdots & \vdots \quad\quad \vdots \quad \vdots & \left.\begin{array}{c}\\\\\\\end{array}\right\} \theta \\
\lambda_m & a_{m1}x_1 + \cdots + a_{mn}x_n + b_m \bowtie_m 0 & \\
\hline
& c_{\ell^*,1}x_1 + \cdots + c_{\ell^*,n}x_n + d_{\ell^*} \geq 0 \leftarrow \eta(\ell^*) \\
& -1 \geq 0 \leftarrow false
\end{array}$$

(a) Initialization Tabular

$$\begin{array}{c|cc}
\mu & c_{\ell,1}x_1 + \cdots + c_{\ell,n}x_n \quad\quad\quad\quad + d_\ell \geq 0 \leftarrow \eta(\ell) \\
\lambda_0 & 1 \geq 0 \\
\lambda_1 & a_{11}x_1 + \cdots + a_{1n}x_n + a'_{11}x'_1 + \cdots + a'_{1n}x'_n + b_1 \bowtie_1 0 \\
\vdots & \vdots \quad\quad \vdots \quad \vdots \quad\quad \vdots \quad \vdots \quad\quad \left.\begin{array}{c}\\\\\end{array}\right\} \rho \\
\lambda_m & a_{m1}x_1 + \cdots + a_{mn}x_n + a'_{m1}x'_1 + \cdots + a'_{mn}x'_n + b_m \bowtie_m 0 \\
\hline
& c_{\ell',1}x'_1 + \cdots + c_{\ell',n}x'_n + d_{\ell'} \geq 0 \leftarrow \eta(\ell')' \\
& -1 \geq 0 \leftarrow false
\end{array}$$

(b) Consecution Tabular

Fig. 3. Tabular for Initialization and Consecution [55]

where, $c_{\ell_i,1}, c_{\ell_i,2}, d_{\ell_i}$ are unknown coefficients to be resolved. Then, we generate the constraints from the initialization as well as consecution conditions via the Farkas' tabular in Fig. 1c and derive the initialization tabular and consecution tabular as shown in Fig. 3. Setting $\varphi = \theta$ and $\psi = \eta(\ell^*)$ in Theorem 1, the initialization of $\theta \models \eta(\ell^*)$ results in linear constraints, while the consecution condition $\eta(\ell) \wedge \rho \models \eta(\ell')'$ by setting $\varphi = \eta(\ell) \wedge \rho$ and $\psi = \eta(\ell')'$ results in quadratic constraints since we have a fresh λ, which we denote as μ, multiplied by $\eta(\ell)$.

In this context, we adopt a unified notation by employing $\eta(\ell)$ to represent both affine expressions and affine inequalities interchangeably throughout the manuscript. To resolve the quadratic constraints from the consecution, the previous approach [69] has considered heuristics to guess the value of the multiplier μ through either practical rules such as factorization or setting μ manually to 0 or 1. These settings relax the original consecution definition into two stronger forms:

- **Local Consecution**: For transition $\tau : \langle \ell_i, \ell_j, \rho \rangle$, $\rho \models \eta(\ell_j)' \geq 0$,
- **Incremental Consecution**: For transition $\tau : \langle \ell_i, \ell_j, \rho \rangle$, $\rho \models \eta(\ell_j)' \geq \eta(\ell_i)$.

Then, we collect the derived constraints to constitute a formula Φ in CNF, which further expands into a DNF Φ'. Note that each disjunctive clause in the DNF Φ' is an affine assertion defining a polyhedral cone, and we solve for the unknown coefficients of the templates by computing the generators of these polyhedral cones. We present a polyhedral cone in the DNF Φ' of the ATS in Fig. 4b,

$$\begin{bmatrix} c_{12} - c_{22} = 0, \\ c_{21} \geq 0, \\ c_{11} + c_{12} \geq 0, \\ 50c_{12} + d_2 \geq 0, \\ 50c_{11} + d_1 - 49c_{21} - d_2 \geq 0 \end{bmatrix}$$

(a) A clause in the DNF

type	c_{11}	c_{12}	d_1	c_{21}	c_{22}	d_2	$\eta(\ell_1)$	$\eta(\ell_2)$
point	0	0	0	0	0	0	$0 \geq 0$	$0 \geq 0$
line	1	-1	0	0	-1	50	$x - y = 0$	$-y + 50 = 0$
ray	0	0	49	1	0	0	$49 \geq 0$	$x \geq 0$
ray	0	0	1	0	0	0	$1 \geq 0$	$0 \geq 0$
ray	1	0	-50	0	0	0	$x - 50 \geq 0$	$0 \geq 0$
ray	0	0	1	0	0	1	$1 \geq 0$	$1 \geq 0$

(b) generators and their invariants

Fig. 4. Example of a clause in the DNF with its generators and invariants

along with corresponding generators, as shown in Fig. 4a, where "point" means a single vector, "ray" means a vector that can be scaled by an arbitrary positive value, and "line" means a vector that can be scaled by any positive or negative value. When putting the generators back to invariants, we obtain the invariants shown in the right part of Fig. 4b. The resulting disjunctive invariant is the disjunction of invariants at all locations over an ATS Γ, corresponding to the invariants required on different loop paths.

Invariant Propagation. In the preceding exposition, the computation of conjunctive affine invariants adheres to existing approaches [55,69]. However, the strategies employed to resolve invariants at each location across the entire ATS result in a significant decline in computational efficiency. To address this limitation, we introduce a propagation technique predicated on existing invariant computation results:

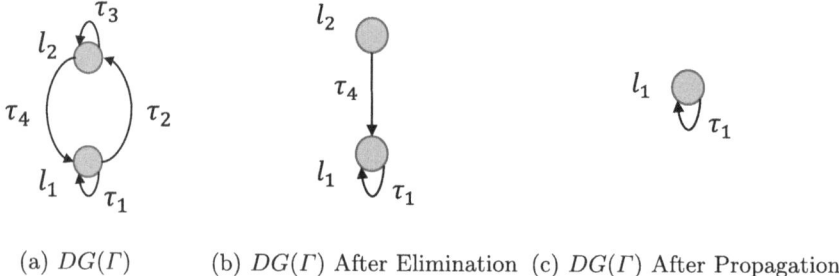

(a) $DG(\Gamma)$ (b) $DG(\Gamma)$ After Elimination   (c) $DG(\Gamma)$ After Propagation

Fig. 5. Procedure of Invariant Propagation for example in Fig. 2

Consider the affine transition system Γ in Fig. 2b. Its underlying directed graph $DG(\Gamma)$ is given in Fig. 5a. We first compute the invariant $\eta(\ell_2) := y = 50 \land 0 \leq x \leq 49$ at the initial location ℓ_2 as above. Then, we can eliminate all transitions pointing to ℓ_2 (the correctness of which is demonstrated in the following section), obtaining the graph in Fig. 5b. Notably, there exists a topological order, where ℓ_2 precedes ℓ_1. Thus, we propagate the invariants at ℓ_2 along the transition τ_4's transition guard ρ to establish the initial condition θ of the ATS over ℓ_1 in Fig. 5c, which is derived from removing ℓ_2 after propagation. Finally, by solving the invariants for the simplified ATS instead of the original ATS, we obtain the affine invariant $\eta(\ell_1) = (x = y \wedge 50 \le x \le 99)$. Note that our invariant propagation is different from abstract interpretation, see Sect. 6 for details.

Moreover, $\eta(\ell_1)$ corresponds to the invariant within the loop, specifically representing the invariant at the first **case**. For the exit state $x = 100$, the disjunctive invariant generated within the loop, in conjunction with the loop exit condition $\neg G$, derives the program state $x = 100$ outside the loop through stepwise deduction.

4 Algorithmic Details in Our Approach

Below we present our approach for generating affine disjunctive invariants over affine programs. We first illustrate our control flow transformation for non-nested loops, then our invariant propagation to reduce invariant computation, and finally the resolution of the infeasible implication and the extension to nested loops.

4.1 Control Flow Transformation

We fix the set of program variables in a loop as $X = \{x_1, \ldots, x_n\}$ and identify it as the set of variables in the ATS to be derived from the loop. We consider the canonical form of a non-nested affine while loop as in Fig. 6 similar to [45], where we have:

- The column vector $\mathbf{x} = (x_1, \ldots, x_n)^\mathrm{T}$ represents the vector of program variables, and G is a disjunction of affine assertions that serves as the loop condition.
- Each $\mathbf{F}_i$ ($1 \le i \le m$) is an affine function, i.e., $\mathbf{F}_i(\mathbf{x}) = \mathbf{A}\mathbf{x} + \mathbf{b}$ where $\mathbf{A}$ (resp. $\mathbf{b}$) is an $n \times n$ square matrix (resp. n-dimensional column vector) that specifies the affine update under the function $\mathbf{F}_i$ in the conditional branch ϕ_i. The assignment $\mathbf{x} := \mathbf{F}_i(\mathbf{x})$ is considered simultaneously for the variables in $\mathbf{x}$ so that in one execution step, the current valuation σ is updated to $\mathbf{F}_i(\sigma)$.
- The statements $\delta_1, \ldots, \delta_m$ specify whether the loop continues after the affine update of the conditional branches $\phi_1, \ldots, \phi_m$. Each statement δ_i is either the **skip** statement that has no effect or the **break** statement that exits.

$$
\begin{aligned}
&\textbf{while } (G) \ \{ \\
&\quad \textbf{case } \phi_1: \ \mathbf{x} := \mathbf{F}_1(\mathbf{x}) \, ; \delta_1 \, ; \\
&\quad \quad \vdots \\
&\quad \textbf{case } \phi_m: \ \mathbf{x} := \mathbf{F}_m(\mathbf{x}) \, ; \delta_m \, ; \\
&\}
\end{aligned}
$$

Fig. 6. The canonical form of a non-nested affine while loop

Any non-nested affine while loop with a break statement can be transformed into the canonical form in Fig. 6 by recursively examining the substructures of the loop body. A detailed recursive transformation process is provided in our extended version [48]. Note that although the transformation into our canonical form may cause exponential growth in the number of conditional branches in the loop body, in practice a loop typically has a small number of conditional branches and further improvement can be carried out by removing invalid branches (i.e., those whose branch condition is unsatisfiable, such as τ_2 in Fig. 2).

Moreover, such a canonical form is often necessary to capture precise disjunctive information in a while loop. Each case corresponds not merely to an individual loop path but also encapsulates the set of states at the entry of a loop. By finely partitioning the incoming program states according to branching conditions and formulating constraints among these states, we endeavor to precisely characterize the dynamics of internal state transitions within loops exhibiting multi-phase behavior. Below we demonstrate our control flow transformation that transforms the canonical form into an ATS.

Formally, the ATS Γ_W for a loop W in our canonical form is given as follows:

- The set of locations is $\{\ell_1, \ldots, \ell_m, \ell_e\}$, where each ℓ_i $(1 \leq i \leq m)$ corresponds to i-th case in the canonical form and ℓ_e is the termination location of the loop.
- For each $1 \leq i \leq m$ such that $\delta_i =$ **break**, we have the transition (we denote $\mathbf{x}' := (x'_1, \ldots, x'_n)^{\mathrm{T}}$)

$$\tau_i = (\ell_i, \ell_e, G \wedge \phi_i \wedge \mathbf{x}' = \mathbf{F}_i(\mathbf{x}))$$

that specifies the one-step jump from ℓ_i to the termination location ℓ_e.
- For each $1 \leq i, j \leq m$ such that $\delta_i \neq$ **break**, we have the transition

$$\tau_{ij} = (\ell_i, \ell_j, G \wedge \phi_i \wedge G[\mathbf{x}'/\mathbf{x}] \wedge \phi_j[\mathbf{x}'/\mathbf{x}] \wedge \mathbf{x}' = \mathbf{F}_i(\mathbf{x}))$$

that specifies the jump from ℓ_i in the current loop iteration to ℓ_j in the next loop iteration.
- For each $1 \leq i \leq m$ such that $\delta_i \neq$ **break**, we have the transition

$$\tau'_i = (\ell_i, \ell_e, G \wedge \phi_i \wedge (\neg G)[\mathbf{x}'/\mathbf{x}] \wedge \mathbf{x}' = \mathbf{F}_i(\mathbf{x}))$$

for the jump from ℓ_i to the termination location ℓ_e.

After the transformation, we remove transitions with an unsatisfiable guard condition to reduce the size of the derived ATS. The transformation for our running example has been given in Fig. 2. The transformed ATS Γ_W enables us to apply existing approaches [55,69] to generate invariants at the locations of the ATS Γ_W. Finally, recall that the overall disjunctive invariant for the ATS Γ_W is the disjunction of the invariants at all the locations.

As for the previous approaches [55,69] that only set locations at the initial location of the loop, our control flow transformation has different locations corresponding to different paths of the original loop, thereby achieving fine-grained piecewise invariants. Moreover, contrary to the granular program translations employed in traditional software model checking, which is similar to the control flow transformation, our motivation is mainly to integrate this transformation with Farkas' Lemma. By analyzing the transition patterns among internal loop paths, we aim to more effectively capture the phase-specific characteristics of multi-stage programs.

As demonstrated by our experiments, considering transitions between any pair of paths (ℓ_i, ℓ_j), rather than partitioning the loop into more complex path regular expressions (using regular expressions to abstract loop behaviors over multiple iterations) allows us to maintain a balance between the precision of invariant generated and the efficiency of constraint solving. To further improve the efficiency, we have designed the invariant propagation algorithm shown below for these directed graphs constructed from paths.

4.2 Invariant Propagation

In the computation of invariants, previous approaches [55,69] require to generate the invariants at all the locations of an ATS. As invariant computation is usually expensive, it is important to explore optimizations that avoid redundant computations. In this section, we propose a novel invariant propagation technique that is applicable to any directed graph of an ATS and achieves maximal efficiency in specific graph structures such as directed cycles. Below we demonstrate the procedure of invariant propagation via Algorithm 1.

The algorithm consists of the following steps: First, we initialize the assertion map and use the classical Tarjan's algorithm [81] to compute a list of strongly connected components (SCCs) in the directed graph $DG(\Gamma)$ (lines 1–2). Depending on whether the graph is decomposable, i.e., whether the size of the SCC list is more than one, we consider two cases:

(i) For a directed graph that can be decomposed into multiple SCCs, we start from the entry SCC and traverse the list of SCCs in breadth-first order, computing invariants for each SCC recursively and integrating them into the final assertion map η (lines 3–19).

(ii) For a directed graph that is a single SCC, we compute the initial invariants at the starting location using the previous method [55], then eliminate the start location ℓ^*, traverse each edge originating from ℓ^*, propagate the invariants to the remaining sub-graph, and disjunctively merge the returned inductive assertion mappings to produce the final disjunctive invariant (lines 20–26).

We formally define the specific functions involved in the algorithm as follows:

1. **Merge**(η_1, η_2) (line 16 and line 24). We extend η to a mapping from the set of locations L to disjunction of affine assertions, specifically representing affine inequalities in DNF. The Merge function is thereby defined as a new mapping such that, for any location $\ell \in L$:

$$\text{Merge}(\eta_1, \eta_2)(\ell) = \eta_1(\ell) \vee \eta_2(\ell)$$

Algorithm 1. $InvProp(\Gamma, DG(\Gamma), \ell^*)$

Require: Γ — ATS, $DG(\Gamma)$ — directed graph of Γ, ℓ^* — initial location of Γ.
Ensure: η — an inductive assertion map for Γ.
 1: *Init assertion map η for Γ.*
 2: SCCs $\leftarrow$ $Tarjan(DG(\Gamma), \Gamma)$ ▷ Find all SCCs in the directed graph
 3: **if** Len(SCCs) $\neq 1$ **then**
 4:    id $\leftarrow$ $FindSCC(\ell^*, \text{SCCs})$ ▷ Find the SCC containing ℓ^*
 5:    stack.$push$(id, ℓ^*)
 6:    **while** $\neg$stack.$isEmpty()$ **do**
 7:      (cur, ℓ_s) $\leftarrow$ stack.$pop()$ ▷ ℓ_s is the initial location of current SCC
 8:      $\Gamma_s \leftarrow$ SCCs[cur] ▷ cur is the index of current traversed SCC
 9:      $\eta_s \leftarrow InvProp(\Gamma_s, DG(\Gamma_s), \ell_s)$ ▷ Process single SCC
10:      **for each** transition τ directed from ℓ_s to ℓ_t **do**
11:         next $\leftarrow$ $FindSCC(\ell_t, \text{SCCs})$
12:         **if** next $\neq$ cur **then**
13:            stack.$push$(next, ℓ_t) ▷ Traverse SCCs in BFS order
14:         **end if**
15:      **end for**
16:      $\eta \leftarrow Merge(\eta, \eta_s)$ ▷ Combine assertion maps disjunctively
17:    **end while**
18:    **return** η
19: **end if**
20: $\eta \leftarrow InitInv(\Gamma, \ell^*)$ ▷ Compute invariant only in initial location
21: $\Gamma_s \leftarrow Project(\Gamma, \ell^*)$ ▷ Derive sub-ATS Γ_s by removing ℓ^*
22: **for each** transition τ directed from ℓ^* to ℓ_t **do**
23:    $\eta_s \leftarrow InvProp(\Gamma_s, DG(\Gamma_s), \ell_t)$
24:    $\eta \leftarrow Merge(\eta, \eta_s)$
25: **end for**
26: **return** η

2. **Project**(Γ, ℓ^*) (line 21). Considering the directed graph DG(Γ) corresponding to the ATS Γ, we remove all edges associated with the node ℓ^*, as well as the node itself. The derived ATS corresponding to the resulting sub-graph is denoted by Project(Γ, ℓ^*).

Note that in Algorithm 1, we omitted the initial condition θ and the propagation effect along the transitions. At each point of our algorithm (line 10 and line 22) that tackles a transition $\langle \ell_s, \ell_t, \rho \rangle$, the propagation effect is computed as the post image of the conjunction of the invariant on ℓ_s and the guard ρ via polyhedral projection onto the primed variables X' and serves as the initial condition θ of the new ATS including ℓ_t.

Example 1. Recall the example in Sect. 3, specifically Fig. 5. Here, Γ is an indivisible SCC. After computing the invariant $\eta(\ell_2)$ of the ATS Γ at the initial location ℓ_2, we consider all transitions (i.e. $\{\tau_4\}$) starting from the initial location ℓ_2, as depicted in the figure. Then, we propagate the invariant through the transition τ_4 to ℓ_1. After project to obtain the remaining sub-ATS Γ_{sub}, composed of ℓ_1 and its self-loop transition, we recursively compute this indivisible SCC to obtain the complete inductive assertion map. □

Our invariant propagation technique applies to all ATS. The main advantage to incorporate this technique is that it allows the generation of invariants only at the initial locations of (sub-)SCCs, thus avoiding the generation of the invariants at all locations as adopted in [55,69]. In the case that the directed graph of the input ATS is a cycle, our invariant propagation reaches the highest efficiency that generates the invariant only at the initial location of the cycle and derives invariants at other locations of the cycle by propagation, since the cycle has an explicit topological order after the removal of the initial location. This advantage becomes more prominent in loops with a non-neglectable amount of conditional branches. The soundness of the invariant propagation is given in the following theorem.

Theorem 2. *The assertion maps generated by the invariant propagation algorithm are inductive.*

Proof. We prove by induction on the number k of locations in the input ATS Γ that the assertion map obtained by our invariant propagation algorithm for the ATS Γ is inductive.

We first consider the base case, i.e., $k = 1$. In this case, $DG(\Gamma)$ has only one location, which is obviously indivisible. Here, the function **InitInv**(), previously mentioned as applying Farkas' Lemma for conjunctive invariant computation, is called. Therefore, the resulting assertion map is inductive, and its correctness is guaranteed by the prior results.

Assuming that the case when the size of Γ equals k holds, we prove that it holds for Γ of size $k + 1$. For an ATS Γ with $k + 1$ locations, if it is divisible, it can be decomposed into several sub-SCCs Γ_{sub} with sizes less than or equal to k. After the call to function **InvProp**() at Line 8, we obtain an inductive assertion mapping by the inductive condition. The **Merge**() function does not affect the inductive condition of the combined mapping. On the other hand, if it is indivisible, then our approach computes the invariant at its initial location and, after projecting away the initial location ℓ^*, obtains a sub-ATS Γ_{sub} of size k. Similarly, the recursive call to invariant propagation at Line 20 and merging the returned results always yields an inductive assertion map by the inductive condition. □

4.3 Other Optimizations

Loop Summary. To address more general control flow, such as nested loops, we use the standard method of loop summary to express the input-output relationship of the inner loops (while adding fresh variables for input values) to handle nested loops, as described in our extended version [48].

Infeasible Implication. In the previous results [55,69], the infeasible implication is not handled in their prototype. Recall the infeasible implication corresponding to $\eta(\ell) \wedge \rho \models -1 \geq 0$ illustrated in Fig. 3b. To fully address this issue, we can simply set $\mu = 1$ in Fig. 3b so that the nonlinear multiplier μ is eliminated. The correctness is given by the following theorem.

Theorem 3. *Let Γ be an ATS. For any AAM η that fulfills the initial and consecution conditions derived from the ATS Γ with the original constraints for the infeasible implication as in each consecution tabular of Fig. 3b (aimed at $-1 >= 0$) with each μ in an infeasible implication instantiated as k for some $k > 0$, it is equivalent to setting all μ's to 1 while preserving the constraints of infeasible implication.*

We present our proof in our extended version [48]. The main idea of the proof is that, for the infeasible implication case, by scaling each λ_i other than μ, the consecution tabular used to generate polyhedra is transformed into an equivalent tabular with scaled lambda variables λ'_i. so that it suffices to choose the multiplier μ to be 1.

5 Experimental Evaluation

In this section, we present the evaluation of the implementation (referred to as DInvG) of our approach to generate disjunctive affine invariants. We focus on the following two questions (**RQ1** and **RQ2**).

- **RQ1:** How competitive is DInvG when compared with other approaches?
- **RQ2:** How effective does invariant propagation enhance our approach?

5.1 Experimental Setup

Implementation. We implement our approach (including the algorithmic techniques in Sect. 4) as a prototype DInvG, dividing the implementation into front-end and back-end. The front-end utilizes Clang Static Analyzer [18] to extract and transform C programs, processing programs into the format required by the back-end. The back-end is an extension of StInG [79] written in C++ and uses PPL 1.2 [6] for polyhedra manipulation (e.g., projection, generator computation, etc.), which generates invariants and propagate them to obtain a disjuntive invariant as the loop invariant.

Environment. All experiments are conducted on a machine equipped with a 12th-generation Intel(R) Core(TM) i7-12800HX CPU, 16 cores, 2304 MHz, 9.5 GB RAM, running Ubuntu 20.04 (LTS). Following the competition settings of SV-COMP, for studies **RQ1** and **RQ2**, we impose a time limit of 900 s.

Benchmarks. We have a total of 114 affine programs, 38.6% of which have disjunctive features, sourced from: 1) 105 benchmarks from the SV-COMP, ReachSafety-Loop track. We excluded those with arrays, pointers, and other non-numeric features, those with modulus, division, polynomial, and other non-linear operations. 2) 9 benchmarks from the recent paper [9], which include complex nested loops and examples with disjunctive features.

Methodology. In **RQ1**, we compare DInvG utilizing invariant propagation techniques with several state-of-the-art software verifiers:

- Veriabs [28] is a state-of-the-art software verifier that is an integration of various strategies such as fuzz testing, k-induction, loop shrinking, loop pruning, full-program induction, explicit state model checking and other invariant generation techniques, which is capable to deal with programs with disjunctive features.
- CPAChecker [25] is a well-developed software verifier that is based on bounded model checking and interpolation and has a comprehensive ability to verify various kinds of properties.
- OOPSLA23 [82] is a recent recurrence analysis tool that handles only loops with the ultimate strict alternation pattern that eventually the loop will alternate between different modes periodically and performs good on such class of programs, which thus excels in the verification of disjunctive programs with alternating modes.
- DIG [59] is an invariant generation tool considering disjunctive features in programs and utilizes front-end CIVL [74] to obtain symbolic execution traces. It employs dynamic analysis along with efficient algorithms from algebra and geometry to solve numerical invariant templates, thereby generating numerical invariants at any position within a program, which is capable of extensively handling the programs with array, nonlinear, linear and disjunctive features.
- IKOS with *Polyset* domain from PPLite [8,11] is a classic abstract interpretation framework with various interface supports. The *Polyset* abstract domain is an efficient implementation of the powerset of polyhedra and serves as an alternative to the trace partitioning strategy implemented in Astree [23].

In **RQ2**, we focus on comparing the impact of the invariant propagation technique on the time efficiency. By contrasting the tool's performance when calculating invariants for each location individually against using invariant propagation, we analyze the role of invariant propagation.

5.2 Tool Comparison (RQ1)

Our work primarily focuses on the generation of disjunctive invariants, whereas tools like CPAChecker and Veriabs are specifically designed as bug finders for verifying assertions. However, by integrating the PPL library [6] and Z3 [89], we use the generated invariants to verify the correctness of assertions and demonstrate the precision of the invariants generated by DInvG.

Table 1. Comparisons Over 114 Benchmarks

Benchmark		DInvG			Veriabs			CPAChecker		
Source	#Num	#Ver.	#Unk.	Time (s)	#Ver.	#Unk.	Time (s)	#Ver.	#Unk.	Time (s)
loop-invariants	5	4	1	0.47	5	0	153.31	4	1	1001.49
loop-new	2	2	0	0.11	0	2	959.74	0	2	1807.20
loop-invgen	5	4	1	0.41	5	0	160.51	0	5	4518.19
loops-crafted-1	25	20	5	4.84	25	0	4010.55	0	25	22607.55
loop-simple	2	1	1	2	1	1	944.69	1	1	919.03
loop-zilu	26	26	0	0.77	25	1	1064.60	26	0	307.18
loops	18	15	3	3.26	17	1	536.33	17	1	1123.35
loop-lit	10	10	0	22.22	10	0	280.87	5	5	5655.33
loop-acceleration	10	9	1	0.32	9	1	493.13	9	1	1030.78
loop-crafted	2	2	0	0.09	2	0	49.59	2	0	27.97
[9]	9	8	1	2.12	9	0	286.24	4	5	4576.98
Total	114	101	13	**34.65**	108	6	8939.56	68	46	43573.42

Benchmark		OOPSLA23			DIG			IKOS + PPLite		
Source	#Num	#Ver.	#Unk.	Time (s)	#Ver.	#Unk.	Time (s)	#Ver.	#Unk.	Time (s)
loop-invariants	5	1	4	14.09	0	5	2344.12	3	2	0.88
loop-new	2	0	2	5.83	0	2	241.68	1	1	988.06
loop-invgen	5	4	1	14.78	1	4	264.34	5	0	1.03
loops-crafted-1	25	22	3	82.03	10	15	6030.28	0	25	8270.92
loop-simple	2	0	2	5.82	0	2	493.24	2	0	10.01
loop-zilu	26	0	26	68.24	19	7	6878.38	17	9	1827.31
loops	18	4	14	48.36	3	15	5241.99	7	11	909.04
loop-lit	10	7	3	29.59	3	7	2024.29	5	5	3993.86
loop-acceleration	10	8	1	27.48	3	7	2263.40	6	4	1.66
loop-crafted	2	2	0	5.63	0	2	503.76	2	0	0.33
[9]	9	6	3	28.98	1	8	497.92	8	1	2.04
Total	114	55	59	330.83	40	74	26783.40	56	58	16005.15

The complete comparison results of DInvG with other tools are presented in Table 1. In the table, *Source* indicates the source category of the benchmark. The term *#Ver.* represents the number of examples correctly verified by the verifier, and *#Unk.* (unknown) mainly arises from the following situations: a) The front-end fails to parse correctly, resulting in program crashes. b) Returns **Unknown.** c) Timeouts. For the benchmarks from [9], which do not contain assertions to be verified, we modify the invariants generated by our DInvG as assertions and test them over the other tools to obtain results.

From the table, it is evident that DInvG typically requires less than 0.3 s on average for verification, and its overall verification accuracy is very close to that of the SV-COMP 2023 Reachability track winner Veriabs, while significantly

Table 2. Experiment for Invariant Propagation

Benchmark	DInvG					
	No PPG			PPG		
Source	#Num	#Ver.	#Unk.	Time (s)	#Ver.	#Unk.	Time (s)
SV-COMP	105	91	14	1825.53	93	12	32.53
paper	9	8	1	10.76	8	1	2.12

outperforming Veriabs in terms of time efficiency by 10X to 1000X. This is mainly because Veriabs employs a rich strategy to assist verification, granting it a stronger verification capability but also requiring more time for most examples. CPAChecker experienced a broad range of timeouts in examples with complex loops that could not be verified within a finite unfolding of loops. This is due to the intrinsic limitations of its bounded model checking approach, and its loop unwinding strategy also results in verification times on the dataset that significantly exceed those of other tools.

Despite the fact that the tool from [82] has the second fewest number of verified benchmarks, it outperforms other tools in examples suitable for recurrence analysis. For DIG, we employ it to generate loop invariants and post conditions, and use Z3 [89] prover to verify the assertion. Nevertheless, the frontend of DIG necessitates CIVL's reliance on extracting symbolic execution paths from the program. When processing loops, it similarly depends on loop unrolling, and if it cannot fully unroll loops within a small bound, it determines that locations after the loop are unreachable. Consequently, it exhibits issues analogous to those of CPAChecker. Additionally, for some randomly assigned variables in SV-COMP, DIG lacks a suitable modeling. We have already reported several bugs via issues on GitHub. As a classical framework for abstract interpretation, IKOS with PPLite did not deliver optimal verification outcomes on the dataset. In some straightforward nested loops and more extensive loop iterations, it either failed to converge to a fixed point, or the precision of the invariants obtained upon convergence was insufficient to verify assertions, thereby causing timeouts or unknown in certain instances.

In summary, we conclude that DInvG significantly outperforms other tools such as Veriabs in time efficiency for affine numerical programs, while its verification capability is not inferior to the SV-COMP winner Veriabs. We also conducted an in-depth analysis of the cases where our DInvG returns **Unknown**. The primary reasons for the issues include: a) the absence of type range constraints at the front end, b) reliance on modular arithmetic, c) the need for more complex loop generalizations, d) exceeding the computational precision of the PPL library, and e) exponential arithmetic that surpasses the modeling capabilities of linear templates. 7–8 of these cases could be further solved by optimizing implementations. In the verifiable cases, the preliminary implementation of DInvG has already far surpassed existing methods in efficiency.

5.3 Ablation Study in Invariant Propagation (RQ2)

In this section, we conduct an ablation study to evaluate the performance of the invariant propagation technique within DInvG. In Table 2, we present the overall results, where we can clearly observe that the use of invariant propagation leads to a 5X–50X improvement in time efficiency.

More specifically, through the scatter plot in Fig. 7, we compared the time performance of individual examples before and after the application of invariant propagation techniques. In some cases, invariant propagation led to significant efficiency improvements (10X-1000X). This is due to the fact that for more complex programs, the size of the ATS Γ is larger, and applying invariant propagation techniques on this basis can maximize performance optimization. Since the tool itself performs efficiently in most examples, the optimization brought by this technique is not apparent in those cases in the graph where the time is below 0.1 s. As the propagation itself, including the projection of sub-ATS, incurs a certain time cost, which dilutes the time optimization brought about by invariant propagation.

In conclusion, invariant propagation significantly enhances the tool's scalability and yields superior optimization results for complex examples. This also reveals that, within our constraint-solving methodology, the cost of computing invariants at any given location constitutes the principal computational bottleneck. By reducing the number of locations that need to be computed and leveraging prior results to avoid redundant polyhedral operations, we can effectively enhance efficiency.

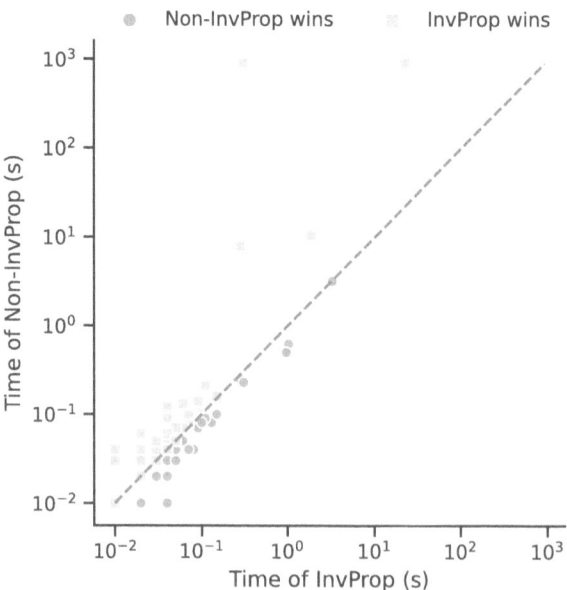

Fig. 7. Comparison for Invariant Propagation

5.4 Caveat to Correctness

This section elucidates configurations that may induce subtle deviations from real-world programs or alternative models during the empirical evaluation of our tool.

- In our current experimental setup, we have not accounted for the behavior of machine integers during overflow conditions. Consequently, our verification process is confined to affine programs that do not encounter overflow errors.
- Within the context of control flow transformations, we introduce uncertainty into conditional statements by adhering to the SV-COMP guidelines. This is achieved by replacing branch conditions with functions that return random Boolean values, thereby emulating the semantics of non-deterministic branches. Nonetheless, we have yet to effectively model uncertainty in variable coefficients, specifically affine inequalities with coefficients represented as intervals.

6 Related Works

Our methodology enhances conjunctive affine invariants by integrating optimizations from prior research [19,45,55,69] and utilizing control flow transformation techniques to extend them to disjunctive forms. A principal contribution of this paper is the mitigation of computational inefficiencies resulting from the exponential state space expansion associated with disjunctive extensions, achieved through invariant propagation. Consequently, this approach distinguishes our work from existing studies. The work [39] generates disjunctive invariants by predefining disjunctive templates, heuristically selecting physical cut points (while we select abstract locations from loop paths) and transforming the quadratic constraints from Farkas' Lemma into SAT solving. Other approaches for conjunctive affine invariant generation include [40,61]. These approaches propose completely different techniques, and thus are orthogonal to our approach.

Polynomial invariant generation [1,15,17,21,43,44,47,53,60,66,68,87] has been widely investigated. Most of these approaches consider conjunctive polynomial invariants only. Compared with conjunctive polynomial invariants, disjunctive affine invariants capture the precise feature of phase and mode changes in affine loops, and therefore are more precise.

The works [54,85] are based on path dependency automata, requiring precise estimates of the number of iterations in loops, which limits their analysis to programs with regular alternation and inductive variables (computable general terms). The work [72] studies the detection of multiphase disjunctive invariants. Multiphase invariants are a special case of our control flow transformation since each phase in a multiphase loop cannot go back to previous phases, while in our control flow transformation, locations can go back and forth via transitions. Thus, we have a wider class of disjunctive invariants as compared with [72].

Our control flow transformation is related to control flow refinement [7,27,38, 75] in the literature. These approaches mostly focus on representing the control flow of multiple loop iterations as regular expressions and refine these regular expressions by various approaches such as abstract domains, simulation relation and even invariant generation to reduce infeasible paths. Our control flow transformation considers the loop body within a single loop iteration, and is dedicated to the application of Farkas' Lemma. Thus, our control flow transformation has a different focus compared with these results. Moreover, the use of Farkas' Lemma can circumvent the issue that finer control flow may not always lead to finer analysis in control flow refinement [27].

Our invariant propagation is related to abstract interpretation [5,9,24,37,42, 76]. The main difference is that it propagates the *already-computed* invariants (via Farkas' Lemma) to yet not computed locations as much as possible to minimize the invariant generation computation, while abstract interpretation usually requires an involved fixed-point iteration to *compute* invariants.

Recurrence analysis [33,49,50] works well over programs with specific structure that ensures closed form solutions. For example, the most related recurrence analysis approach [82] (that also targets disjunctive invariants) solves the exact invariant over the class of loops with (ultimate) strict alternation between different modes. Compared with recurrence analysis, our approach does not require specific program structure to ensure closed form solution, but is less precise over programs that can be solved exactly by recurrence analysis.

Finally, we compare our approach with other methods such as machine learning, inference and data-driven approaches. Unlike constraint solving that can have an accuracy guarantee for the generated invariants based on the constraints, these methods cannot have an accuracy guarantee. Furthermore, machine learning and data-driven approaches themselves cannot guarantee that the generated assertions are indeed invariants. Moreover, our approach can generate invariants *without* the need of a goal property, while several approaches (such as IC3 [77], CLN2INV [64,67]) usually requires a goal property. Note that the invariant generation without a given goal property is a classical setting (see e.g. [19,24]), and has applications in loop summary and probabilistic program verification (see e.g. [13,83]).

LLM-based invariant generation methods [84] performs poorly on certain complex programs exhibiting disjunctive features. Those large-scale models have been unable to precisely comprehend the disjunctive properties inherent in these programs, and the invariants they produce often necessitate iterative interaction with Frama-C until an invariant that can be successfully verified by Frama-C is generated.

7 Conclusion

In this work, we propose a novel approach to generate affine disjunctive invariants over affine loops. Our novelty lies in combining a control flow transformation to extract the interleaving relationships between loop paths and employing Farkas'

Lemma to solve the disjunctive invariants of loops. Additionally, we apply invariant propagation techniques to mitigate the computational costs of exponential explosion. A thorough resolution of the infeasible implication in the application of Farkas' Lemma and an extension to nested loops through loop summary are proposed as optimizations for practical program verification. Experimental results show that our approach is competitive with state-of-the-art software verifiers in affine disjunctive invariant generation over affine loops.

Acknowledgments. We thank anonymous reviewers for constructive comments. This work is supported by the National Key R&D Program of China (No. 2022YFA1005101) and the National Natural Science Foundation of China (Nos. 61872232, 62172271).

References

1. Adjé, A., Garoche, P.-L., Magron, V.: Property-based polynomial invariant generation using sums-of-squares optimization. In: Blazy, S., Jensen, T. (eds.) SAS 2015. LNCS, vol. 9291, pp. 235–251. Springer, Heidelberg (2015). https://doi.org/10.1007/978-3-662-48288-9_14
2. Albarghouthi, A., Li, Y., Gurfinkel, A., Chechik, M.: Ufo: a framework for abstraction- and interpolation-based software verification. In: Madhusudan, P., Seshia, S.A. (eds.) CAV 2012. LNCS, vol. 7358, pp. 672–678. Springer, Heidelberg (2012). https://doi.org/10.1007/978-3-642-31424-7_48
3. Alias, C., Darte, A., Feautrier, P., Gonnord, L.: Multi-dimensional rankings, program termination, and complexity bounds of flowchart programs. In: Cousot, R., Martel, M. (eds.) SAS 2010. LNCS, vol. 6337, pp. 117–133. Springer, Heidelberg (2010). https://doi.org/10.1007/978-3-642-15769-1_8
4. Asadi, A., Chatterjee, K., Fu, H., Goharshady, A.K., Mahdavi, M.: Polynomial reachability witnesses via stellensätze. In: PLDI, pp. 772–787. ACM (2021). https://doi.org/10.1145/3453483.3454076
5. Bagnara, R., Hill, P.M., Ricci, E., Zaffanella, E.: Precise widening operators for convex polyhedra. In: Cousot, R. (ed.) SAS 2003. LNCS, vol. 2694, pp. 337–354. Springer, Heidelberg (2003). https://doi.org/10.1007/3-540-44898-5_19
6. Bagnara, R., Ricci, E., Zaffanella, E., Hill, P.M.: Possibly not closed convex polyhedra and the parma polyhedra library. In: Hermenegildo, M.V., Puebla, G. (eds.) SAS 2002. LNCS, vol. 2477, pp. 213–229. Springer, Heidelberg (2002). https://doi.org/10.1007/3-540-45789-5_17
7. Balakrishnan, G., Sankaranarayanan, S., Ivancic, F., Gupta, A.: Refining the control structure of loops using static analysis. In: Chakraborty, S., Halbwachs, N. (eds.) Proceedings of the 9th ACM & IEEE International conference on Embedded software, EMSOFT 2009, Grenoble, France, 12–16 October 2009, pp. 49–58. ACM (2009). https://doi.org/10.1145/1629335.1629343
8. Becchi, A., Zaffanella, E.: Pplite: zero-overhead encoding of nnc polyhedra. Inf. Comput. **275**, 104620 (2020)
9. Boutonnet, R., Halbwachs, N.: Disjunctive relational abstract interpretation for interprocedural program analysis. In: Enea, C., Piskac, R. (eds.) VMCAI 2019. LNCS, vol. 11388, pp. 136–159. Springer, Cham (2019). https://doi.org/10.1007/978-3-030-11245-5_7

10. Bradley, A.R., Manna, Z., Sipma, H.B.: Linear ranking with reachability. In: Etessami, K., Rajamani, S.K. (eds.) CAV 2005. LNCS, vol. 3576, pp. 491–504. Springer, Heidelberg (2005). https://doi.org/10.1007/11513988_48
11. Brat, G., Navas, J.A., Shi, N., Venet, A.: IKOS: a framework for static analysis based on abstract interpretation. In: Giannakopoulou, D., Salaün, G. (eds.) SEFM 2014. LNCS, vol. 8702, pp. 271–277. Springer, Cham (2014). https://doi.org/10.1007/978-3-319-10431-7_20
12. Calcagno, C., Distefano, D., O'Hearn, P.W., Yang, H.: Compositional shape analysis by means of bi-abduction. J. ACM **58**(6), 26:1–26:66 (2011). https://doi.org/10.1145/2049697.2049700
13. Chakarov, A., Sankaranarayanan, S.: Probabilistic program analysis with martingales. In: Sharygina, N., Veith, H. (eds.) CAV 2013. LNCS, vol. 8044, pp. 511–526. Springer, Heidelberg (2013). https://doi.org/10.1007/978-3-642-39799-8_34
14. Chatterjee, K., Fu, H., Goharshady, A.K.: Non-polynomial worst-case analysis of recursive programs. ACM Trans. Program. Lang. Syst. **41**(4), 20:1–20:52 (2019). https://doi.org/10.1145/3339984
15. Chatterjee, K., Fu, H., Goharshady, A.K., Goharshady, E.K.: Polynomial invariant generation for non-deterministic recursive programs. In: PLDI, pp. 672–687. ACM (2020). https://doi.org/10.1145/3385412.3385969
16. Chen, Y., Xia, B., Yang, L., Zhan, N., Zhou, C.: Discovering non-linear ranking functions by solving semi-algebraic systems. In: Jones, C.B., Liu, Z., Woodcock, J. (eds.) ICTAC 2007. LNCS, vol. 4711, pp. 34–49. Springer, Heidelberg (2007). https://doi.org/10.1007/978-3-540-75292-9_3
17. Chen, Y.-F., Hong, C.-D., Wang, B.-Y., Zhang, L.: Counterexample-guided polynomial loop invariant generation by lagrange interpolation. In: Kroening, D., Păsăreanu, C.S. (eds.) CAV 2015. LNCS, vol. 9206, pp. 658–674. Springer, Cham (2015). https://doi.org/10.1007/978-3-319-21690-4_44
18. Clang static analyzer: a source code analysis tool that finds bugs in c, c++, and objective-c programs (2022). https://clang-analyzer.llvm.org/
19. Colón, M.A., Sankaranarayanan, S., Sipma, H.B.: Linear invariant generation using non-linear constraint solving. In: Hunt, W.A., Somenzi, F. (eds.) CAV 2003. LNCS, vol. 2725, pp. 420–432. Springer, Heidelberg (2003). https://doi.org/10.1007/978-3-540-45069-6_39
20. Colóon, M.A., Sipma, H.B.: Synthesis of linear ranking functions. In: Margaria, T., Yi, W. (eds.) TACAS 2001. LNCS, vol. 2031, pp. 67–81. Springer, Heidelberg (2001). https://doi.org/10.1007/3-540-45319-9_6
21. Cousot, P.: Proving program invariance and termination by parametric abstraction, lagrangian relaxation and semidefinite programming. In: Cousot, R. (ed.) VMCAI 2005. LNCS, vol. 3385, pp. 1–24. Springer, Heidelberg (2005). https://doi.org/10.1007/978-3-540-30579-8_1
22. Cousot, P., Cousot, R.: Abstract interpretation: a unified lattice model for static analysis of programs by construction or approximation of fixpoints. In: POPL, pp. 238–252. ACM (1977). https://doi.org/10.1145/512950.512973
23. Cousot, P., Cousot, R., Feret, J., Mauborgne, L., Miné, A., Monniaux, D., Rival, X.: The astrée analyzer. In: Programming Languages and Systems: 14th European Symposium on Programming, ESOP 2005, Held as Part of the Joint European Conferences on Theory and Practice of Software, ETAPS 2005, Edinburgh, UK, 4–8 April 2005. Proceedings 14, pp. 21–30. Springer, Heidelberg (2005)
24. Cousot, P., Halbwachs, N.: Automatic discovery of linear restraints among variables of a program. In: POPL, pp. 84–96. ACM Press (1978). https://doi.org/10.1145/512760.512770

25. Cpachecker: The configurable software-verification platform (2022). https://cpachecker.sosy-lab.org
26. Csallner, C., Tillmann, N., Smaragdakis, Y.: Dysy: dynamic symbolic execution for invariant inference. In: ICSE, pp. 281–290. ACM (2008).https://doi.org/10.1145/1368088.1368127
27. Cyphert, J., Breck, J., Kincaid, Z., Reps, T.W.: Refinement of path expressions for static analysis. Proc. ACM Program. Lang. **3**(POPL), 45:1–45:29 (2019). https://doi.org/10.1145/3290358
28. Darke, P., Agrawal, S., Venkatesh, R.: Veriabs: a tool for scalable verification by abstraction (competition contribution). In: Tools and Algorithms for the Construction and Analysis of Systems: 27th International Conference, TACAS 2021, Held as Part of the European Joint Conferences on Theory and Practice of Software, ETAPS 2021, Luxembourg City, Luxembourg, 27 March–1 April 2021, Proceedings, Part II 27, pp. 458–462. Springer, Heidelberg (2021)
29. David, C., Kesseli, P., Kroening, D., Lewis, M.: Danger invariants. In: Fitzgerald, J., Heitmeyer, C., Gnesi, S., Philippou, A. (eds.) FM 2016. LNCS, vol. 9995, pp. 182–198. Springer, Cham (2016). https://doi.org/10.1007/978-3-319-48989-6_12
30. Dillig, I., Dillig, T., Li, B., McMillan, K.L.: Inductive invariant generation via abductive inference. In: OOPSLA, pp. 443–456. ACM (2013). https://doi.org/10.1145/2509136.2509511
31. Donaldson, A.F., Haller, L., Kroening, D., Rümmer, P.: Software verification using k-induction. In: Yahav, E. (ed.) SAS 2011. LNCS, vol. 6887, pp. 351–368. Springer, Heidelberg (2011). https://doi.org/10.1007/978-3-642-23702-7_26
32. Farkas, J.: A fourier-féle mechanikai elv alkalmazásai (Hungarian). Mathematikaiés Természettudományi Értesitö **12**, 457–472 (1894)
33. Farzan, A., Kincaid, Z.: Compositional recurrence analysis. In: FMCAD, pp. 57–64. IEEE (2015)
34. Gan, T., Xia, B., Xue, B., Zhan, N., Dai, L.: Nonlinear craig interpolant generation. In: Lahiri, S.K., Wang, C. (eds.) CAV 2020. LNCS, vol. 12224, pp. 415–438. Springer, Cham (2020). https://doi.org/10.1007/978-3-030-53288-8_20
35. Garg, P., Löding, C., Madhusudan, P., Neider, D.: ICE: a robust framework for learning invariants. In: Biere, A., Bloem, R. (eds.) CAV 2014. LNCS, vol. 8559, pp. 69–87. Springer, Cham (2014). https://doi.org/10.1007/978-3-319-08867-9_5
36. Garg, P., Neider, D., Madhusudan, P., Roth, D.: Learning invariants using decision trees and implication counterexamples. In: POPL, pp. 499–512. ACM (2016). https://doi.org/10.1145/2837614.2837664
37. Gopan, D., Reps, T.: Guided static analysis. In: Nielson, H.R., Filé, G. (eds.) SAS 2007. LNCS, vol. 4634, pp. 349–365. Springer, Heidelberg (2007). https://doi.org/10.1007/978-3-540-74061-2_22
38. Gulwani, S., Jain, S., Koskinen, E.: Control-flow refinement and progress invariants for bound analysis. In: Hind, M., Diwan, A. (eds.) Proceedings of the 2009 ACM SIGPLAN Conference on Programming Language Design and Implementation, PLDI 2009, Dublin, Ireland, 15–21 June 2009, pp. 375–385. ACM (2009). https://doi.org/10.1145/1542476.1542518
39. Gulwani, S., Srivastava, S., Venkatesan, R.: Program analysis as constraint solving. In: PLDI, pp. 281–292. ACM (2008). https://doi.org/10.1145/1375581.1375616
40. Gupta, A., Rybalchenko, A.: InvGen: an efficient invariant generator. In: Bouajjani, A., Maler, O. (eds.) CAV 2009. LNCS, vol. 5643, pp. 634–640. Springer, Heidelberg (2009). https://doi.org/10.1007/978-3-642-02658-4_48

41. He, J., Singh, G., Püschel, M., Vechev, M.T.: Learning fast and precise numerical analysis. In: PLDI, pp. 1112–1127. ACM (2020). https://doi.org/10.1145/3385412.3386016
42. Henry, J., Monniaux, D., Moy, M.: PAGAI: a path sensitive static analyser. Electron. Notes Theor. Comput. Sci. **289**, 15–25 (2012). https://doi.org/10.1016/j.entcs.2012.11.003
43. Hrushovski, E., Ouaknine, J., Pouly, A., Worrell, J.: Polynomial invariants for affine programs. In: LICS, pp. 530–539. ACM (2018). https://doi.org/10.1145/3209108.3209142
44. Humenberger, A., Jaroschek, M., Kovács, L.: Automated generation of non-linear loop invariants utilizing hypergeometric sequences. In: ISSAC, pp. 221–228. ACM (2017). https://doi.org/10.1145/3087604.3087623
45. Ji, Y., Fu, H., Fang, B., Chen, H.: Affine loop invariant generation via matrix algebra. In: Shoham, S., Vizel, Y. (eds.) Computer Aided Verification - 34th International Conference, CAV 2022, Haifa, Israel, 7–10 August 2022, Proceedings, Part I. Lecture Notes in Computer Science, vol. 13371, pp. 257–281. Springer, Heidelberg (2022). https://doi.org/10.1007/978-3-031-13185-1_13
46. K., H.G.V., Shoham, S., Gurfinkel, A.: Solving constrained horn clauses modulo algebraic data types and recursive functions. Proc. ACM Program. Lang. **6**(POPL), 1–29 (2022). https://doi.org/10.1145/3498722
47. Kapur, D.: Automatically generating loop invariants using quantifier elimination. In: Deduction and Applications. Dagstuhl Seminar Proceedings, vol. 05431. Internationales Begegnungs- und Forschungszentrum für Informatik (IBFI), Schloss Dagstuhl, Germany (2005). http://drops.dagstuhl.de/opus/volltexte/2006/511
48. Ke, J., Fu, H., Liu, H., Chen, L., Li, G.: Affine disjunctive invariant generation with farkas' lemma. arXiv preprint arXiv:2307.13318 (2023)
49. Kincaid, Z., Breck, J., Boroujeni, A.F., Reps, T.W.: Compositional recurrence analysis revisited. In: PLDI, pp. 248–262. ACM (2017). https://doi.org/10.1145/3062341.3062373
50. Kincaid, Z., Cyphert, J., Breck, J., Reps, T.W.: Non-linear reasoning for invariant synthesis. Proc. ACM Program. Lang. **2**(POPL), 54:1–54:33 (2018). https://doi.org/10.1145/3158142
51. Larraz, D., Rodríguez-Carbonell, E., Rubio, A.: SMT-based array invariant generation. In: Giacobazzi, R., Berdine, J., Mastroeni, I. (eds.) VMCAI 2013. LNCS, vol. 7737, pp. 169–188. Springer, Heidelberg (2013). https://doi.org/10.1007/978-3-642-35873-9_12
52. Le, T.C., Zheng, G., Nguyen, T.: SLING: using dynamic analysis to infer program invariants in separation logic. In: McKinley, K.S., Fisher, K. (eds.) Proceedings of the 40th ACM SIGPLAN Conference on Programming Language Design and Implementation, PLDI 2019, Phoenix, AZ, USA, 22–26 June 2019, pp. 788–801. ACM (2019). https://doi.org/10.1145/3314221.3314634
53. Lin, W., Wu, M., Yang, Z., Zeng, Z.: Proving total correctness and generating preconditions for loop programs via symbolic-numeric computation methods. Front. Comp. Sci. **8**(2), 192–202 (2014). https://doi.org/10.1007/s11704-014-3150-6
54. Lin, Y., et al.: Inferring loop invariants for multi-path loops. In: International Symposium on Theoretical Aspects of Software Engineering, TASE 2021, Shanghai, China, 25–27 August 2021, pp. 63–70. IEEE (2021). https://doi.org/10.1109/TASE52547.2021.00030
55. Liu, H., Fu, H., Yu, Z., Song, J., Li, G.: Scalable linear invariant generation with Farkas' lemma. Proc. ACM Program. Lang. **6**(OOPSLA2) (2022). https://doi.org/10.1145/3563295

56. Manna, Z., Pnueli, A.: Temporal Verification of Reactive Systems - Safety. Springer, Heidelberg (1995)
57. McMillan, K.L.: Quantified invariant generation using an interpolating saturation prover. In: Ramakrishnan, C.R., Rehof, J. (eds.) TACAS 2008. LNCS, vol. 4963, pp. 413–427. Springer, Heidelberg (2008). https://doi.org/10.1007/978-3-540-78800-3_31
58. Nguyen, T., Kapur, D., Weimer, W., Forrest, S.: Using dynamic analysis to discover polynomial and array invariants. In: ICSE, pp. 683–693. IEEE Computer Society (2012). https://doi.org/10.1109/ICSE.2012.6227149
59. Nguyen, T., Nguyen, K., Duong, H.: Syminfer: inferring numerical invariants using symbolic states. In: Proceedings of the ACM/IEEE 44th International Conference on Software Engineering: Companion Proceedings, pp. 197–201 (2022)
60. de Oliveira, S., Bensalem, S., Prevosto, V.: Polynomial invariants by linear algebra. In: Artho, C., Legay, A., Peled, D. (eds.) ATVA 2016. LNCS, vol. 9938, pp. 479–494. Springer, Cham (2016). https://doi.org/10.1007/978-3-319-46520-3_30
61. de Oliveira, S., Bensalem, S., Prevosto, V.: Synthesizing invariants by solving solvable loops. In: D'Souza, D., Narayan Kumar, K. (eds.) ATVA 2017. LNCS, vol. 10482, pp. 327–343. Springer, Cham (2017). https://doi.org/10.1007/978-3-319-68167-2_22
62. Padon, O., McMillan, K.L., Panda, A., Sagiv, M., Shoham, S.: Ivy: safety verification by interactive generalization. In: PLDI, pp. 614–630. ACM (2016). https://doi.org/10.1145/2908080.2908118
63. Podelski, A., Rybalchenko, A.: A complete method for the synthesis of linear ranking functions. In: Steffen, B., Levi, G. (eds.) VMCAI 2004. LNCS, vol. 2937, pp. 239–251. Springer, Heidelberg (2004). https://doi.org/10.1007/978-3-540-24622-0_20
64. Riley, D., Fedyukovich, G.: Multi-phase invariant synthesis. In: Proceedings of the 30th ACM Joint European Software Engineering Conference and Symposium on the Foundations of Software Engineering, pp. 607–619 (2022)
65. Rodríguez-Carbonell, E., Kapur, D.: An abstract interpretation approach for automatic generation of polynomial invariants. In: Giacobazzi, R. (ed.) SAS 2004. LNCS, vol. 3148, pp. 280–295. Springer, Heidelberg (2004). https://doi.org/10.1007/978-3-540-27864-1_21
66. Rodríguez Carbonell, E., Kapur, D.: Automatic generation of polynomial loop invariants: algebraic foundations. In: ISSAC, pp. 266–273. ACM (2004). https://doi.org/10.1145/1005285.1005324
67. Ryan, G., Wong, J., Yao, J., Gu, R., Jana, S.: CLN2INV: learning loop invariants with continuous logic networks. In: 8th International Conference on Learning Representations, ICLR 2020, Addis Ababa, Ethiopia, 26–30 April 2020. OpenReview.net (2020). https://openreview.net/forum?id=HJlfuTEtvB
68. Sankaranarayanan, S., Sipma, H., Manna, Z.: Non-linear loop invariant generation using gröbner bases. In: POPL, pp. 318–329. ACM (2004). https://doi.org/10.1145/964001.964028
69. Sankaranarayanan, S., Sipma, H.B., Manna, Z.: Constraint-based linear-relations analysis. In: Giacobazzi, R. (ed.) SAS 2004. LNCS, vol. 3148, pp. 53–68. Springer, Heidelberg (2004). https://doi.org/10.1007/978-3-540-27864-1_7
70. Schrijver, A.: Theory of linear and integer programming. Wiley-Interscience series in discrete mathematics and optimization, Wiley (1999)
71. Sharma, R., Aiken, A.: From invariant checking to invariant inference using randomized search. Formal Methods Syst. Des. **48**(3), 235–256 (2016). https://doi.org/10.1007/s10703-016-0248-5

72. Sharma, R., Dillig, I., Dillig, T., Aiken, A.: Simplifying loop invariant generation using splitter predicates. In: Gopalakrishnan, G., Qadeer, S. (eds.) CAV 2011. LNCS, vol. 6806, pp. 703–719. Springer, Heidelberg (2011). https://doi.org/10.1007/978-3-642-22110-1_57
73. Sharma, R., Gupta, S., Hariharan, B., Aiken, A., Liang, P., Nori, A.V.: A data driven approach for algebraic loop invariants. In: Felleisen, M., Gardner, P. (eds.) ESOP 2013. LNCS, vol. 7792, pp. 574–592. Springer, Heidelberg (2013). https://doi.org/10.1007/978-3-642-37036-6_31
74. Siegel, S.F., et al.: Civl: the concurrency intermediate verification language. In: Proceedings of the International Conference for High Performance Computing, Networking, Storage and Analysis, pp. 1–12 (2015)
75. Silverman, J., Kincaid, Z.: Loop summarization with rational vector addition systems. In: Dillig, I., Tasiran, S. (eds.) CAV 2019. LNCS, vol. 11562, pp. 97–115. Springer, Cham (2019). https://doi.org/10.1007/978-3-030-25543-5_7
76. Singh, G., Püschel, M., Vechev, M.T.: Fast polyhedra abstract domain. In: Castagna, G., Gordon, A.D. (eds.) Proceedings of the 44th ACM SIGPLAN Symposium on Principles of Programming Languages, POPL 2017, Paris, France, 18–20 January 2017, pp. 46–59. ACM (2017)
77. Somenzi, F., Bradley, A.R.: IC3: where monolithic and incremental meet. In: Bjesse, P., Slobodová, A. (eds.) International Conference on Formal Methods in Computer-Aided Design, FMCAD '11, Austin, TX, USA, 30 October–02 November 2011, pp. 3–8. FMCAD Inc. (2011). http://dl.acm.org/citation.cfm?id=2157657
78. Srivastava, S., Gulwani, S.: Program verification using templates over predicate abstraction. In: Hind, M., Diwan, A. (eds.) Proceedings of the 2009 ACM SIGPLAN Conference on Programming Language Design and Implementation, PLDI 2009, Dublin, Ireland, 15–21 June 2009, pp. 223–234. ACM (2009). https://doi.org/10.1145/1542476.1542501
79. Sting: Stanford invariant generator (2006). http://theory.stanford.edu/~srirams/Software/sting.html
80. Software verification competition (2023). https://sv-comp.sosy-lab.org
81. Tarjan, R.: Depth-first search and linear graph algorithms. SIAM J. Comput. **1**(2), 146–160 (1972)
82. Wang, C., Lin, F.: Solving conditional linear recurrences for program verification: the periodic case. In: OOPSLA. ACM (2023)
83. Wang, J., Sun, Y., Fu, H., Chatterjee, K., Goharshady, A.K.: Quantitative analysis of assertion violations in probabilistic programs. In: PLD, pp. 1171–1186. ACM (2021). https://doi.org/10.1145/3453483.3454102
84. Wen, C., et al.: Enchanting program specification synthesis by large language models using static analysis and program verification. In: International Conference on Computer Aided Verification, pp. 302–328. Springer, Heidelberg (2024). https://doi.org/10.1007/978-3-031-65630-9_16
85. Xie, X., Chen, B., Liu, Y., Le, W., Li, X.: Proteus: computing disjunctive loop summary via path dependency analysis. In: Zimmermann, T., Cleland-Huang, J., Su, Z. (eds.) Proceedings of the 24th ACM SIGSOFT International Symposium on Foundations of Software Engineering, FSE 2016, Seattle, WA, USA, 13–18 November 2016, pp. 61–72. ACM (2016). https://doi.org/10.1145/2950290.2950340
86. Xu, R., He, F., Wang, B.: Interval counterexamples for loop invariant learning. In: ESEC/FSE, pp. 111–122. ACM (2020). https://doi.org/10.1145/3368089.3409752
87. Yang, L., Zhou, C., Zhan, N., Xia, B.: Recent advances in program verification through computer algebra. Front. Comput. Sci. China **4**(1), 1–16 (2010). https://doi.org/10.1007/s11704-009-0074-7

88. Yao, J., Ryan, G., Wong, J., Jana, S., Gu, R.: Learning nonlinear loop invariants with gated continuous logic networks. In: PLDI, pp. 106–120. ACM (2020). https://doi.org/10.1145/3385412.3385986
89. Z3 (2023). https://github.com/Z3Prover/z3

Automatic Inference of Relational Object Invariants

Yusen Su[1](✉), Jorge A. Navas[2], Arie Gurfinkel[1], and Isabel Garcia-Contreras[1,3]

[1] Department of Electrical and Computer Engineering, University of Waterloo, Waterloo, ON, Canada
{y256su,arie.gurfinkel,igarciac}@uwaterloo.ca
[2] Certora Inc., Seattle, WA, USA
jorge@certora.com
[3] Black Duck Software, Inc., Burlington, MA, USA

Abstract. Relational object invariants (or representation invariants) are relational properties held by the fields of a (memory) object throughout its lifetime. For example, the length of a buffer never exceeds its capacity. Automatic inference of these invariants is particularly challenging because they are often broken temporarily during field updates.

In this paper, we present an Abstract Interpretation-based solution to infer object invariants. Our key insight is a new object abstraction for memory objects, where memory is divided into multiple *memory banks*, each containing several objects. Within each bank, objects are abstracted by separating the *most recently used* (MRU) object, represented precisely with strong updates, while the rest are summarized. For an effective implementation of this approach, we introduce a new composite abstract domain, which forms a reduced product of numerical and equality sub-domains. This design efficiently expresses relationships between a small number of variables (e.g., fields of the same abstract object).

We implement the new domain in the CRAB abstract interpreter and evaluate it on several benchmarks for memory safety. We show that our approach is significantly more scalable for relational properties than the existing implementation of CRAB. To evaluate precision, we have integrated our analysis as a pre-processing step to SEABMC bounded model checker, and show that it is effective at both discharging assertions during pre-processing, and significantly improving the run-time of SEABMC.

Keywords: Static Analysis · Abstract Interpretation · Object Invariants · Abstract Domains

1 Introduction

Program invariants are crucial to capture properties that persist during runtime. Verifying programs with classes or data structures requires determining *repre-*

```
1  #define N 100
2  struct byte_buf {
3      int len;
4      int cap;
5      char *buf;
6  };
7  int main() {
8      struct byte_buf *ary[N];
9      for (int i = 0; i < N; ++i) {
10         struct byte_buf *p =
                 malloc(sizeof(struct byte_buf));
11         int sz = i + 1;
12         p->len = i; p->cap = sz;
13         p->buf = malloc(sz);
14         ary[i] = p;
15     }
16     char *new_buf = malloc(20);
17     ary[0]->len = 15;
18     ary[0]->cap = 20;
19     ary[0]->buf = new_buf;
20     assert(ary[0]->len <= ary[0]->cap);
21     ary[0]->buf[ary[0]->len] = '\0';
22 }
```

Fig. 1. A simple C program.

Fig. 2. Abstract memory state on line 17 of Fig. 1.

sentation invariants [19] that express *consistency* properties (e.g., the length of a vector never exceeds its capacity) of those data types. For memory objects, representation invariants as *object invariants* describe relational properties among object fields that hold across all program states where these objects are alive. These invariants are essential for proving memory safety and functional correctness of a program. However, the invariants become imprecise when the static analyzer is uncertain about which memory objects are affected by field updates, typically represented as *weak* updates.

Consider a C program in Fig. 1 that uses a `byte_buf` to represent a resizable byte buffer with length and capacity. The program keeps an array `ary` of byte buffers. Each initialized element of `ary` satisfies an invariant: `len <= cap`. Discovering this invariant is crucial for establishing memory safety (e.g., proving safe access on line 21), yet, notoriously hard for abstract interpreters. Note that *recency* [1] does not help here because all memory stores after the `for` loop are modeled as weak updates. For instance, Mopsa [22] with recency does not prove the assertion on line 20, since the inferred invariant is $len > 0 \land cap > 1$.

In this paper, we present a new technique for inferring object invariants. We capture field updates *strongly* in a separate temporary object abstraction and join it with previously established invariants only when necessary. While preserving soundness, our approach produces more precise analysis results by not weakening inferred invariants with intermediate object states between updates.

First, we introduce a new concrete memory model that organizes memory as a collection of *memory banks*, each containing certain memory objects. The partitioning is achieved by a parameterized function that assigns each memory object in the program a corresponding bank. Each bank has two components:

storage, holding objects, and *cache*, storing the object being read from or written to. For example, all byte buffers in Fig. 1 are placed into the storage of the same bank. The field updates on line 12 require loading the byte buffer referred by pointer `p` into the cache before updates. The cache singles out the object being modified. For brevity, we specify this usage pattern with a size of one as *most recently used* (MRU) and denote the object in the cache as the MRU object.

Second, we follow a standard summarization-based abstraction with a single summary object with its invariants representing properties common to all the objects stored in each bank. Similar to the concrete model, all memory updates are handled through the MRU object. This avoids temporarily breaking the invariants of the (abstract) summary object, as changes to the MRU object do not impact the summarized invariants until it is merged back. Figure 2 presents the changes in the abstract memory state at line 17. The memory bank for byte buffers includes one MRU object and one summary object. Before evaluating line 17, as shown in Fig. 2(a), `p` refers to the MRU object, since the last two field updates (line 12) happened on this object. Following the initialization loop, `len <= cap` is kept for both MRU and summary objects.

The cache may *miss* if the cached object is no longer the MRU. For example, the field update, `ary[0]->len = 15` , on line 17 requires access to the byte buffer referenced by `ary[0]` , while the cache still holds the object referred by `p` . In this case, the cached object is *packed* back to the summary (see Fig. 2(a)) and the new MRU object is *unpacked* from the summary (Fig. 2(b)). We track pointer alias information to decide when to pack and unpack. Before each memory access, if the dereferenced pointer does not alias with the pointer accessed to the MRU object, packing and unpacking occur. In this example, after the loop computation, `p` does not alias `ary[0]` .

After the cache is replaced, the field update, `ary[0]->len = 15` , breaks the invariant `len <= cap` , but our solution (Fig. 2(c)) ensures that we update the content of the MRU object properly without affecting the invariants in the summary object. Then, the invariant is restored at line 18, thus proving the assertion on line 20 and memory safety on line 21 through our invariants in the cache.

Third, we introduce a new abstract domain, called *MRUD*, that infers automatically object invariants based on our new memory model. This domain requires combining heap (memory abstraction), must alias (flow-sensitive points-to information) and value (numerical relational invariants) analyses. Using a monolithic numerical domain is highly inefficient because of the large number of dimensions required to model all program variables and their ghost versions that keep track of base addresses, offsets, etc. However, a key insight is that each transfer function typically affects a small subset of variables (e.g., reading a field only updates the corresponding integer/pointer value). Based on this observation, *MRUD* is designed as a composite abstract domain where each memory bank is modeled separately and the propagation of facts between them is carefully limited to a small set of shared variables. This modular design is what makes MRUD both scalable for large code bases and capable of preserving precise object invariants.

We implemented MRUD in the CRAB analyzer [13] and evaluated both its scalability and precision. For scalability, we compare it to the summarization-based abstract domain implemented in CRAB. Our approach shows improved scalability, with 75X faster performance than the state-of-the-art. For precision, we compare it to the recency domain implemented on Mopsa using a small set of benchmarks. The results show that our approach successfully proves all assertions in the programs and achieves better precision by preserving object invariants. Additionally, we use MRUD in a case study with the bounded model checker SEABMC, where it effectively proves and discharges memory safety checks to reduce the verification cost of SEABMC.

In summary, the contributions of this paper are: (1) We introduce a new memory model designed for object abstraction as an alternative to the C memory model, and describe the concrete semantics of an intermediate representation based on the new model (Sect. 3); (2) We describe the MRUD and corresponding abstract transfer functions, and introduce a domain reduction for invariant refinement (Sect. 4); (3) We detail our implementation (Sect. 5) and evaluate it in the CRAB analyzer (Sect. 6).

2 Preliminaries

Without loss of generality, we assume that the input program is in CrabIR [13] intermediate representation. The syntax of CrabIR is shown in Fig. 3. We assume that each memory object is a collection of integer and pointer fields. A pointer is a pair of a base address and an offset, where an offset is given by a number num and an optional field name fld. All named fields have fixed offsets. That is, field names are redundant – they are use to simplify the abstraction function in the abstract semantics. In our implementation, the field names are automatically discovered by a whole-program pointer analysis during compilation from the source language to CrabIR.

We write $\mathcal{V}$ for the set of all program variables. The set $\mathcal{V}$ is partitioned into: integers $\mathcal{V}_{int}$, pointers $\mathcal{V}_{ptr}$, and fields $\mathcal{V}_{fld}$. The union of $\mathcal{V}_{int}$ and $\mathcal{V}_{ptr}$ is called *scalars*. The statements in CrabIR consist of gotos, assumptions, assertions, and arithmetic and memory operations. All statements are strongly typed. Allocation of memory objects is performed by alloc (allocate). Pointer arithmetic is handled by the gep instruction that computes a destination address using the base pointer and an integer offset. Memory reads and writes are done by load and store, respectively. As usual, a program $\mathcal{P}$ is a control flow graph (CFG) whose basic blocks are annotated with statements from Fig. 3. CrabIR also supports C-like memory objects and it does not require them to be partitioned into fields. These are handled as in prior work [13]. We omit such objects in the theoretical exposition in the paper, but handle them as in [13] in our implementation.

$$P ::= F^+$$
$$F ::= \text{declare } fun(v^*)\{\ BB^+\ \}$$
$$BB ::= l : S^* \text{ goto } l^+ \ |$$
$$\quad\quad l : S^* \text{ return } v^*$$
$$S ::= \text{assert}(E_{\text{cond}}) \ | \ \text{assume}(E_{\text{cond}}) \ |$$
$$\quad\quad \text{num} := E_{\text{int}} \ | \ S_{\text{ptr}}$$

$$S_{\text{ptr}} ::= \text{ptr} := \text{alloc}(\text{fld}, \text{num}) \ |$$
$$\quad\quad \text{ptr2}, \text{fld2} := \text{gep}(\text{ptr1}, \text{fld1}, \text{num}) \ |$$
$$\quad\quad \text{scl} := \text{load}(\text{ptr}, \text{fld}) \ | \ \text{store}(\text{ptr}, \text{fld}, \text{scl})$$
$$E_{\text{int}} ::= \text{Const} \ | \ \text{num} \ | \ E_{\text{int}} \ op_{int} \ E_{\text{int}}$$
$$E_{\text{cond}} ::= E_{\text{int}} \ op_{cmp} \ E_{\text{int}}$$

Fig. 3. The syntax of CrabIR.

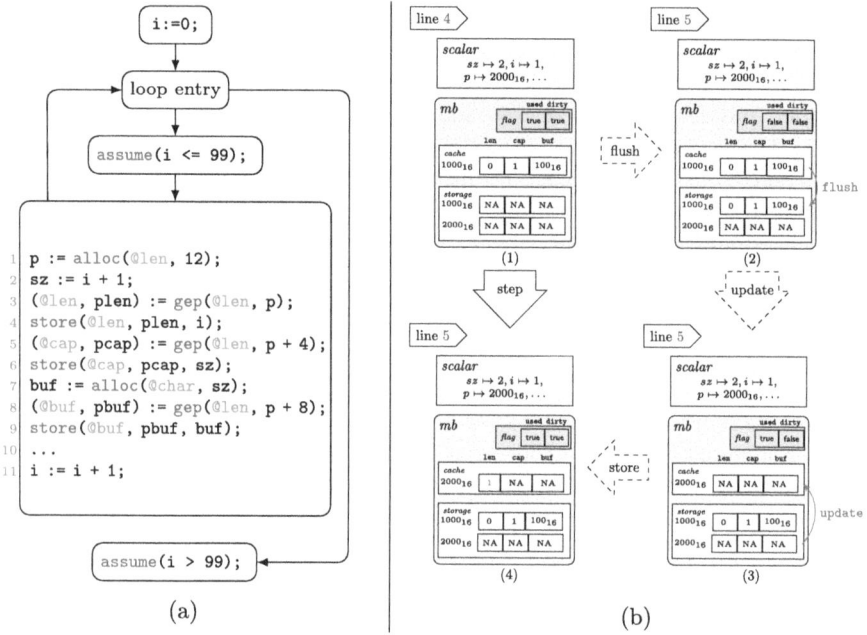

Fig. 4. (a) A program, and (b) an execution of line 4 under RUMM.

We assume the reader is familiar with a standard numerical abstract domain that provides the following operations: join ($\sqcup$), meet ($\sqcap$), widen (∇), projection ($\text{project}(d, \mathcal{V})$) that projects an abstract value d to the variable set $\mathcal{V}$, forget ($\text{forget}(d, v)$) that removes a variable v from an abstract value d, and constrain ($\text{addCons}(d, c)$) that restricts an abstract value d by a linear constraint c.

We use an equality domain over variable sets $\mathcal{V}$ to express equivalence relations such as $x \approx y$. The equality domain can be implemented using weakly relational numerical domains (e.g., [17,20,21]). We assume the equality domain has the following special operations: addEqual for adding an equality, equals for testing whether an abstract value entails an equality, and toCons for computing closed form of all implied equalities.

3 Recent-Use Memory Model

A memory model defines how memory is structured and accessed in the operational semantics (i.e., execution) of the program. The standard C memory model (CMM) treats each allocation as a blob of bytes. Specifically, each memory object is a blob of bytes (logically sub-divided into fields). A pointer is a pair (b, o) of an object identifier b (a.k.a., the *base* address) and a numeric offset o within that object. At allocation, an object occupies a blob in memory at an address determined by the memory allocator. Each memory operation is performed through a pointer to access the object's content. In practice, CMM is typically implemented by a flat memory model of the underlying architecture. However, non-flat memory models with multiple address spaces are common, especially in embedded systems [14].

In this paper, we introduce a new memory model, called *recent-use memory model* (RUMM), that differentiates between the most recently used (MRU) object and other memory objects. RUMM partitions memory into multiple *banks*, each with (a) a *storage* – a blob of bytes that permanently stores memory objects, and (b) a *cache* – a blob of bytes that temporarily holds the MRU object of that bank. The notion of objects and pointers in RUMM is exactly as in CMM. Furthermore, RUMM is parameterized by a function findmb that maps allocation sites to specific memory banks of RUMM. This is similar to a pool allocation, where objects are allocated in different pools [18]. Each object is allocated as a blob in the selected bank's storage, with each bank managing its allocations.

What makes RUMM special is its handling of read and write operations. To access an object x from a given bank, x is first loaded into the cache and then accessed from there. If a different object y currently occupies the cache, y is flushed back to its place in its memory bank before x is loaded. Thus, multiple read and write operations that work on the same object only use the cache, until the cache is flushed when a new object, from the same bank, is accessed.

Figure 4a shows a CrabIR for the for loop in Fig. 1. Variables prefixed with @ are the fields of byte_buf . The loop starts at the *entry block* and checks whether the counter i meets the enter/exit condition. In CrabIR, assume is used to enforce this condition. The loop initializes a memory object, increments the counter, and loops back to the loop entry. Figure 4b illustrates the execution of line 4 during the *second* iteration of the loop. Figure 4b(1) shows the state at line 4, where scalar variables map to their values as *scalar* and a memory bank mb is provided to store memory objects allocated at line 1. We assume the first two iterations allocate objects at addresses 1000_{16} and 2000_{16}, respectively. The fields of each object are visually represented as slots, with either concrete values or marked as not available (NA) if uninitialized. The storage keeps two uninitialized objects, while the cache holds the MRU object. The object at address 1000_{16} is the MRU since its last access is at line 9 during the first iteration. The cache status is indicated by two flags: *used*, indicating the cache is active, and *dirty*, meaning the cache value has been updated. When store at line 4 accesses the object with address 2000_{16}, the cache flushes the object (1000_{16}) back to the

$[\![\text{ptr} := \text{alloc}(\text{fld}, \text{num})]\!]^{\text{RUMM}}(\sigma) \equiv$
　let $\langle scalar, mem \rangle = \sigma$ **in**
　let $mb = \text{findmb}(\text{fld}, mem)$ **in**
　let $\langle cache, storage, flag \rangle = mb$ **in**
　let $\langle \_, sz \rangle = scalar[\text{num}]$ **in**
　let $\langle \text{ptr}^{base}, storage' \rangle =$
　　　$\text{allocator}_{mb}(storage, sz)$ **in**
　let $scalar' =$
　　　$scalar[\text{ptr} \mapsto \langle \text{ptr}^{base}, 0 \rangle]$ **in**
　let $mb' = \langle cache, storage', flag \rangle$ **in**
　$\langle scalar', mem \setminus \{mb\} \cup \{mb'\} \rangle$

$[\![\text{scl} := \text{load}(\text{ptr}, \text{fld})]\!]^{\text{RUMM}}(\sigma) \equiv$
　let $\langle scalar, mem \rangle = \sigma$ **in**
　let $mb = \text{findmb}(\text{fld}, mem)$ **in**
　let $\langle \text{ptr}^{base}, \_ \rangle = scalar[\text{ptr}]$ **in**
　let $mb' = \text{cacheSync}(mb, \text{ptr}^{base})$ **in**
　let $\langle cache, \_, \_ \rangle = mb'$ **in**
　let $\langle \_, fields \rangle = cache$ **in**
　let $scalar' =$
　　　$scalar[\text{scl} \mapsto fields[\text{fld}]]$ **in**
　$\langle scalar', mem \setminus \{mb\} \cup \{mb'\} \rangle$

$[\![\text{ptr2}, \text{fld2} := \text{gep}(\text{ptr1}, \text{fld1}, \text{num})]\!]^{\text{RUMM}}(\sigma) \equiv$
　let $\langle scalar, mem \rangle = \sigma$ **in**
　let $\langle \text{ptr1}^{base}, offset \rangle = scalar[\text{ptr}]$ **in**
　let $\langle \_, val \rangle = scalar[\text{num}]$ **in**
　let $offset' = offset + val$ **in**
　let $scalar' =$
　　　$scalar[\text{ptr2} \mapsto \langle \text{ptr1}^{base}, offset' \rangle]$ **in**
　$\langle scalar', mem \rangle$

$[\![\text{store}(\text{ptr}, \text{fld}, \text{scl})]\!]^{\text{RUMM}}(\sigma) \equiv$
　let $\langle scalar, mem \rangle = \sigma$ **in**
　let $mb = \text{findmb}(\text{fld}, mem)$ **in**
　let $\langle \text{ptr}^{base}, \_ \rangle = scalar[\text{ptr}]$ **in**
　let $mb' = \text{cacheSync}(mb, \text{ptr}^{base})$ **in**
　let $\langle cache, storage, \_ \rangle = mb'$ **in**
　let $\langle cache^{base}, fields \rangle = cache$ **in**
　let $cache' = \langle cache^{base},$
　　　$fields[\text{fld} \mapsto scalar[\text{scl}]] \rangle$ **in**
　let $mb'' =$
　　　$\langle cache', storage, \langle \text{true}, \text{true} \rangle \rangle$ **in**
　$\langle scalar, mem \setminus \{mb\} \cup \{mb''\} \rangle$

Fig. 5. CrabIR statements operating under RUMM.

storage (Fig. 4b(2)) and updates with the uninitialized object from the storage (Fig. 4b(3)). The cache is then ready to write @len with a value of 1 (Fig. 4b(4)).

We argue that RUMM is compatible with CMM. This follows from: (1) RUMM organizes memory objects into separate, non-overlapping memory banks; (2) The usage of cache is an extra step that does not invalidate the properties of each object. The semantics of CrabIR are the same under both memory models. In the following, we formalize the concrete semantics of CrabIR under RUMM.

A CrabIR program has scalars (i.e., integers $\mathcal{V}_{int}$ and pointers $\mathcal{V}_{ptr}$) whose values are represented as *cells*. A cell, $cell \in \text{Cell} : \mathbb{N} \times \mathbb{Z}$, represents either a pointer's base address and offset, denoted as $\langle baddr, offset \rangle$, or an integer value: $\langle 0, val \rangle$. Formally, a scalar state is $scalar \in \text{Scalar} : \mathcal{V}_{scl} \mapsto \text{Cell}$. To avoid redundancy, we explicitly associate the base address of a ptr with a ghost variable $\text{ptr}^{base} \in \mathcal{V}_{ptr}^{base}$. For example, if a pointer p is $\langle 100_{16}, 8 \rangle$, then p^{base} is 100_{16}.

The memory is modeled as a set of memory banks, $mem \in \text{Memory} : \{mb \mid mb \in \text{MB}\}$. Each bank, $mb \in \text{MB} : \text{Cache} \times \text{Storage} \times \text{Flag}$, holds memory values for cache, storage, and boolean flags. The cache, $cache \in \text{Cache} : \mathbb{N} \times \text{FldVal}$, includes the cached object's base address (as $cache^{base}$) and field values. The field values (as cells) are kept in an environment $fields \in \text{FldVal} : \mathcal{V}_{fld} \mapsto \text{Cell}$. The storage, $storage \in \text{Storage} : \mathbb{N} \mapsto \text{FldVal}$, maps base addresses of memory objects to the corresponding field environment. The cache boolean flags, *flag*, indicate if it is occupied (*used*) and overwritten (*dirty*). Overall, a concrete program state

$$\begin{aligned}
&\texttt{cacheSync}(mb, \texttt{ptr}^{base}) \equiv \\
&\quad \textbf{let } \langle cache, storage, \langle used, dirty \rangle \rangle = mb \textbf{ in} \\
&\quad \textbf{let } \langle cache^{base}, \_ \rangle = cache \textbf{ in} \\
&\quad \textbf{let } mb' = \\
&\qquad \textbf{if } \neg used \land \texttt{ptr}^{base} \neq cache^{base} \textbf{ then} \\
&\qquad\quad \textbf{let } storage' = \textbf{if } dirty \textbf{ then} \\
&\qquad\qquad \texttt{flush}(cache, storage) \textbf{ else } storage \textbf{ in} \\
&\qquad\quad \textbf{let } cache' = \texttt{refresh}(storage', \texttt{ptr}^{base}) \textbf{ in} \\
&\qquad\quad \textbf{let } mb' = \\
&\qquad\qquad \langle cache', storage', \langle \textsf{true}, \textsf{false} \rangle \rangle \textbf{ in} \\
&\qquad\qquad mb' \\
&\qquad \textbf{else } mb \\
&\quad \textbf{in } mb'
\end{aligned}$$

$$\begin{aligned}
&\texttt{flush}(cache, storage) \equiv \\
&\quad \textbf{let } \langle cache^{base}, fields \rangle = \\
&\qquad cache \textbf{ in} \\
&\quad storage[cache^{base} \mapsto fields] \\
\\
&\texttt{refresh}(storage, \texttt{ptr}^{base}) \equiv \\
&\quad \langle \texttt{ptr}^{base}, storage[\texttt{ptr}^{base}] \rangle
\end{aligned}$$

Fig. 6. Cache operations.

$\sigma \in \textsf{State}$ is a tuple: $\langle scalar, mem \rangle$. We assume findmb maps a field variable and memory state to a memory bank, indicating in which bank the field is stored.

Figures 5 and 6 describe the changes to a program state at each memory and pointer arithmetic statements in CrabIR. The function $[\![\cdot]\!]^{\text{RUMM}}(\cdot)$ takes a statement and a program state and returns the computed state under RUMM. The initial state's *scalar* is an empty map. Each bank *mb* contains an empty *cache*, an empty map *storage*, and a $\langle \textsf{false}, \textsf{false} \rangle$ cache flags.

The alloc statement creates a new memory object of size num, assigns it to a specific bank's storage. The bank is determined by fld through findmb, and its allocator constructs the object and returns its base address assigned to ptr.

The gep computes a new pointer value for ptr2 by adding an offset num to the pointer value of ptr1. Earlier, we assume all pointer arithmetic stays inbounds, so the ptr2 and ptr1 have the same base address but (presumably) different offsets.

The load operation accesses the object pointed by ptr from the cache associated with the corresponding memory bank. To ensure the object is cached, we use the cacheSync function to check if the cache is missed. If so, we flush the cache back to the storage with flush if the cache is modified, and then load the new MRU object by calling refresh . The flush function moves the currently cached object into *storage*, while refresh refreshes the cache with the object pointed by ptr. After that, the object at ptr is in the cache, so the flag *used* is set to true . The value of scl in *scalar* gets updated by the cached field fld. Similarly, store updates the field for the object, using cacheSync to ensure it is in the cache. The flag *dirty* is set to true , indicating the object has been modified.

Overall, RUMM offers a different way to organize C memory by partitioning it into multiple banks, with additional space (i.e., the cache) to temporarily hold a memory object for reads and writes. This setup is very convenient for two reasons: first, it allows strong updates on the cache; second, it provides a straightforward memory abstraction by summarizing all objects from the same bank into one and simplifies the design of MRUD, as described in Sect. 4.

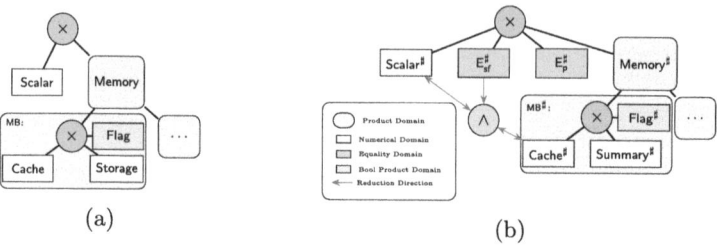

Fig. 7. (a) Concrete domain and (b) MRUD hierarchy.

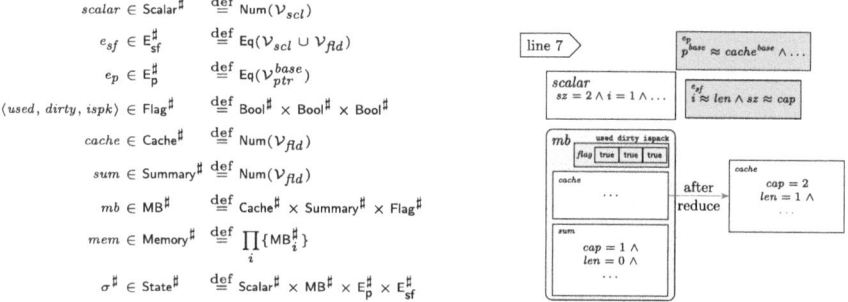

Fig. 8. Abstract semantic domains.   **Fig. 9.** State at line 7, 2nd iteration.

4   An Abstract Domain for Inferring Object Invariants

In this section, we introduce MRUD, a new abstract domain that is a (partially) reduced product of the domains for scalars, pointers, and objects. After setting up the domain, we detail key transfer functions and the reduction procedure.

Similar to the concrete domain in Fig. 7a, the MRUD is shown in Fig. 7b. It is a reduced product of four domains: (a) a numerical domain $\mathsf{Scalar}^\sharp$, (b) an equality domain $\mathsf{E}^\sharp_p$, (c) an equality domain $\mathsf{E}^\sharp_{sf}$, and (d) a collection of product domains $\mathsf{Memory}^\sharp : \{\mathsf{MB}^\sharp\}$. $\mathsf{MB}^\sharp$ is a product of two numerical domains, and three Boolean domains: $\mathsf{Cache}^\sharp \times \mathsf{Summary}^\sharp \times \mathsf{Flag}^\sharp$. Figure 8 shows the abstract semantic domains where variables are mapped to unique *dimensions* of each abstract domain. Most domains correspond to those in concrete semantics, except for a few that provide additional information. Specifically, $\mathsf{E}^\sharp_{sf}$ represents the value equivalence of fields and scalars, which enables information propagation between $\mathsf{Scalar}^\sharp$ and $\mathsf{Cache}^\sharp$ for domain reduction. $\mathsf{E}^\sharp_p$ captures the aliasing properties of pointers, indicating which pointer refers to which object. The added Boolean domain in $\mathsf{Flag}^\sharp$ is a flag for later use. All domains are parameterized by

$[\![\mathsf{ptr} := \mathsf{alloc}(\mathsf{fld}, \mathsf{num})]\!]^{\mathrm{RUMM}}(\sigma^\sharp) \equiv$
 let $\langle scalar, e_{sf}, e_p, mem \rangle = \sigma^\sharp$ in
 let $scalar' = \mathsf{forget}(scalar, \mathsf{ptr})$ in
 let $scalar'' = \mathsf{addCons}($
   $scalar', \mathsf{ptr} \neq 0)$ in
 let $e_p' = \mathsf{forget}(e_p, \mathsf{ptr}^{base})$ in
 $\langle scalar'', e_{sf}, e_p', mem \rangle$

$[\![\mathsf{store}(\mathsf{ptr}, \mathsf{fld}, \mathsf{scl})]\!]^{\mathrm{RUMM}}(\sigma^\sharp) \equiv$
 let $\langle scalar, e_{sf}, e_p, mem \rangle = \sigma^\sharp$ in
 let $mb = \mathsf{findmb}^\sharp(\mathsf{fld}, mem)$ in
 let $\langle e_p', mb' \rangle = \mathsf{cacheSync}^\sharp($
   $mb, e_p, \mathsf{ptr})$ in
 let $\langle cache, sum, \langle \_, \_, ispk \rangle \rangle = mb'$ in
 let $cache' = \mathsf{forget}(cache, \mathsf{fld})$ in
 let $e_{sf}' = \mathsf{forget}(e_{sf}, \mathsf{fld})$ in
 let $e_{sf}'' = \mathsf{addEqual}(e_{sf}', \mathsf{scl}, \mathsf{fld})$ in
 let $flag = \langle \mathsf{true}, \mathsf{true}, ispk \rangle$ in
 let $mb'' = \langle cache', sum, flag \rangle$ in
 $\langle scalar, e_{sf}'', e_p',$
   $mem \setminus \{mb\} \cup \{mb''\} \rangle$

$[\![\mathsf{ptr2}, \mathsf{fld2} := \mathsf{gep}(\mathsf{ptr1}, \mathsf{fld1}, \mathsf{num})]\!]^{\mathrm{RUMM}}(\sigma^\sharp) \equiv$
 let $\langle scalar, e_{sf}, e_p, mem \rangle = \sigma^\sharp$ in
 let $scalar' = \mathsf{forget}(scalar, \mathsf{ptr2})$ in
 let $scalar'' = \mathsf{addCons}($
   $scalar', \mathsf{ptr2} = \mathsf{ptr1} + \mathsf{num})$ in
 let $e_p' = \mathsf{forget}(e_p, \mathsf{ptr2}^{base})$ in
 let $e_p'' = \mathsf{addEqual}($
   $e_p', \mathsf{ptr2}^{base}, \mathsf{ptr1}^{base})$ in
 $\langle scalar'', e_{sf}, e_p'', mem \rangle$

$[\![\mathsf{scl} := \mathsf{load}(\mathsf{ptr}, \mathsf{fld})]\!]^{\mathrm{RUMM}}(\sigma^\sharp) \equiv$
 let $\langle scalar, e_{sf}, e_p, mem \rangle = \sigma^\sharp$ in
 let $mb = \mathsf{findmb}^\sharp(\mathsf{fld}, mem)$ in
 let $\langle e_p', mb' \rangle = \mathsf{cacheSync}^\sharp($
   $mb, e_p, \mathsf{ptr})$ in
 let $scalar' = \mathsf{forget}(scalar, \mathsf{scl})$ in
 let $e_{sf}' = \mathsf{forget}(e_{sf}, \mathsf{scl})$ in
 let $e_{sf}'' = \mathsf{addEqual}(e_{sf}', \mathsf{fld}, \mathsf{scl})$ in
 $\langle scalar', e_{sf}'', e_p'',$
   $mem \setminus \{mb\} \cup \{mb'\} \rangle$

Fig. 10. Abstract transformers for memory operations.

relational abstract domains like Zones [20]. An abstract state $\sigma^\sharp$ is represented by lattice elements within the MRUD.

Figure 9 shows the abstract state at line 7 during the second iteration of the CrabIR example from Fig. 4a. We assume that the Zones domain is used for equality and numerical domains. We only show the invariants for scalars i and sz, and fields len and cap. *scalar* shows invariants for the scalars i and sz. The sole memory bank mb represents the objects of type byte_buf. The *cache* shows the invariants for the MRU byte_buf object referenced by pointer p. This follows from the equality $p^{base} \approx cache^{base}$ in e_p. The *cache* does not have any explicit invariants for fields. However, the fields invariants are *implicitly* represented through the invariants in *scalar* and the equalities in e_{sf}, $i \approx len$ and $sz \approx cap$, that connect fields and scalars. These equalities are established during field writes. For instance, $i \approx len$ is there because instruction store(@len, plen, i) was used to update the field len with scalar i. Finally, *sum* shows the object invariants for the objects initialized at the first iteration. Specifically, the fields of that object satisfy len <= cap.

The most relevant transfer functions for inferring object invariants are shown in Fig. 10. For the initial state of analysis, we assign all subdomain elements with $\top$, except for *flag* in each memory bank as $\langle \mathsf{false}, \mathsf{false}, \mathsf{false} \rangle$. The third flag, $ispk$, is false to indicate the *sum* does not represent any concrete objects.

For alloc, the transformer assigns a ptr as not NULL in *scalar* indicating the valid address of the allocated object that ptr refers to. For gep, the transformer computes the address for ptr2 by addition in *scalar* and establishes an equiva-

$\text{cacheSync}^\sharp(mb, e_p, \text{ptr}) \equiv$
  let $\langle cache, sum, \langle used, dirty, ispk \rangle \rangle = mb$ **in**
  let $\langle e_p', mb' \rangle =$
    if $\neg used \land \neg\text{equals}(e_p, \text{ptr}^{base}, cache^{base})$ **then**
      let $sum', ispk' = $ **if** $dirty$ **then** $\text{pack}^\sharp(cache, sum, ispk)$ **else** $sum, ispk$ **in**
      let $cache' = \text{unpack}^\sharp(sum')$ **in**
      let $e_p' = \text{forget}(e_p, cache^{base})$ **in**
      let $e_p'' = \text{addEqual}(e_p', \text{ptr}^{base}, cache^{base})$ **in**
      $\langle e_p'', \langle cache', sum', \langle \text{true}, \text{false}, ispk' \rangle \rangle \rangle$
    else $\langle e_p, mb \rangle$
  in $\langle e_p', mb' \rangle$
$\text{pack}^\sharp(cache, sum, ispk) \equiv$ **if** $\neg ispk$ **then** $\langle \text{copy}(cache), \text{true} \rangle$ **else** $\langle sum \sqcup cache, ispk \rangle$
$\text{unpack}^\sharp(sum) \equiv \text{copy}(sum)$

Fig. 11. Abstract cache operations.

lence between ptr2 and ptr1 in e_p, denoting that the two pointers refer to the same memory object. For load/store, the transformer requires that the object referred by ptr is in the cache before it is accessed. The function $\text{cacheSync}^\sharp$ in Fig. 11 checks for a cache miss and handles operations when a miss happens. It tests whether ptr refers to the cached object by comparing ptr^{base} with $cache^{base}$ in e_p. When the cache is missed, the function performs $\text{pack}^\sharp$ and $\text{unpack}^\sharp$. The $\text{pack}^\sharp$ operation merges $cache$ into sum. The invariants of the first cached object are copied to sum because, initially, sum does not represent any concrete objects. We change the flag $ispk$ to true since the sum now holds the invariants for that object. Any subsequent packs use the join operation. The $\text{unpack}^\sharp$ is achieved by copying the sum as the new $cache$. The $\text{pack}^\sharp$ and $\text{unpack}^\sharp$ operations are similar to the *fold* and *expand* in [12] but simpler because $cache$ and sum are two domain values underlying the same field dimensions. After unpacking, $cache^{base}$ equals ptr^{base}, signifying the $cache$ is for the new MRU object. The transformer then performs a strong read/update in $cache$ without changing any invariant stored in sum. The read/update creates an equivalence relation between fld and scl in e_{sf} through addEqual. For field read, the transformer discards the information in scl before adding the equality. For field update, the transformer forgets information about fld ahead of setting the equality and sets *dirty* to true afterward.

Other abstract operators, including join, meet, widening, and narrowing, are computed pointwise over subdomains with an additional caching step: packing the dirty cache for each memory bank and resetting it as unused. The full definition for applying domain operators is available in the extended version of the paper [27].

We argue that the abstract semantics is sound as it is systematically derived from the concrete semantics. At each program point, the scalar abstraction overapproximates the set of numeric values or addresses of each scalar variable. For memory objects, the abstraction collapses concrete objects in each memory bank into one summary (abstract) object, also as an over-approximation. The sound-

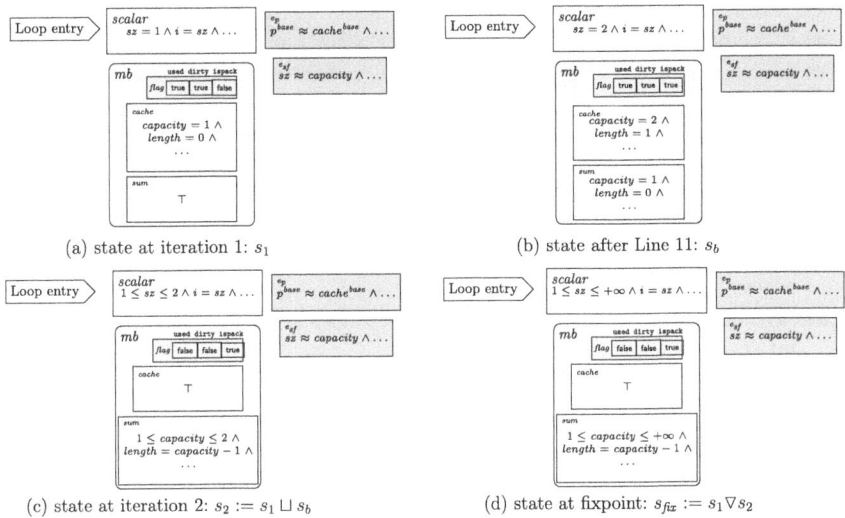

Fig. 12. Fixpoint computation for the entry state of the loop in Fig. 4a.

ness argument follows from our design of abstraction and Galois connections. We omit it here since the abstraction is straightforward.

Figure 12 illustrates the computation of abstract states at the loop entry of Fig. 4a. In Fig. 12a, state s_1 represents the an abstract state at the loop entry after the first iteration opf the loop. Since during the first iteration only one byte_buf object is initialized, the cache in s_1 has the invariants only of that object: $len = 0$ and $cap = 1$, while the summary has no objects (i.e., $ispk$ flag is unset). The next abstract state is s_b (Fig. 12b) after line 11. During the second iteration, the cache is flushed for the new byte_buf object and the summary only maintains the invariants for the flushed object. Then, s_1 and s_b are joined at the loop entry, resulting in s_2 (Fig. 12c). The join is pairwise across subdomains after the caches of both states are flushed. Finally, the widening operator is applied to reach a fixpoint, as shown in Fig. 12d.

As the memory and scalar properties are kept separately, we configure a domain reduction step to exchange information between each bank's *cache* and *scalar* through the equalities that are introduced during load and store. We use a bidirectional reduction (see red arrows on the right of Fig. 7): one direction flows from the Cache$^\sharp$ of each memory bank to Scalar$^\sharp$; the other is in the opposite. The domain reduction follows Fig. 13 which reduces an abstract state as $\sigma^\sharp$ in two steps by propagates numerical properties (1) from each *cache* into *scalar*, and (2) from the *scalar* back to each *cache*. The algorithm computes the iterated pairwise reduction through reduce which operates on each bank's *cache* and *scalar*. For example, Fig. 9 shows the *cache* after applying the reduction whose values are refined for cap and len based on equalities generated for field updates through scalars sz and i in *scalar*. The *cache* is reduced through the step (2) which involves reduce converting equalities ($len \approx i$ and $cap \approx sz$) into linear

$\text{reduce}(base_{src}, base_{dst}, e) \equiv$
  $\text{let } e' = \text{project}(e, \mathcal{V}_{src} \cup \mathcal{V}_{dst}) \text{ and } cons = \text{toCons}(e') \text{ in}$
  $\text{let } base'_{dst} = \text{project}((base_{dst} \sqcap \text{addCons}(base_{src}, cons)), \mathcal{V}_{dst}) \text{ in}$
  $base'_{dst}$

$\text{reduction}(\sigma^\sharp) \equiv$
  $\text{let } \langle scalar, e_{sf}, e_p, mem \rangle = \sigma^\sharp \text{ in}$
  $\text{for all } mb \in mem \text{ do}$ ▷ Step 1: reduce from caches to base
    $\text{let } \langle cache, \_, \_ \rangle = mb \text{ in}$
    $scalar' := \text{reduce}(cache, scalar, e_{sf})$
  $\text{for all } mb \in mem \text{ do}$ ▷ Step 2: reduce from base to caches
    $\text{let } \langle cache, sum, flag \rangle = mb \text{ in}$
    $\text{let } cache' = \text{reduce}(scalar', cache, e_{sf}) \text{ in}$
    $mb := \langle cache', sum, flag \rangle$ ▷ Update mb directly
  $\langle scalar', e_{sf}, e_p, mem \rangle$

Fig. 13. Domain reduction.

constraints and adding them to *scalar*. Then, it performs a meet with *cache* to propagate numerical information from *scalar*. Finally, it projects the result of the meet to the field variables, and obtains the new *cache*.

When reduction is executed once, it refines the abstract values in each bank's *cache* and *scalar* in the state. It adds numerical properties and preserves equalities. This ensures that it is both reductive and sound. We terminate the reduction after one iteration for each of the two directions.

In summary, we introduce MRUD, a composite abstract domain and its corresponding transformer for inferring object invariants. As a reduced product of domains for scalars and objects, MRUD is effective for scalable analysis. The reduction algorithm leverages equalities between variables to avoid precision loss.

5 Implementation

We have implemented the MRUD[1] in CRAB [13], a library for building abstract interpretation-based analyses. The Memory$^\sharp$ is implemented using a Patricia tree [24] for *structural sharing* among multiple abstract elements during analysis. This approach prevents redundant copying of domain values when computing the outputs of domain operators and transfer functions, allowing efficient memory sharing for parts of the abstract state that remain unchanged after an operation. For example, two domain elements of Memory$^\sharp$ share memory banks if they are unchanged during computation.

We have developed a custom equality domain based on a union-find data structure to represent variable equivalence (e.g., $x \approx y$). The details of this domain are available in the extended version of the paper [27]. Each equivalence

---

[1] Publicly available at https://github.com/LinerSu/crab/tree/VMCAI-2025.

class corresponds to a set of variables (e.g., $\{p^{base}, cache^{base}\}$ as $p^{base} \approx cache^{base}$ in Fig. 9). This structure fits the representation of equivalence relations and efficiently supports domain operation. Our implementation also partitions $\mathsf{E}^{\sharp}_{sf}$ into reduced product of smaller domains for better alignment with variable packing [3]. Specifically, we use an equality domain $\mathsf{E}^{\sharp}_{s}$ for scalars and $\mathsf{E}^{\sharp}_{f}$, in each memory bank, for fields. The domain value of $\mathsf{E}^{\sharp}_{sf}$ is the union of these smaller domain values. For example, $i \approx len \wedge sz \approx cap$ is maintained as two classes $e_{sf} := \{i, len\}, \{sz, cap\}$ which are equivalent to splitted classes as $e_s := \{i, \tilde{a}\}, \{sz, \tilde{b}\}$ and $e_f := \{len, \tilde{a}\}, \{cap, \tilde{b}\}$ with special representatives $\tilde{a}, \tilde{b}$.

For memory partitioning, we use SEADSA [10] to divide the memory used by the program into memory banks, with each bank containing objects from the same allocation site. As mentioned earlier, in CrabIR, a field variable represents an offset to access an object field. The findmb function of RUMM is defined by mapping fields to their corresponding bank. However, in practice, not all field offsets can be determined statically. We over-approximate the values of such field by $\top$. Improving this is left for future work.

For effective and efficient domain reduction, we use heuristics to balance precision and performance. MRUD tracks which direction needs reduction. For example, if equalities between fields and scalars only affect memory reads, there is no need to apply a reduction to refine the corresponding cache. We also allow reduction to be performed on demand. For instance, reduction is applied when an assertion is present in the program.

6 Evaluation

We performed three kinds of experiments: **scale**, **precision**, and **case study**. All experiments were conducted on a desktop computer with an Intel Xeon E5-2680 @2.50GHz, with 256 GB RAM, and are available at https://doi.org/10.5281/zenodo.13849174.

First, the **scale** experiment compares the performance of MRUD ($\mathcal{D}_\mathcal{O}$) with the summarization-based [13] domain ($\mathcal{D}_\mathcal{S}$) from CRAB by timing analysis of 114 programs: 5 from [13], and 109 from GNU Coreutils [11]. We used the Zones[2] [10] abstract domain for its simplicity and sufficiency in expressing (relational) memory safety invariants. The primary goal is to show that $\mathcal{D}_\mathcal{O}$ scales better than $\mathcal{D}_\mathcal{S}$ due to the effect of variable packing [3] in $\mathcal{D}_\mathcal{O}$ that follows from representing each partition with a different DBM, while $\mathcal{D}_\mathcal{S}$ relies on a single DBM for expressing all scalars (included ghost ones) and summary variables. Another goal is to measure the overhead introduced by domain reduction, which incurs extra costs. To evaluate this, we provide two additional strategies: FULL, which applies reduction at each transfer function, and NONE, where no reduction is applied, and compare them with the heuristic strategy, OPT. These three strategies highlight the different costs of reduction.

---

[2] The Zones domain represents all the binary relationships between two-variable difference (including zero), stored in a Difference-Bound Matrix (DBM).

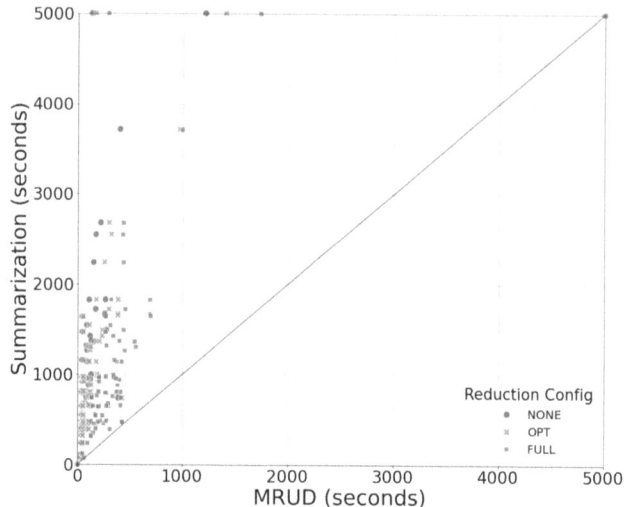

Fig. 14. Scalability results. Summarization refers to $\mathcal{D}_\mathcal{S}$ and MRUD to $\mathcal{D}_\mathcal{O}$.

Figure 14 shows the timing results, with a timeout of 5 000 seconds per program. Both domains time out on 6 cases, while $\mathcal{D}_\mathcal{S}$ times out on 2 more cases. Excluding timeout cases, $\mathcal{D}_\mathcal{O}$ outperforms $\mathcal{D}_\mathcal{S}$ on nearly every benchmark. On average, $\mathcal{D}_\mathcal{O}$ with NONE, OPT, and FULL configurations is 81x, 76x, and 57x faster than $\mathcal{D}_\mathcal{S}$, respectively. This demonstrates the advantage of composite abstract domains for inferring object invariants in large and complex programs, regardless of the domain reduction strategy used.

We analyze ginstall from GNU Coreutils to understand why $\mathcal{D}_\mathcal{O}$ is faster. The running time for $\mathcal{D}_\mathcal{S}$ is 1 846s, while for $\mathcal{D}_\mathcal{O}$, it takes 273s. Most of the time in both domains is spent on join operations, where $\mathcal{D}_\mathcal{S}$ spends 600s, while $\mathcal{D}_\mathcal{O}$ takes 95s. Joining in $\mathcal{D}_\mathcal{O}$ is also efficient because it allows to share DBMs across memory banks from other states (structural sharing for Memory$^\sharp$ domain). Another reason is that most DBMs in $\mathcal{D}_\mathcal{O}$ are small, making their joins less costly compared to $\mathcal{D}_\mathcal{S}$, where large DBMs are involved. This efficiency is also reflected in the time to copy DBMs: $\mathcal{D}_\mathcal{S}$ takes 260s, while $\mathcal{D}_\mathcal{O}$ takes 20s.

As for domain reduction, applying it at each transfer function is inefficient, as FULL takes 144 (177) seconds longer than OPT (NONE) on average. The heuristics strategy (OPT) effectively handles complex programs without significant performance loss.

Second, the **precision** experiment compares $\mathcal{D}_\mathcal{O}$ against existing heap abstract domains: $\mathcal{D}_\mathcal{S}$ and Mopsa with recency abstraction ($\mathcal{D}_\mathcal{R}$). Since all three domains follow allocation-site abstraction, which summarizes multiple objects into one and treats them indistinguishable, it becomes challenging to precisely track field updates on individual concrete objects. Specifically, $\mathcal{D}_\mathcal{S}$ cannot overcome this limitation. $\mathcal{D}_\mathcal{R}$ improves precision by differentiating the most recently allocated object at the same site. $\mathcal{D}_\mathcal{O}$ provides a more general strategy by dis-

```
1  void foo(){
2    char ary1[1], ary2[2];
3    struct byte_buf o1 = {.len = 0,
         .cap = 1, .buf=ary1};
4    struct byte_buf o2 = {.len = 1,
         .cap = 2, .buf=ary2};
5    struct byte_buf *p;
6    if (/*some conditions*/) {
7      p = &o1;
8    } else {
9      p = &o2;
10   }
11   p->len = 15; p->cap = 20;
12   ...
13 }
```

Fig. 15. Another C program.

Table 1. Precision results.

Program	#A	$\mathcal{D}_\mathcal{O}$ safe	$\mathcal{D}_\mathcal{S}$ safe	$\mathcal{D}_\mathcal{S}$ warn	$\mathcal{D}_\mathcal{R}$ safe	$\mathcal{D}_\mathcal{R}$ warn
bytebuf	3	3	0	3	0	3
bytebuf_memcpy	3	3	0	3	0	3
bytebuf_path	3	3	1	2	1	2
ipc_handler	3	3	2	1	2	1
mult_bytebuf	3	3	0	3	0	3
object	1	1	0	1	0	1
range	2	2	1	1	0	2

tinguishing the most recently used object. As a result, $\mathcal{D}_\mathcal{O}$ still precisely models field updates after object initialization, such as field updates on lines 17 and 19 in Fig. 1, which either $\mathcal{D}_\mathcal{R}$ or $\mathcal{D}_\mathcal{S}$ cannot handle.

Another challenge is path sensitivity since unclear pointer aliasing leads to imprecise modeling of field updates. For example, in Fig. 15, two byte_buf objects, o1 and o2 , are allocated separately, and a pointer p is referred to either o1 or o2 . Modeling strong field updates in line 11 requires knowing which object is being updated, but it is unknown which object the pointer p refers to. Both $\mathcal{D}_\mathcal{R}$ and $\mathcal{D}_\mathcal{S}$ can track field updates precisely, but they need more accurate points-to information. $\mathcal{D}_\mathcal{O}$, however, allows strong updates by placing o1 and o2 in the same memory bank. When updating a field on either object, we load it into the cache and perform strong updates without precise pointer aliasing.

We provide a set of 7 benchmarks[3] with similar code pattern like examples in Figs. 1 and 15 for evaluation and configure all three domains using the octagon domain. Table 1 shows that $\mathcal{D}_\mathcal{O}$ successfully proves all assertions, showing the effectiveness of our methodology in providing a more precise memory abstraction. Conversely, $\mathcal{D}_\mathcal{S}$ and $\mathcal{D}_\mathcal{R}$ largely fail due to weak updates, as discussed above.

Third, we present a **case study** which integrates an Abstract Interpreter (AbsInt) into a Bounded Model Checker (BMC) pipeline for memory safety verification. This new pipeline, AI4BMC, uses AbsInt to verify and remove a number of assertions before passing the problem to the SMT solver.

The AI4BMC pipeline, shown in Fig. 16, starts by compiling and instrumenting the input program with buffer overflow checks. Next, AbsInt is applied to remove as many of these checks as possible. Now, the program still keeps the original loops. Then, the loops are unrolled using a user-supplied bound for BMC. Later, we run another AbsInt round to eliminate buffer overflow checks in the simplified program with unrolled loops. Last, we continue with the BMC pipeline, as in SeaBMC [25], that generates a Verification Condition (VC) in

---

[3] Available at: https://github.com/LinerSu/MRU-Domain-Benchmarks.

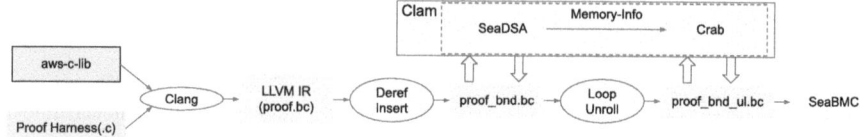

Fig. 16. The AI4BMC pipeline.

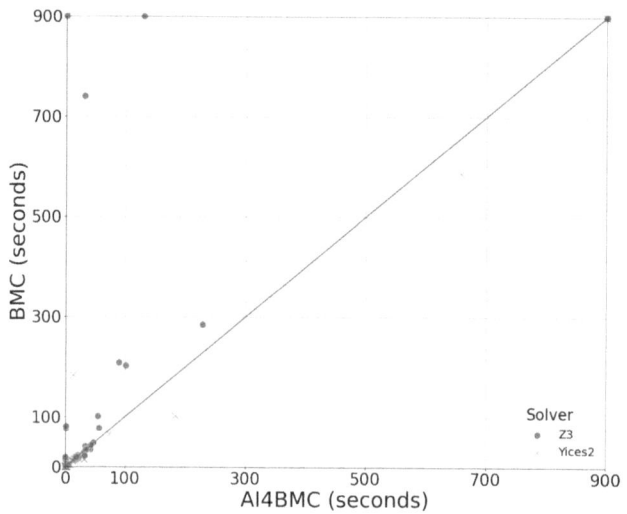

Fig. 17. AI4BMC vs. BMC.

SMT-LIB and uses an SMT-solver to check the VC's satisfiability such that the original program is safe if and only if SMT-LIB formula is unsatisfiable.

The motivation for AI4BMC is that many memory safety arguments are simple and are established independently of loop bounds. We expect AbsInt to verify those, leaving less work for BMC. Thus, we consider AI4BMC pipeline successful if (a) AbsInt discharges some buffer overflow checks before loop unwinding, and (b) AI4BMC requires less overall runtime than the BMC pipeline.

We developed two benchmark suites from industrial code. The first is based on `aws-c-commons` verification tasks, where we reduce assertions only to memory safety. The second is based on a more complex code from AWS C SDK in C99 implementation. Together, there are 109 verification tasks. The benchmarks[4] have been adapted to simplify control flow since proving all memory safety checks requires path-sensitivity.

We evaluate the effectiveness AI4BMC by comparing it with SEABMC which was previously compared against other state-of-the-art tools in [25]. Our performance evaluation focuses on these metrics: (1) *Faster* indicates AI4BMC outperforms BMC; (2) *Slower* means AI4BMC is slower than BMC; (3) *AbsInt*

---

[4] Available at https://github.com/LinerSu/verify-c-common/tree/VMCAI-2025.

Table 2. AI4BMC vs. BMC details.

Category	Metric	% Metric	Number of Cases	
			AI4BMC (Z3)	AI4BMC (Y2)
Performance Comparison	*Faster* (Time Difference > 5s)	> 95%	10	2
		others	6	1
	Slower (Time Difference > 5s)	≤ 50%	4	2
		others	0	4
AbsInt Performance	*AbsInt Time* in AI4BMC time	> 40%	65	74
		≤ 40%	39	29
Precision	*AbsInt Solving Rate* before LU	100%	37	37
		> 50%	52	52
	AbsInt Solving Rate after LU	100%	6	6
		> 50%	1	1

Time expresses the run-time of AbsInt in the AI4BMC pipeline. For precision, we provide the *AbsInt Solving Rate*, showing how many checks are solved before or after loop unrolling (LU). We used MRUD for CRAB (AbsInt) and chose two SMT-solvers for SEABMC: Z3[5] [23], and YICES2 [9]. Experiments were conducted under 900 seconds timeout and all results are summarized in Fig. 17 and Table 2.

First, comparing performance between AI4BMC and BMC. With Z3, AI4BMC timed out in 5 cases, while BMC timed out in 7 cases; AbsInt helped solving 2 more cases. Excluding timeouts, AI4BMC is at least 5 s *faster* than BMC in 16 cases. The speed-up comes from AbsInt proving and discharging assertions checks. In 10 of these 16 cases, the speed-up exceeds over 95%, with AbsInt completely solving the checks in 9 cases. The other 6 cases show at least a 20% speed-up. AbsInt takes under one second on average in all 16 cases. There are 4 cases in which AI4BMC is at least 5 s *slower* than BMC. In two of these, the slowdowns are due to Z3 taking 6s extra solving time on average, which is not surprising since the SMT performance is not always deterministic. In the other two, although Z3 solving time is decreased, AbsInt slows down by taking around 11s, roughly a third of the total run-time.

The results with YICES2 are similar, but YICES2 is faster and exhibits better stability. Both AI4BMC and BMC timed out in 4 cases and 5 cases individually, with 1 case where AbsInt improves performance. AI4BMC outperforms BMC in 3 cases with at least a 93% improvement. However, AI4BMC is slower in 6 cases, 4 of which are affected by the slowdown of AbsInt. The other 2 cases are due to the slows down of SEABMC and YICES2. The SEABMC experiences a slowdown due to lambda-encoding, where the beta-reduction simplification time is not deterministic. While switching to array-encoding shows the effectiveness of AbsInt, this slows overall performance for both AI4BMC and BMC.

---

[5] We fixed the performance issue on Z3. The one we used is available at: https://github.com/LinerSu/z3/tree/fix-performance.

Overall, the performance results show that AbsInt improves the overall performance of using BMC regardless of the solver used.

Second, in evaluating the performance of AbsInt, runtime ratios depend on the total running time of AI4BMC and the solver selected. With Z3, AbsInt takes over 40% of the time on 65 cases, but these cases terminate within 50s, with AbsInt averaging only 0.1s and maxing at 1.2s. For the rest of the 39 cases, AbsInt takes 40% or less, with 5 cases exceeding 50s and 34 cases under 50s. For these 5 longer cases, AbsInt accounts for under 2%, averaging 1s with a maximum of 1.5s. For the 34 shorter cases, AbsInt contribution was below 36%. With Yices2, the runtime percentage of AbsInt increases because Yices2 is efficient, with more cases where AbsInt accounts for a significant portion of the runtime. In summary, using AbsInt has no big cost, compared with the solving time of SMT solver.

Last, for assertion rate, AbsInt solved more than 50% of assertions in 89 cases before LU, completely solving 37 cases, and in 7 cases after LU, fully solving 6 cases. We only have 8 cases where AbsInt solves less than half of the checks. The reasons are: (1) the widening operation produces too imprecise invariants that cannot be recovered by narrowing. AbsInt needs more precise widening techniques to prove more checks; (2) Some memory safety invariants cannot be expressed by Zones or Octagons, and instead require more complex abstract domains such as Polyhedra; (3) Memory safety checks for C string require tracking the length of strings that our implementation does not support. We believe using [15] to determine the null character of each string will improve overall precision.

In this case study, we demonstrate the effectiveness of using AbsInt in the BMC pipeline. By using the Zones, it proves most memory safety checks in this industry project and reduces the number of checks BMC handles. This speeds up both BMC encoding and SMT solver performance.

7 Related Works

To deal with a potentially unbounded number of memory objects, most abstract analysis frameworks group memory objects together into *summary objects* (e.g., [12]). A summary object represents properties that are common to all objects it stands for. The most common summarization is *Allocation Site Abstraction* (ASA) [5] that groups objects by their allocation site. In ASA, all concrete objects allocated at a certain line of a program are represented by one abstract summary object. Since each summary object represents a set of objects, it supports only *weak* updates – an assignment to the field of an object does not override previous value, but rather adds to it, to capture that the field update may modify only one object out of the summary. This significantly degrades analysis precision.

The loss of precision is specifically important during object creation, when an object is first allocated and then initialized field-by-field. In ASA, because of weak updates, this results in all properties of the summary being lost since the newly allocated object has no properties in common with already summarized objects. A common solution, e.g., used by Mopsa, is *recency abstraction* [1] that refines ASA into: (a) the most recently allocated object, and (b) the rest. Since most recent object is a singleton, it can be updated *strongly*, i.e., field updates overwrite previous values. Our approach is a further refinement that separates objects not by recency of *creation*, but by recency of *use*. In principle, other extensions of recency, such as [2] can be combined with our technique for further precision improvement.

The temporary isolation of recently-used objects avoids invariant violations in summarized objects during individual field updates. Our pack and unpack methods communicate changes between these two types of objects. This is similar to corresponding methods in [4], where the annotated *pack/unpack* statements manage transitions of mutable objects during class method calls, allowing temporary updates while maintaining class invariants (i.e., invariants for all instances of a given class). Similarly, JayHorn [16] uses *push/pull* statements for encoding each memory access. Each *pull* statement reads fields of an object to make invariants available, while a following *push* statement updates fields to ensure modifications preserve invariants. The concept of *pack/unpack* has been used in refinement types [26], where the inference algorithm obtains predicates with *fold/unfold* operations to prevent temporary invariant violations of objects from the same allocation site. Unlike our work, all prior work uses heuristics to manage placement of *fold/unfold* operations. In contrast, our analysis automatically processes these during analysis.

The domain hierarchy in our MRUD uses two strategies. First, variable packing [3] is used to pack program variables for fields of memory objects in each memory bank. With two numerical domains per pack, our approach allows for the independent updating of invariants for each bank. The packing is rarely used in computing memory properties, but Toubhans et al. [28] introduced a product of memory domains that pack variables used for lists, trees, and other fixed-size structures. Second, domain reduction [6] helps exchange equivalences between scalars and object fields. This is commonly used when abstract domains are organized modularly. Astrée [7] combines various abstract domains in a sequence, using reduction steps for forward and backward propagation of information between them. [8] interprets the Nelson-Oppen procedure as a domain reduction, propagating (dis)equalities across domains.

8 Conclusion

In this work, we present a new methodology for inferring object invariants that avoids temporarily breaking invariants following the concept of caching. Our new abstract domain, parameterized by numerical and equality domains, organizes a structured hierarchy, enabling scalable analysis of complex programs. We design a reduction algorithm following equalities introduced across numerical domains to avoid significant precision loss. Our results demonstrate that MRUD enhances both precision and scalability and can be effectively integrated with other verification techniques for memory safety.

References

1. Balakrishnan, G., Reps, T.: Recency-abstraction for heap-allocated storage. In: Yi, K. (ed.) SAS 2006. LNCS, vol. 4134, pp. 221–239. Springer, Heidelberg (2006). https://doi.org/10.1007/11823230_15
2. Balatsouras, G., Smaragdakis, Y.: Structure-sensitive points-to analysis for C and C++. In: Rival, X. (ed.) SAS 2016. LNCS, vol. 9837, pp. 84–104. Springer, Heidelberg (2016). https://doi.org/10.1007/978-3-662-53413-7_5
3. Blanchet, B., et al.: A static analyzer for large safety-critical software. In: Cytron, R., Gupta, R. (eds.) Proceedings of the ACM SIGPLAN 2003 Conference on Programming Language Design and Implementation 2003, San Diego, California, USA, 9–11 June 2003, pp. 196–207. ACM (2003). https://doi.org/10.1145/781131.781153
4. Chang, B.E., Leino, K.R.M.: Inferring object invariants: extended abstract. In: Cortesi, A., Logozzo, F. (eds.) Proceedings of the First International Workshop on Abstract Interpretation of Object-oriented Languages, AIOOL@VMCAI 2005, Paris, France, 21 January 2005. Electronic Notes in Theoretical Computer Science, vol. 131, pp. 63–74. Elsevier (2005). https://doi.org/10.1016/J.ENTCS.2005.01.023
5. Chase, D.R., Wegman, M.N., Zadeck, F.K.: Analysis of pointers and structures. In: Fischer, B.N. (ed.) Proceedings of the ACM SIGPLAN'90 Conference on Programming Language Design and Implementation (PLDI), White Plains, New York, USA, 20–22 June 1990, pp. 296–310. ACM (1990). https://doi.org/10.1145/93542.93585
6. Cousot, P., Cousot, R.: Systematic design of program analysis frameworks. In: Aho, A.V., Zilles, S.N., Rosen, B.K. (eds.) Conference Record of the Sixth Annual ACM Symposium on Principles of Programming Languages, San Antonio, Texas, USA, January 1979, pp. 269–282. ACM Press (1979). https://doi.org/10.1145/567752.567778
7. Cousot, P., et al.: Combination of abstractions in the ASTRÉE static analyzer. In: Okada, M., Satoh, I. (eds.) ASIAN 2006. LNCS, vol. 4435, pp. 272–300. Springer, Heidelberg (2007). https://doi.org/10.1007/978-3-540-77505-8_23
8. Cousot, P., Cousot, R., Mauborgne, L.: The reduced product of abstract domains and the combination of decision procedures. In: Hofmann, M. (ed.) FoSSaCS 2011. LNCS, vol. 6604, pp. 456–472. Springer, Heidelberg (2011). https://doi.org/10.1007/978-3-642-19805-2_31
9. Dutertre, B.: Yices 2.2. In: Biere, A., Bloem, R. (eds.) CAV 2014. LNCS, vol. 8559, pp. 737–744. Springer, Cham (2014). https://doi.org/10.1007/978-3-319-08867-9_49
10. Gange, G., Navas, J.A., Schachte, P., Søndergaard, H., Stuckey, P.J.: Exploiting sparsity in difference-bound matrices. In: Rival, X. (ed.) SAS 2016. LNCS, vol. 9837, pp. 189–211. Springer, Heidelberg (2016). https://doi.org/10.1007/978-3-662-53413-7_10
11. GNU Project: GNU core utilities official page. https://www.gnu.org/software/coreutils/
12. Gopan, D., DiMaio, F., Dor, N., Reps, T., Sagiv, M.: Numeric domains with summarized dimensions. In: Jensen, K., Podelski, A. (eds.) TACAS 2004. LNCS, vol. 2988, pp. 512–529. Springer, Heidelberg (2004). https://doi.org/10.1007/978-3-540-24730-2_38
13. Gurfinkel, A., Navas, J.A.: Abstract interpretation of LLVM with a region-based memory model. In: Bloem, R., Dimitrova, R., Fan, C., Sharygina, N. (eds.) Software Verification - 13th International Conference, VSTTE 2021, New Haven, CT, USA,

18–19 October 2021, and 14th International Workshop, NSV 2021, Los Angeles, CA, USA, July 18-19, 2021, Revised Selected Papers. Lecture Notes in Computer Science, vol. 13124, pp. 122–144. Springer, Cham (2021). https://doi.org/10.1007/978-3-030-95561-8_8

14. Huston, B.: Single-chip microcomputers can be easy to program. In: American Federation of Information Processing Societies: 1982 National Computer Conference, 7–10 June 1982, Houston, Texas, USA. AFIPS Conference Proceedings, vol. 51, pp. 85–93. AFIPS Press (1982). https://doi.org/10.1145/1500774.1500786

15. Journault, M., Miné, A., Ouadjaout, A.: Modular static analysis of string manipulations in C programs. In: Podelski, A. (ed.) SAS 2018. LNCS, vol. 11002, pp. 243–262. Springer, Cham (2018). https://doi.org/10.1007/978-3-319-99725-4_16

16. Kahsai, T., Kersten, R., Rümmer, P., Schäf, M.: Quantified heap invariants for object-oriented programs. In: Eiter, T., Sands, D. (eds.) LPAR-21, 21st International Conference on Logic for Programming, Artificial Intelligence and Reasoning, Maun, Botswana, 7–12 May 2017. EPiC Series in Computing, vol. 46, pp. 368–384. EasyChair (2017). https://doi.org/10.29007/ZRCT

17. Karr, M.: Affine relationships among variables of a program. Acta Informatica **6**, 133–151 (1976). https://doi.org/10.1007/BF00268497

18. Lattner, C., Adve, V.S.: Automatic pool allocation: improving performance by controlling data structure layout in the heap. In: Sarkar, V., Hall, M.W. (eds.) Proceedings of the ACM SIGPLAN 2005 Conference on Programming Language Design and Implementation, Chicago, IL, USA, 12–15 June 2005, pp. 129–142. ACM (2005). https://doi.org/10.1145/1065010.1065027

19. Meyer, B.: Object-Oriented Software Construction, 2nd edn. Prentice-Hall Inc, USA (1997)

20. Miné, A.: A new numerical abstract domain based on difference-bound matrices. In: Danvy, O., Filinski, A. (eds.) PADO 2001. LNCS, vol. 2053, pp. 155–172. Springer, Heidelberg (2001). https://doi.org/10.1007/3-540-44978-7_10

21. Miné, A.: The octagon abstract domain. In: Burd, E., Aiken, P., Koschke, R. (eds.) Proceedings of the Eighth Working Conference on Reverse Engineering, WCRE 2001, Stuttgart, Germany, 2–5 October 2001, p. 310. IEEE Computer Society (2001). https://doi.org/10.1109/WCRE.2001.957836

22. Monat, R., Ouadjaout, A., Miné, A.: Mopsa-c: modular domains and relational abstract interpretation for C programs (competition contribution). In: Sankaranarayanan, S., Sharygina, N. (eds.) Tools and Algorithms for the Construction and Analysis of Systems - 29th International Conference, TACAS 2023, Held as Part of the European Joint Conferences on Theory and Practice of Software, ETAPS 2022, Paris, France, 22–27 April 2023, Proceedings, Part II. Lecture Notes in Computer Science, vol. 13994, pp. 565–570. Springer, Cham (2023). https://doi.org/10.1007/978-3-031-30820-8_37

23. de Moura, L., Bjørner, N.: Z3: an efficient SMT solver. In: Ramakrishnan, C.R., Rehof, J. (eds.) TACAS 2008. LNCS, vol. 4963, pp. 337–340. Springer, Heidelberg (2008). https://doi.org/10.1007/978-3-540-78800-3_24

24. Okasaki, C., Gill, A.: Fast mergeable integer maps. In: Notes of the ACM SIGPLAN Workshop on ML, pp. 77–86 (1998)

25. Priya, S., Su, Y., Bao, Y., Zhou, X., Vizel, Y., Gurfinkel, A.: Bounded model checking for LLVM. In: Griggio, A., Rungta, N. (eds.) 22nd Formal Methods in Computer-Aided Design, FMCAD 2022, Trento, Italy, 17–21 October 2022, pp. 214–224. IEEE (2022). https://doi.org/10.34727/2022/ISBN.978-3-85448-053-2_28

26. Rondon, P.M., Kawaguchi, M., Jhala, R.: Low-level liquid types. In: Hermenegildo, M.V., Palsberg, J. (eds.) Proceedings of the 37th ACM SIGPLAN-SIGACT Symposium on Principles of Programming Languages, POPL 2010, Madrid, Spain, 17–23 January 2010, pp. 131–144. ACM (2010). https://doi.org/10.1145/1706299.1706316
27. Su, Y., Navas, J.A., Gurfinkel, A., Garcia-Contreras, I.: Automatic inference of relational object invariants (2024). https://arxiv.org/abs/2411.14735
28. Toubhans, A., Chang, B.-Y.E., Rival, X.: An abstract domain combinator for separately conjoining memory abstractions. In: Müller-Olm, M., Seidl, H. (eds.) SAS 2014. LNCS, vol. 8723, pp. 285–301. Springer, Cham (2014). https://doi.org/10.1007/978-3-319-10936-7_18

Author Index

A
Assolini, Nicola II-50
Azeem, Muqsit II-97

B
Bentele, Manuel I-74
Bloem, Roderick II-137
Bora, Utpal II-121
Bruhns, Gustav S. II-208
Bruse, Florian I-29
Bucev, Mario II-185

C
Chait-Roth, Devora II-231
Chakraborty, Debraj II-97
Chassot, Samuel II-185
Chen, Liqian I-187
Chen, Xin I-125

D
Deshmukh, Jyotirmoy V. I-3
Di Pierro, Alessandra II-50
Dross, Claire I-163

E
Eichler, Paul I-101
Erhard, Julian I-74

F
Fang, Dongliang II-163
Felix, Simon II-185
Fu, Hongfei I-187

G
Garcia-Contreras, Isabel I-214
Gurfinkel, Arie I-214

H
Hansen, Martin P. II-208
Hebsgaard, Rasmus II-208
Heizmann, Matthias I-74
Huguet, Joffrey I-163
Hyldgaard, Frederik M. W. II-208

I
Ishii, Daisuke I-148

J
Jacobs, Swen I-101
Joshi, Saurabh II-121

K
Kanav, Sudeep II-97
Kanig, Johannes I-163
Ke, Jingyu I-187
Klumpp, Dominik I-74
Koskinen, Eric II-26
Křetínský, Jan II-97
Kunčak, Viktor II-185

L
Larrauri, Alberto II-137
Li, Guoqiang I-187
Li, Qin I-125
Liu, Hongming I-187
Liu, Puzhuo II-163

M
Maderbacher, Benedikt II-137
Mastroeni, Isabella II-3, II-50
Mercer, Eric I-52
Mohagheghi, Mohammadsadegh II-97
Mohr, Stefanie II-97
Muduganti, Gautam II-121

N
Namjoshi, Kedar S. II-75, II-231
Navas, Jorge A. I-214

P
Pan, Zhiwen II-163
Pincus, Jared II-26
Pingle, Aabha Shailesh I-3

R
Raghothaman, Mukund I-3
Ravi, Srivatsan I-3

S
Saan, Simmo I-74
Schramka, Filip II-185
Schüssele, Frank I-74
Schwarz, Michael I-74
Seidl, Helmut I-74
Si, Shuaizong II-163
Srba, Jiří II-208
Storey, Kyle I-52
Su, Yusen I-214
Sun, Limin II-163
Sun, Zhouyue I-187
Sur, Deepayan I-3

T
Tilscher, Sarah I-74
Trefler, Richard II-75

U
Upadrasta, Ramakrishna II-121

V
Vojdani, Vesal I-74

W
Wang, Yuncheng II-163
Weil-Kennedy, Chana I-101
Weininger, Maximilian II-97
Windisch, Felix II-137

X
Xia, Yuan I-3

Y
Yang, Junfeng I-125

Z
Zhang, Min I-125
Zhang, Ruoxi II-75
Zhang, Weidong II-163
Zheng, Yaowen II-163

The manufacturer's authorised representative in the EU is Springer Nature Customer Service Centre GmbH, Europaplatz 3, 69115 Heidelberg, Germany. If you have any concerns regarding our products, please contact ProductSafety@springernature.com

Printed and bound by CPI Group (UK) Ltd, Croydon, CR0 4YY

26/03/2026

02078973-0003